I0070006

STATE OF CALIFORNIA
DEPARTMENT OF NATURAL RESOURCES
WARREN T. HANNUM, Director

DIVISION OF MINES
Ferry Building, San Francisco

W. BURLING TUCKER   State Mineralogist

San Francisco]   Bulletin 135   [October 1946

PLACER MINING FOR GOLD IN CALIFORNIA

By

CHARLES VOLNEY AVERILL

Contributing Authors

F. W. COLLINS   P. MALOZEMOFF
L. L. HUELSDONK   C. M. ROMANOWITZ
OLAF P. JENKINS   H. A. SAWIN
          H. H. SYMONS

printed in CALIFORNIA STATE PRINTING OFFICE

LETTER OF TRANSMITTAL

To His Excellency,
THE HONORABLE EARL WARREN,
Governor of the State of California

Sir: I have the honor to transmit herewith for reprinting, Bulletin
135, *Placer Mining for Gold in California,* of the Division of Mines. This
volume was originally prepared under the direction of former State
Mineralogist, Walter W. Bradley, and published after his retirement,
in October 1946 when W. Burling Tucker held the same position. Since
the issue of the book is now exhausted and there is still a continued
demand for it by industry, a second printing is recommended by Olaf P.
Jenkins, one of its authors, who is now Chief of the Division of Mines and
State Mineralogist.

The volume is divided into four sections and an appendix: Placer
mining methods, Geology of placer deposits, Prospecting and sampling
placer deposits, Placer mines by counties, and Laws affecting placer
mining. Eight authors have contributed to the book: Charles Volney
Averill, F. W. Collins, L. L. Huelsdonk, Olaf P. Jenkins, P. Malozemoff,
C. M. Romanowitz, H. A. Sawin, and H. H. Symons. Assembling and
coordinating the volume were done by C. V. Averill.

Respectfully submitted,

WARREN T. HANNUM, Director
Department of Natural Resources

October 13, 1949.

CONTENTS

CONTENTS—Continued

ILLUSTRATIONS

CONTENTS—Continued

ILLUSTRATIONS—Continued

ABBREVIATIONS

a-c	alternating current (electric)
ARS	abrasion resisting steel (U. S. Steel Corporation)
C	carbon
cfs.	cubic feet per second
cu. ft.	cubic foot or feet
cu. yd.	cubic yard(s)
d-c	direct current (electric)
E.	east
ft.	foot or feet
gpm.	gallons per minute
H.	Humboldt Meridian
hp.	horsepower
hr.	hour(s)
Inf. Circ.	Information Circular (U. S. Bureau of Mines)
in.	inch(es)
kva.	kilovolt-amperes
lb.	pound(s)
M.	1000
M.D.	Mount Diablo Meridian
mg.	milligram(s)
min.	minute(s)
N.	north
no.	number
oz.	ounce(s)
R.	range
rpm.	revolutions per minute
S.	south
S.B.	San Bernardino Meridian
sec.	section(s)
sq.	square
T.	township
vol.	volume
W.	west

SECTION 1

PLACER MINING METHODS

Introduction

The following paragraphs describing small-scale placer mining during the depression of the 1930's are abstracted from a report[1] by the Federal Works Agency, Work Projects Administration, and from the letter of transmittal of that report by Corrington Gill, Assistant Commissioner.

This report shows that hand placering for gold is a vanishing frontier enterprise from which it is now next to impossible to extract a living. Soon after the depression set in thousands of unemployed with their families attempted small-scale placer mining as a source of livelihood. During the early years of the rush to the creeks the number of would-be miners failing to find gold was 20 times greater than the number of miners who had been successful in recovering an amount sufficient for even one sale. Disillusionment was rapid, and by 1933, a year in which at least 100,000 men tried their hand at placer mining, departures greatly exceeded arrivals at the diggings. By 1937 the number seeking gold had dropped to approximately 22,000, of whom a fifth recovered no gold at all. Moreover, small-scale placer mining has generally offered employment only for a very short time even to those who had some success. About half of those who found any gold gave up the effort within a month, and three-quarters within two months. Because climate and stream conditions frequently limit the work-year, and because seasonal jobs in other industries sometimes are available at higher wages, even the comparatively small number of full-time miners worked only eight months out of the year.

The average gross earnings for the miners who found gold in California, where most of the hand placering is carried on, were $6.02 per week for the three years 1935-37, and weekly income of nearly a third of the placer operators did not exceed $3.50. These figures represent gross earnings for a full week's work; returns per calendar week are lower because of broken working time; net returns are still smaller because of commissions paid to bullion buyers and necessary expenses incidental to mining.

When the low level of weekly earnings and the short periods of work are known it is not surprising that yearly returns from gold placering by hand methods are found to be pitifully small. Gross average annual earnings per miner for California ranged from $44 to $59 in the years 1935-37.

The survey did reveal one small group of miners to whom placering is important. These are the men to whom placering offers an opportunity for work in off seasons and to earn something between jobs. When lumber camps are idle, when no harvests are offering work, when shops are closed for repairs or waiting for orders, placering provides

---

[1] Newcomb, R., Merrill, C. W., and Kiessling, R. L., Employment and income from gold placering by hand methods: Work Projects Administration, National Research Project, Rept No. E-14, 1940.

FIG. 1.   Primitive gold recovery methods—pan, rocker, arrastre. *Cut by courtesy of The Argonaut; reprinted from California Journal of Mines and Geology, April-July 1934, p. 262.*

something to do even though the returns are small.   In certain limited areas, therefore, placering may yield enough to men with irregular jobs to be of marked aid to them even though it does not yield enough for support in the absence of other sources of income.

If placering is thus looked upon only as a supplemental source of income for residents of the areas with placer deposits, it can be made to fill a definite but very minor place in the economy of the few communities in which gold-bearing gravels are found and to help a few hundred men at most.

The needs of the unemployed in the early 1930's caused many persons to grasp at any possible source of income, no matter how small or temporary it might be.   Moreover, perplexed local relief officials, who were as yet receiving no aid from Federal sources, welcomed any possible source of help for the long lines of unemployed that gathered at their offices.

At that critical time fabulous tales of rich gold strikes came to the unemployed and the relief officials.   The reports were listened to eagerly by many, and the farther from the gold streams they spread, the more fantastic they became and the more readily they were believed and acted upon.   The greater the distance, the greater was the urge to get to the streams.   Many local relief officials even "staked" families to gasoline and food for a one-way trip to the new Eldorados.

The increase in the price of gold from $20.67 in 1932 to $25.56 in 1933 and to $35.00 in 1934 made such stories seem even more plausible and helped further to stimulate the migration of men to the creeks, despite the unfortunate experiences related by most of those drifting back from the gold-bearing areas.   Stories of those who succeeded in making a living and of the very few who made strikes continued to be magnified out of all proportion, both in passing from mouth to mouth and in the press, and brought new recruits to the streams as late as 1937.

It is obvious to those versed in gold mining that the facts are greatly exaggerated in these stories. To the hard-pressed unemployed, however, these accounts sounded like the answer to their need. How could they know that for every one who made a strike in placer mining, tens of thousands would find little or nothing, that not more than a few score at most could possibly expect to develop a profitable lode mine, and that large amounts of capital would be required for most of these mines? The experience of the thousands who are unsuccessful in placering does not make news; the story of the man here and there who is lucky does. Most of the accounts were stories of success, stories which were news but which were misleading to the unemployed.

Number of Small-Scale Placer Miners

Many thousands of unemployed and their families joined in the gold rush that followed the spread of such success stories. Creeks that later had only one or two placer miners per mile sometimes harbored 100 men or more per mile searching for precious metal in 1932-33. Of course no count was ever made of those who flocked to the gold-bearing streams, but 100,000 would seem a conservative estimate for 1932 and 1933. The number probably did not drop much until after 1933, for new men kept coming in considerable numbers until 1934. They came from greater and greater distances as the stories spread eastward, and they came rapidly enough to replace the disillusioned families which were leaving. If there was only one turnover from 1932 through 1933, it would mean that 200,000 men tried their hand at placering, and that there was one would-be miner for every 10 men who were at least 21 years of age in California in 1930.

The 12,422 small-scale miners recorded by the United States Mint as selling gold in California in 1937 sold metal valued at only $542,186, compared with gold worth $1,033,093 sold by 19,463 miners in 1935. It might be pointed out also that the greatest productivity was not reached until 1936, after the crowds had left and when those who knew the business were able to work unhindered by scores of would-be placer miners.

Summary of Findings

Small-scale placer mining has certain advantages for the able-bodied unemployed. It provides a meager income to a few without requiring much in the way of training or capital. It enables them to work at any time without going through the sometimes hopeless process of finding an employer. And, in addition, mining has given many who took it up seriously a new sense of self-reliance, of independence, and of initiative. Such results have had a salutary psychological effect on many unemployed during hard times.

To a few who have mined only intermittently and who have relied on the creeks to augment their incomes from other sources rather than to provide them with a living, placering has proved particularly helpful. It has enabled many men, together with their families, to have some occupation between jobs, and it has contributed more to the welfare of these individuals than the small financial returns might suggest. And to a very small proportion of the few who have stuck to the creeks fairly steadily, placer mining has proved profitable.

To some who dislike discipline and authority, placer mining has proved preferable to other ways of making a living. There are men who

Value of gold produced in California, 1848 to 1944*

Period	Placer								Total lode	Total placer and lode value
	Surface placers						Underground placers drift	Total placer		
	Connected-bucket dredges	Dragline dredges	Non-floating[14] washing plants	Hydraulic	Small-scale wet	Hand method dry[14]				
1848-1850								$51,669,767		$51,669,767
1851-1860								581,561,868	$5,874,362	[2]587,436,230
1861-1870								211,388,439	23,487,604	[3]234,876,043
1871-1880								120,910,077	51,818,604	[4]172,728,681
1881-1890								37,881,328	113,643,986	[5]151,525,314
1891-1900	$426,118							29,416,001	117,664,005	[7]147,080,006
1901	471,762			$1,698,720	$719,078		$1,061,489	3,951,049	13,037,995	16,989,044
1902	867,665			1,256,222	1,234,554		[9]889,161	4,247,602	12,662,718	16,910,320
1903	1,475,749			872,812	798,521		905,679	4,052,761	[10]12,247,892	16,300,653
1904	2,187,038			1,028,183	836,115		933,954	4,985,290	13,648,386	18,633,676
1905	3,276,141			975,140	825,555		815,240	5,892,076	13,006,469	18,898,545
1906	5,098,359			1,054,172	617,577		605,817	7,375,925	11,356,527	18,732,452
1907	5,065,437			909,011	302,864		563,383	6,840,695	9,887,233	16,727,928
1908	6,536,189			743,797	560,656		390,545	8,231,187	10,530,372	18,761,559
1909	7,382,950			605,608	376,078		739,797	9,104,433	11,133,437	20,237,870
1910	7,550,254			635,498	186,114		516,929	8,888,795	10,826,645	19,715,440
1901-1910	39,911,544			9,779,163	6,457,112		7,421,994	63,569,813	118,337,674	181,907,487
1911	7,666,461			675,486	164,680		479,900	8,986,527	10,752,381	19,738,908
1912	7,429,955			689,682	138,034		387,992	8,645,663	11,067,815	19,713,478
1913	8,090,294			329,300	224,045		192,538	8,836,177	11,570,781	20,406,958
1914	7,783,394			702,884	264,623		329,948	9,080,849	11,572,647	20,653,496
1915	7,796,465			420,770	118,427		272,955	8,608,617	13,833,679	22,442,296
1916	7,769,227			390,015	165,118		251,297	8,575,657	12,835,084	21,410,741
1917	8,313,527			267,103	126,641		366,759	9,074,030	11,013,474	20,087,504
1918	7,431,927			213,229	85,203		108,420	7,638,779	8,690,174	16,528,953
1919	7,716,919			184,832	45,984		85,341	8,033,076	8,662,879	16,695,955
1920	6,900,366			66,233	31,089		62,925	7,060,613	7,250,430	14,311,043
1911-1920	76,898,635			3,939,634	1,363,844		2,538,075	84,739,988	107,249,344	191,989,332

Note previous tables

Year	1	2	3	4	5	6	7	8	9	10
1921	7,756,787			162,808	107,818		127,411	8,154,824	7,549,998	15,704,822
1922	4,999,215			158,275	89,739		252,626	5,499,855	9,170,491	14,670,346
1923	6,065,735			111,828	160,249		184,771	6,522,583	6,856,430	13,379,013
1924	4,305,521			60,195	124,088		98,568	4,588,372	8,561,803	13,150,175
1925	4,750,842			175,345	103,434		66,523	5,096,144	7,969,186	13,065,330
1926	4,950,545			69,139	97,483		111,236	5,228,403	6,695,078	11,923,481
1927	5,461,929			120,832	112,623		141,929	5,837,313	5,833,705	11,671,018
1928	4,430,913			91,512	174,818		153,386	4,850,629	5,934,686	10,785,315
1929	3,589,259			59,732	136,948		84,668	3,870,607	4,656,096	8,526,703
1930	3,451,801			89,403	151,324		62,615	3,755,143	5,696,019	9,451,162
1921-1930	49,762,547			1,099,069	1,258,524		1,283,733	53,403,873	68,923,492	122,327,365
1931	3,619,355			62,556	227,636		111,199	4,020,746	6,793,416	10,814,162
1932	3,903,481			122,876	533,238		205,880	4,765,475	7,000,251	11,765,726
1933[11]	5,155,716	$1,924	$40,442	114,890	928,098	$5,737	434,036	6,680,843	9,002,232	15,683,075
1934[12]	6,772,380	121,138	203,810	324,397	1,694,919	6,426	454,098	9,577,168	15,554,116	25,131,284
1935[13]	8,274,130	776,701	416,240	476,809	1,545,153	4,494	599,883	12,093,410	19,071,640	31,165,050
1936	9,671,347	1,748,864	422,079	268,450	1,369,620	11,827	837,618	14,329,805	23,380,665	37,710,470
1937	11,303,635	3,294,970	597,765	161,980	896,420	17,010	258,930	16,530,710	24,579,520	41,110,230
1938	13,135,360	4,133,780	806,610	247,135	1,459,010	6,020	250,040	20,037,955	25,851,560	45,889,515
1939	12,959,240	6,038,165	1,450,290	212,065	1,358,525	5,915	228,375	22,261,575	27,972,665	50,234,240
1940	14,523,810	7,181,335	988,120	422,065	1,374,030	7,385	176,575	24,673,320	26,275,165	50,948,485
1931-1940	89,318,454	23,296,877	4,934,356	2,413,223	11,386,649	64,814	3,556,634	134,971,007	185,481,230	320,452,237
1941	14,639,870	7,875,665	1,004,605	355,075	1,086,645	7,700	160,895	25,130,455	24,177,300	49,307,755
1942	10,882,795	4,126,710	351,540	213,710	613,235	2,555	81,865	16,272,410	13,407,485	29,679,895
1943	2,344,965	496,860	104,895	60,305	88,760	140	33,950	3,129,875	2,061,605	5,191,480
1944	2,272,375	218,435	42,350	29,330	49,280	105	14,840	2,626,715	1,481,340	4,108,055
Grand totals	$286,457,203	$36,014,547	$6,437,746 *	$17,889,409	$22,304,049	$75,314	$15,091,986	$1,416,671,616	$833,608,031	$2,250,279,647

* Compiled by Henry H. Symons from publications of the U. S. Bureau of Mines.

[1] Estimated as 100 percent from placer mines.

[2] Estimated as 1 percent from gold-lode mines and 99 percent from placer mines.

[3] Estimated as 10 percent from gold-lode mines and 90 percent from placer mines.

[4] Estimated as 30 percent from gold-lode mines and 70 percent from placer mines.

[5] Estimated as 75 percent from gold-lode mines and 25 percent from placer mines.

[6] Dredge production first recorded in 1898, $18,887; 1899, $206,302; 1900, $200,929. See U. S. Geological Survey Mineral Resources, Pt. 1, 1914, p. 357, for table to date. Note previous tables gave no production for 1898.

[7] Estimated as 80 percent from gold-lode mines and 20 percent from placer mines.

[8] Estimated distribution from information in Annual Report of Director of Mint, 1901, p. 90.

[9] Estimated distribution from information in Annual Report of Director of Mint, 1902, p. 76.

[10] From U. S. Geological Survey tabulation sheets.

[11] Value calculated at an average weighted price of $25.56 a fine ounce; previously $20.6718.

[12] Value calculated at an average weighted price of $34.95 a fine ounce.

[13] Value calculated at $35 a fine ounce.

[14] Prior to 1933 was included with small-scale hand methods wet.

2

prefer to live on 25 cents a day which they themselves earn through placer mining rather than to work for wages or to accept public aid. Placer mining has enabled such men to live their own lives to some degree at least.

Another small group to whom small-scale placering has been helpful includes men with outside incomes or pensions. These men would have had nothing to do if they had lived in the cities, but they can work as hard or as easily as they will on the creeks. Knowing that their pensions will enable them to live, they work at their own convenience and at their own rate of speed on the placer gravels, adding a little to their income and taking advantage of the fact that living costs are lower on the creeks than in town. Placer mining has enabled many retired or pensioned persons to enjoy healthful work in moderation, to increase their small incomes, and to dream of making a rich strike some day.

Men who have shown that they can live within their means and build up their equipment out of an income of a dollar or two a day can sometimes secure backing for larger placer projects that require more capital and will return at least a living wage. Each year a few men demonstrate unusual ability to placer and to conserve their resources and are able to lease good bars and equipment. Only a very few succeed in this way, but they prove that it can be done.

All the men in these groups do not add up to 5 percent of the small-scale placer miners of the country. For 95 percent of those who try to depend on small-scale placer mining for a living, it has turned out to be a delusion and a snare, primarily because earnings are tragically low. The output per man-hour from hand methods of placering on the lean bars still available is too low to support life in modest comfort. Less than half of the men who try it find enough gold to hold them at the streams over a month, and half of those who stay over a month do not remain over 2 months. Even among the better full-time miners, half appear to net less than $7 per week. The result is that most miners follow placering only casually in the hope of having a "lucky break" or in an effort to earn an income to tide them over between other jobs.

Earnings from small-scale placer mining, which are too low to support individuals, are far too low to support a well-rounded family life. Even the more successful miners can make no provision for medical attention, good clothing, social life, reserves for emergencies, facilities for recreation, and other such needs. The small-scale placer miner's family lives at a bare subsistence level and from day to day. The uncertain nature of the work—owing to the fact that the gravels at any particular point may give out at any time and force the family to move—has the further disadvantage of discouraging provision for suitable or permanent dwellings and the making or purchasing of furniture or household equipment. This aspect of placering also makes it particularly difficult for children to be educated satisfactorily.

Children are given very limited educational facilities in the mountain counties at best. When they are reared in tents and shacks and are moved from creek to creek, they have access to poor schools only and cannot hope to receive an education equivalent to that given children of more settled families in the more populous sections. They are handicapped in many other ways. Diets are unbalanced, medical facilities hard to secure, and social contacts scarce.

Finally, families find conditions discouraging because the community life is so unsatisfactory. It is quite different from that of the original pioneers or even of farm families. Pioneers and farmers feel that they own the land and are developing it; they are the people who count; they are the community, and they are able to make a community life of their own even with very limited physical facilities. But the placer miners are temporary interlopers. They own no land and are not developing the area; they are living off, or at best, in the community, not as part of it, and they do not have the resources with which to make a life of their own nor with which to purchase an entree to the life of the community in which they are living. Family and social life are very circumscribed.

Not only was the life of the small-scale placer miner unsatisfactory, particularly if he had a family with him, but the future probably will bring a declining level. Small-scale placer mining in the United States provided fewer than 6,000 men with an average recovery above $3.50 per week gross for more than 1 month out of 12, and it supplied fewer than 350 men with that recovery for more than 6 months out of 12. Unless there is a sharp upward change in the price of gold, it probably will provide fewer and fewer men with even this much income and for shorter and shorter periods each year.

The reason for such unsatisfactory incomes may possibly be better understood when it is recalled that small-scale placer mining by hand methods is an attempt to extract a living from a parsimonious Nature by human muscle, with very little aid from tools. The only energy provided by other than human exertion is a little free water power and power drawn upon by about a third of the full-time miners who utilize gasoline engines to pump water. But even these miners, more fortunate than the rest, shovel gravel themselves.

Wherever human muscle, unaided by power equipment works against nature, it is an almost universal result that returns are very low unless the work requires great skill. This holds true for placer mining. If bars are exceedingly rich, as many of them were for a time in the late 1840's and 1850's, muscle power may extract returns for a time comparable with those won by skilled labor in urban centers. But when the bars are small, lean, and uncertain in their distribution and erratic in their content, as they are in most known auriferous areas available to small-scale miners in the United States today, hand labor expended on them generally cannot yield earnings comparable with wages.

Mechanized mining can still yield good returns in many areas, even on beds with a lower gold content per yard than those being worked by hand, because the gold content is certain, the yardage is extensive, and the amount handled per man-day with the aid of power machinery is many times the yardage one man can handle unaided by machinery. But even when beds are worked by power, they must be extensive and must give a constant yield to be profitable. If they yielded well one day, little the next, and nothing the third, as do many bars worked by hand, they could not be made to pay no matter how much machinery each man could put to work.

In view of the character of the work and its low returns, the question naturally arises as to why and how men adapt themselves to this pioneer type of life and its exceedingly low earnings. The adaptation of those who stick to the work is not so difficult as it might appear, for the selective process quickly weeds out those who cannot adjust themselves readily and leaves those to whom the life does not seem strange and to whom it may even seem attractive. Men who cannot live on a steady diet of canned foods, flapjacks, and beans; who cannot repair their own equipment or fix the roof when it leaks; and who dislike solitude cannot long survive the life at the creeks.

Phrasing it differently, the probability that a miner will adapt himself to placer mining may vary directly with his self-sufficiency. If he can live alone, take care of his own needs, work without supervision, and live on a few cents a day, he may become a full-time, small-scale placer miner. Men to whom such a life appeals, or men to whom it is not unattractive, can adapt themselves to placer mining, and some of them thrive physically on it. But the proportion of workers in California, or even in the country, who can meet such qualifications is very small; so the number who can make a success of or even last at placer mining is very limited. Men who can fix the roof if it leaks, or build it from scratch if necessary, can readily be found; but not many men can both fix the roof and stand living alone under it after working alone all day. So the process of adapting themselves to the creeks is primarily one of selection; most of those who try it cannot adapt themselves, and leave.

Some idea of the difficulties facing a would-be miner entering gold-bearing terrain may be realized when it is recalled that many of the forty-niners failed on the creeks of California when gold was much more plentiful than it is now, and when it is further recalled that in the nearly 100 years during which gold has been actively mined, all the profitable areas have long since been patented or at least taken up as mining claims, or have been purchased for farming or other nonmineral purposes. Consequently, a miner who has been successful in locating a place that looks promising will ordinarily find that someone else has established ownership to it a long time before.

About half of the miners interviewed who gave information on this point (102 out of 201 miners) were working without making any effort to secure permission; 63 were working with permission; 24 owned the claims they were working (mostly claims that were so poor that others had passed them by, but that did yield something); six paid royalties of 10 to 20 percent; and two were supposed to pay royalties above fixed earnings. The rest worked under various sorts of agreements, such as acting as caretaker for property in return for the right to mine. Owners of rich bars of course will not freely permit unrestricted mining, but many private owners of low-grade gravel that will not pay wages make no objection to its being mined without royalties provided the operation does not become a nuisance.

The situation is sometimes different when the men attempt to work on the public domain, for it is the duty of Government officials to protect public property, and they have not been enthusiastic over the invasion of public lands by miners. The Forest Service, for instance, has a very useful policy of keeping a strip of land a quarter of a mile to half a mile wide, on either side of major scenic highways, in its primitive state. Its officials naturally object to the building of hovels within this protected area, though they sympathize with the men and allow them to build half a mile back from the road. But this means that the miners must maintain their own drives to their shacks, which is a real hardship in muddy weather. The danger of forest fires is ever present, and the Forest Service also must be very careful that careless miners do not become a fire hazard. Game wardens may object to the presence of the small-scale placer miners, who sometimes muddy waters and hunt or fish without regard to game laws. The muddying of water used for irrigation purposes may also create difficulties at times. River pollution is another problem where miners work on streams whose waters are used by towns or cities, and restrictions imposed by sanitary districts sometimes add to the miners' difficulties. One of the first adjustments the miners must make, consequently, is that of accommodating themselves to property rights which deprive them of the chance to work the best bars which already are privately owned, and to laws and regulations which interfere with operations on the poorer bars on the public domain.

Those persons who insist on trying small-scale placer mining in spite of the above warnings will find methods described by Boericke.[2] Numerous practical suggestions by a man who states that he has personally made a living from small-scale placer mining over a period of years are contained in a recent book by Douglas.[3] Small-scale devices described below are suitable for sampling large gravel deposits to determine whether the gold-content is sufficient to justify working by machinery on a large scale. Descriptions of the pan, rocker, dip-box, and sluice-box are reprinted with minor changes and additions from an article by Symons.[4]

Pan, Rocker, Dip-Box, and Sluice-Box

The equipment and operations described herein are among the simplest, and have been used in California to recover gold from placers since the days of '49. They are used not only for gold, but any heavy materials may be separated from lighter ones in this way. They are adaptable for the separation of cassiterite (stream tin), tungsten ore, cinnabar, platinum metals, and gem stones.

Gold-Pan and Batea

The gold-pan is used in prospecting for gold, in cleaning gold-bearing concentrates, and in the hand-working of very rich deposits. It is a shallow pan which varies from 15 inches to 18 inches in diameter at the top, and from 2 inches to 2½ inches in depth, the sides having a slope of about 30°. It weighs from 2 to 3 pounds. It is made of a heavy-gauge steel with the rim turned back over a heavy wire to stiffen it. Where amalgamating is to be done in the pan, it is either made of copper or has a copper bottom. When used by a skilled operator, it has a capacity of from half a yard to 1 yard in 10 hours.

The object of panning is to concentrate the heavier materials by washing away the lighter. To do this most efficiently, all material

[2] Boericke, William F., Prospecting and operating small gold placers, 2d ed., New York, John Wiley & Sons, Inc., 1941.

[3] Douglas, Jack, Gold in placer: published by Jack Douglas, Box 21, Dutch Flat, California, 1944.

[4] Symons, Henry H., The pan, rocker, dip-box, and sluice-box: California Jour. Mines and Geology, vol. 30, pp. 126-135, 1934.

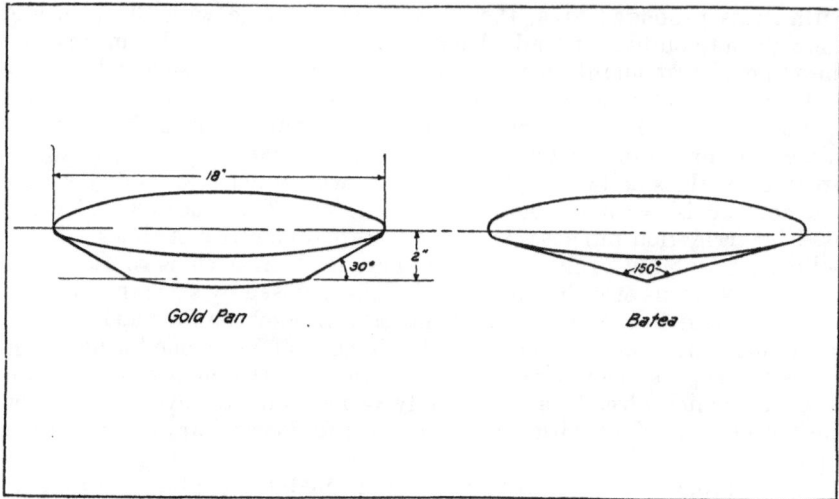

FIG. 2.   Gold pan and batea.   *Reprinted from California Journal of Mines and Geology, April 1932, p. 205.*

should be of as even a size as possible.   The pan is filled about three-quarters full of gravel to be washed, then it is submerged in water. First the large gravel is picked out by hand, then the clay is broken up, after which the operator raises the pan to the edge of the water, inclining it slightly away from him, moving it with a circular motion combined with a slight jerk, thus stirring up the mud and light sand and allowing it to float off.

This is continued until only the heavier materials remain, such as the gold, black sand and other minerals having a high specific gravity. These concentrates are saved until a large quantity accumulates, after which the gold is separated from them.   It may be picked out by hand, amalgamated with quicksilver, sometimes in a copper-bottomed pan. In some cases where the separation is extremely difficult and the quality and quantity justifies the concentrates are shipped to a smelter.   Panning may be best learned by watching an old-timer or experienced operator at work, learning certain tricks in the trade from him.   A clean 6- or 8-inch frying pan makes an excellent prospecting or clean-up pan. It is well to burn out an iron pan after having used quicksilver in it, and then polish it with a soft rock or piece of brick, otherwise it may be impossible to see small colors or flakes of gold.

The batea is cone-shaped and is the equivalent of the pan.   It is made of wood or sheet metal.   It varies from 15 to 24 inches in diameter and has an angle from 150° to 155° at the apex.   Many persons claim that wood will hold fine gold better than metal.   The batea is in common use in Mexico, Central and South America, and Asia.   A shallow wooden chopping bowl may be utilized as a substitute for the batea.   This would be used in the same manner as a pan.

Rocker

The rocker is a machine to save gold from auriferous sands and gravels by concentration (sometimes in conjunction with amalgamation).

Rockers vary greatly in size, shape, and general construction depending on ideas of builders in different localities and on their experience. Designs vary also because of different materials being available and because of variations in the sizes of the particles of gold to be recovered. Rockers vary in length from 24 to 60 inches or more, in width from 12 to 24 inches, and in height from 6 to 24 inches. Some have a single apron, and others two aprons and screens with holes as much as half an inch in diameter. A great variety of devices to recover the gold is found: riffles of all kinds, blanket, carpet, rubber mat, cocoa mat, canvas, cowhide, burlap, and amalgamated copper plate. The writer would suggest as a fairly efficient and easy construction of riffles for a rocker, to clamp ⅜-inch metal lath over a double thickness of blanket so that it can be easily removed for cleaning. Of all wet placer methods for saving gold, the rocker is one of the most economical on water for the amount of material handled. The average rocker when operated by two men has a capacity of about 3 to 5 yards in 10 hours, using 100 to 800 gallons of water.

Construction. Rockers are built in three distinct parts, consisting of a body or sluice box, a screen, and an apron. The floor of the body holds the riffles in which the gold is caught. The screen catches the coarser materials and is a place where clay can be broken up to free it of all small particles of gold. The apron is to carry all material to the head of the rocker, and is made of canvas stretched loosely over a frame. It has a pocket or low place on which coarse gold and black sands can be collected.

The accompanying drawing (fig. 3) gives a suggestion for a knockdown rocker that can be built by any one. The six bolts are removed to dismantle the rocker for easy transportation. The material required to construct it is given in the following tabulation with dimensions in inches:

A End, one piece 1" x 14", 16" long
B Sides, two pieces 1" x 14", 48" long
C Bottom, one piece, 1" x 14", 44" long
D Middle spreader, one piece 1" x 6", 16" long
E End spreader, one piece 1" x 4", 15" long
F Rockers, two pieces 2" x 5", 17" long
H Screen, about 16" square outside dimensions with screen bottom. Four pieces of 1" x 4", 15¼" long and one piece of screen 16" square with ¼" or ½" openings or sheet metal perforated with similar sized openings.
K Apron, made of 1" x 2" strips covered loosely with canvas. For cleats and apron, etc., 27 feet of 1" x 2" is needed. Six pieces of ⅜" iron rod 19" long threaded 2" on each end and fitted with nuts and washers.
L The handle, in the drawing is placed on the screen, although some miners prefer it on the body. When on the screen, it helps in lifting the screen from the body.

If 1- by 14-inch boards cannot be obtained, clear flooring tightly fitted will serve, in which case about 12 feet of 1- by 2-inch cleats in addition to that above mentioned will be needed.

A dipper made of a tomato can (no. 2½) and 30 inches of broom handle is also necessary. Through the center of each of the rockers a spike is placed to prevent slipping during operation. In constructing riffles, it is advisable to build them in such a way that they may be easily removed, so that clean-ups can be made more readily. Two planks about 2 by 8 by 24 inches with a hole in the center to hold the spike in the rockers, are also required. These are used as a bed for the rockers to work on and to adjust the slope of the bed of the rocker.

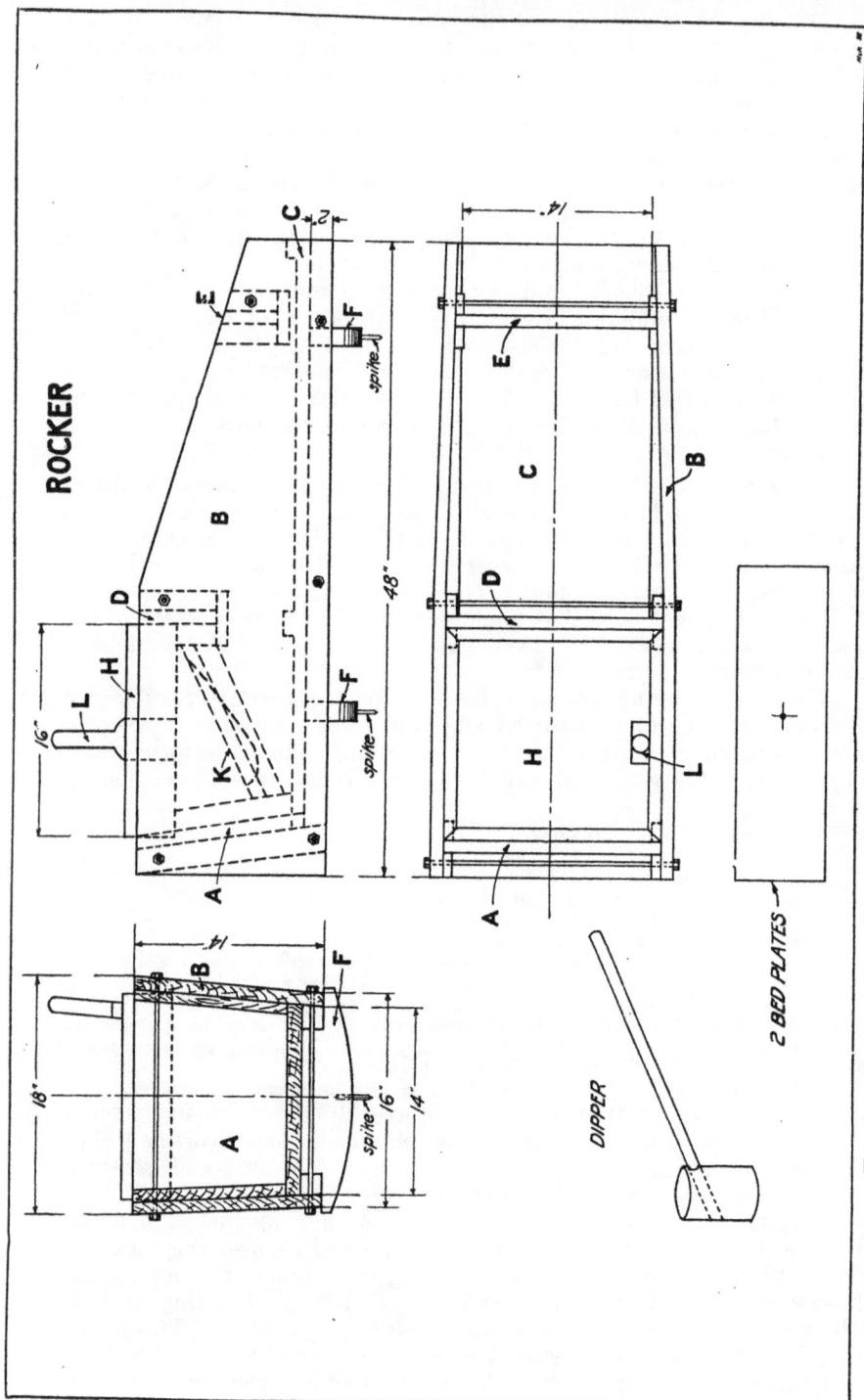

Fig. 3. Construction of knock-down rocker. *Reprinted from California Journal of Mines and Geology, April 1932, p. 206.*

ROCKER PARTS

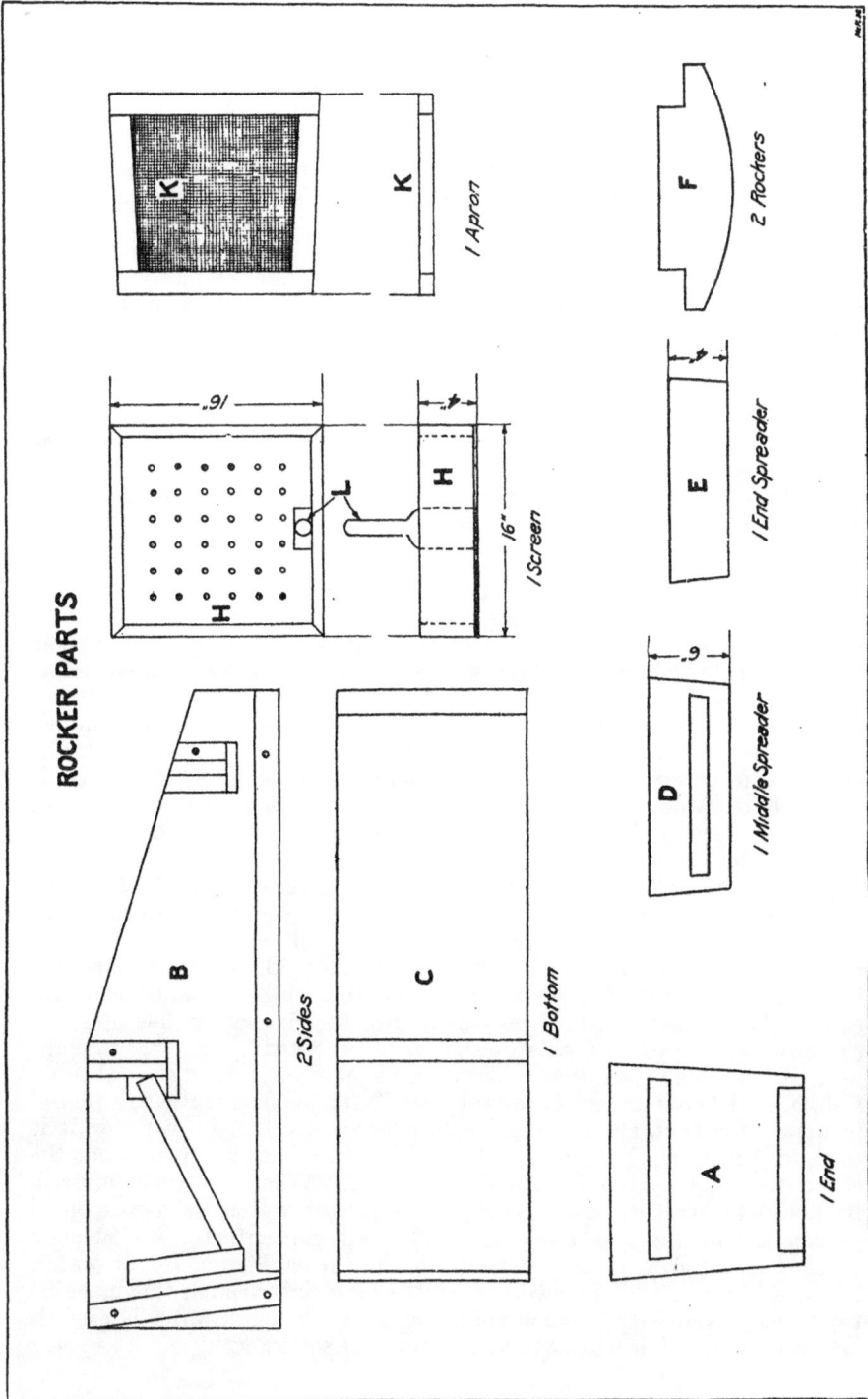

Fig. 4. Rocker parts. *Reprinted from California Journal of Mines and Geology, April 1932, p. 208.*

Assembly. The parts are cut to size as shown on the drawing, figure 4. The cleats on parts A, B, C, and D are of 1- by 2-inch material and are fastened with nails or preferably screws. The screen (H) is nailed together and the handle (L) is bolted to one side. Corners of the screen should be reinforced with pieces of sheetmetal because the screen is being continually pounded by the fall of rocks when the rocker is in use. The apron (K) is a frame nailed together, and canvas is fastened to the bottom. Joints at the corners should be strengthened with strips of tin or other metal.

Parts are assembled as follows: bottom (C), end (A) with cleats inside, middle spreader (D) with cleat toward A, and end spreader (E) are placed in position between the two sides (B) as shown in figure 3. The six bolts are inserted and the nuts are fastened. Rockers (F) should be fastened to bottom (C) with screws. Apron (K) and screen (H) are set in place, and the rocker is ready for use.

If one-quarter-inch lag screws are driven into the bottom of each rocker about 5 inches to each side of the spike, and if the head is allowed to protrude from the wood, a slight bump will be caused as the machine is worked back and forth. This additional vibration will help to concentrate the gold. If these are used, metal strips should be fastened to the bed-plates to protect the wood from wearing.

Operation. When the ground to be worked has been found, the miner picks a place near his source of water for his rocker. The first thing to do here is to set the bed-plates so that the spikes in the rocker fit in the hole in the plate and so that the floor has the proper slope. This slope is decided according to the ground to be worked. Where most of the gold is coarse and there is no clay, the head bed-plate should be 2 inches to 4 inches higher than the tail bed-plate; where most of the gold is fine or clay is present or a combination of both, this slope is lessened sometimes to only an inch. It is hard to save very fine gold if very muddy water is used, as the operation does not let the fine gold settle out but rather floats it off.

After the rocker is placed in position, the screen box is filled with gravel, which is washed off by pouring water over this material with the dipper. The larger gravel, when clean, is either picked out with a fork or by hand and all clay is broken up into a mud. Next the machine is rocked vigorously for several minutes, and water is added continuously. If all material that will pass through the screen has done so, the box is dumped and this operation is repeated until it is thought necessary to clean the apron. The apron should be cleaned several times a shift, as all coarse gold is caught there. The concentrates are placed in a pile for further cleaning. The riffles are cleaned whenever it is thought necessary, but not nearly as frequently as the apron, and the concentrates are saved for further cleaning. When a blanket is used, it should be washed out carefully in a tub of water, as here a good percentage of the fine gold is found. All concentrates are cleaned further in a pan. It is important to use the right amount of water. The use of too much water will carry the material through this machine too quickly, and with it much gold. When not enough water is used, it makes a mud which will not let the fine gold settle.

FIG. 5. A dip-box or short sluice, with iron screen for riffles.
*Photo by Thos. White; reprinted from California Journal of
Mines and Geology, April-July 1934, p. 124.*

Dip-Box

This is a modification of the sluice-box and may be used where
water is scarce and there is not enough grade for an ordinary sluice.
It is portable and may be carried in an automobile. It will permit
handling about as much dirt in a day as a rocker, though the larger
stones will have to be thrown out by hand.

The dip-box is simply a short sluice with a bottom of 1- by 12-inch
lumber to which are nailed sides of three-quarter-inch or 1 inch by 6
inches. The back end piece may also be of 1- by 6-inch stuff and the
lower end piece 1 or $1\frac{1}{2}$ inches high. To catch the gold, the bottom of
the box may be covered with burlap, canvas or thin carpet. Over this,
beginning 1 foot below the back end of the box, may be laid a strip
of heavy wire screen of quarter-inch mesh (made from no. 13 or no. 14
wire) 1 foot wide by 3 feet long. Burlap and screen may be held in
place by cleats along the sides of the box. The dip-box may be made
6 to 8 feet long. Those who use it often claim that practically all the
gold will be saved in the first 3 feet. The box is given a steep grade
by being set on small trestles, the one near the head end being about

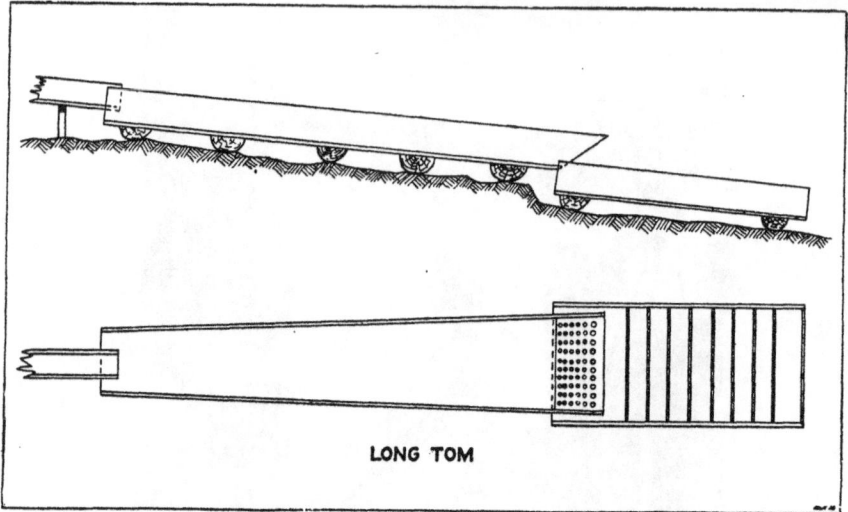

FIG. 6.   Long tom.   *Reprinted from California Journal of Mines and Geology, April-July 1934, p. 132.*

waist high and 6 inches to a foot higher than the lower one, near discharge end.   The dip-box is used by dumping the sand and gravel, a small bucketful at a time, into the back end of box, then pouring water from a dipper, bucket or hose onto it until it is washed through the box, discharging over the lower end.   The gold will lodge mostly in the screen.   Riffles may be put in the lower part of the box to stop gold passing the screen.   Water should not be poured too violently into the box.   The larger stones must be thrown out by hand, unless the box is fitted with a hopper, or a screen.

Puddling Box

Where muddy or clayey material is to be sluiced, the first box of the string can be made into a "puddling box."   This can be 3 feet wide by 6 feet long, or any convenient dimensions, with 6-inch or 8-inch sides, and no riffles.   The clayey material can be shoveled into this box and broken up with a hoe or rake before it passes into the main sluice. Lumps of clay in a sluice may pick up and carry away the gold particles.

Long Tom

A long tom is an inclined trough used to concentrate auriferous earths and gravels.   It has a greater capacity than a rocker, but also uses more water in operation, because the water is the carrying agent of the finer materials.   The long tom is usually of crude construction, being built in two sections, the sluice-box, and the riffle-box (fig. 6). The slope is generally 1 inch to each foot in length, but this is varied as conditions warrant.   The sluice-box is usually about 12 feet long and about 15 to 24 inches wide at the head or upper end, and 24 to 36 inches wide at the tail or lower end, and sides are about 8 inches high at the tail.   A screen or piece of perforated sheet metal prevents the coarse material from going to the riffle-box, and at the head end is a flume or iron pipe from which the water is fed.   The riffle-box is usually shorter than the sluice-box, and slightly wider than the latter at the tail end. It begins just below the first opening in the screen, some-times has a more gentle slope. Here the riffles are placed to catch the gold. The box is often lined with canvas as in the rocker, and it is best to build detachable riffles. The sluice-box should be made of 2-inch lumber to withstand the abrasion of the gravel. The capacity of a long tom is from 4 to 6 yards in a 10-hour day, per man, two to four men working.

Operation. The ground to be worked is shoveled into the sluice-box and washed by the water coming from the head end. One of the men will work the material in the trough with a fork, taking the coarser gravel out when washed clean and keeping the screen from clogging. Clean-ups are made when necessary, usually at the end of the day, but experience might show that they should be made oftener.

Sluicing

Sluicing is a method of working auriferous gravels in a flume called a sluice-box, or in a ditch, and the method is then called ground sluicing. The sluice-box is a crude sloping flume or trough, having riffles on the bottom to catch the gold. Dimensions vary greatly and are governed by the amount of material to be washed through the sluice. The slope varies from 5 to 18 inches in a 12-foot length. The riffles also vary, sometimes there are several kinds in a single sluice, some of which are quite elaborate and require considerable work in laying.

In the rocker and long tom, all the coarse materials are removed, but in the sluice all is allowed to pass through, or in some cases a grizzly is placed at the head of the sluice-box to catch the very coarsest of material, allowing much heavier gravel to enter than in the other devices previously described. This coarse material serves to grind and polish the gold, thereby cleaning it and making it easier to amalgamate and possibly freeing some material mechanically held.

In sluicing, much of the manual labor done in the preceding methods is eliminated, as the water does all of the carrying of the material. In some cases the mining is done by hydraulicking or a stream of water is allowed to fall over a bank and in that way wash the material to the sluice. Sluicing requires more water than the methods previously described, the amount depending upon the material to be washed, and varying from 20 to 80 cubic feet of water to move 1 cubic foot of gravel. Coarse gravel requires more water than fine, but as the grade is increased the amount of water required is lessened. The capacity of the sluice box is governed by its grade and amount of water available as well as its dimensions. In ground sluicing a ditch is dug along bed-rock and natural irregularities in the bottom furnish pockets which catch the gold.

Riffles. Riffles are obstacles placed along the bottom of a sluice which form pockets to catch gold by concentrating the heavier materials. Numerous forms of riffles with innumerable modifications have been devised. Some of the best known are described in the following para-graphs:

Common riffles or slat riffles are strips of wood, iron, or steel extend-ing across the sluice box. The abrasion is so great on wooden riffles that replacement is required often and therefore other types of riffles are preferred in large-scale operations.

Pole riffles are frequently used. These are 2- to 4-inch peeled poles placed either across or lengthwise of the sluice box. This type is used

FIG. 7. Bodinson sampling machine. *Photo by courtesy of Bodinson Manufacturing Company, San Francisco.*

with coarse material and is efficient in concentrating both coarse and fine gold.

Block riffles are made by paving the floor of the sluice-box with wood blocks cut across the grain, and four inches or more high depending on the depth and width of the sluice. They are nailed to narrow slats on the end that is to rest on the bottom of the sluice. The slats are nailed to the sides of the blocks, so that a space is left between the rows of blocks at the top. Spaces between the rows of blocks form the riffles. The blocks may be made either square or round. This method is good for both coarse and fine materials.

Rock or stone riffles are made by paving the floor of the sluice with rock, either stream pebbles or flat stones quarried for the purpose. They are held in place by strips of wood nailed across the bottom at intervals. This method is good for both fine and coarse material, and **extra** good for cemented gravel.

Zig-zag riffles are slats placed part way across the floor of the sluice box alternately from the sides.. This type is good for fine material and concentrates in a similar way to panning.

An undercurrent is a wide flat sluice placed beneath the main sluice box and is used for the purpose of saving the fine gold. It is usually 5 to 20 times as wide as the main sluice and from 10 to 50 feet long. It receives its feed from a grizzly or screen placed in the floor of the main sluice-box, from which the fine material drops into a trough, which distributes the feed evenly across the whole width. Undercurrents usually have a greater slope than the main sluice, because the shallow stream is retarded more by friction.

Small-Scale Placer Machines

A few small-scale machines for recovery of gold from placer gravels are described below because they are believed to be valuable for sampling large deposits to determine whether the gold-content is great enough to justify the use of a dragline dredge, bucket-ladder dredge, hydraulic equipment or other expensive machinery. The trend in sampling is toward the use of such labor-saving machines instead of the rocker and other hand-operated devices described above. Additional machines, some of them much different in design from those mentioned below, have been described in the California Journal of Mines and Geology.[5]

Bodinson Manufacturing Company, 2401 Bayshore Boulevard, San Francisco, California, has made, on special order, machines for sampling placer gravels. The following equipment was included: revolving screen 30 inches in diameter by 50 inches long perforated with three-eighths-inch holes; 2-inch or 2½-inch heavy duty pump; and 2½-hp., 1400 rpm. Novo single-cylinder gasoline engine. The engine was arranged to drive the other machines through belts, chains, and sprockets. All machines were mounted on a steel frame, and some were mounted on rubber tires to make a two-wheel automobile trailer.

Denver Mechanical Gold Pan is made by Denver Equipment Company, 1400 Seventeenth Street, Denver, Colorado. This machine has a motion somewhat similar to that used in hand-panning, imparted by an eccentric which makes 240 complete oscillations per minute. Power for both the eccentric and a pump to furnish water is supplied by a gasoline engine of ¾-hp. Gravel is placed in a hopper 2¼ feet above the base of the machine and passes over an upper screen with water from a spray-pipe. The upper screen is a heavy punched plate which passes ¼-inch material. The hopper is provided with a lip to hold back large nuggets. The ¼-inch material passing through the coarse screen goes to a fine screen below, and fine sizes pass through to three concentrating pans below. The top pan is of copper and is used to amalgamate fine gold with quicksilver. Overflow passes to the two lower pans, which are provided with rubber mats covered with heavy wire screen of 1-inch mesh to act as riffles. Manufacturers state that the machine is very efficient in recovering both coarse and fine gold. Capacity of a single machine is 1½ to 2 cubic yards bank run per hour. The manufacturer supplies small trommels to mount over either one or two of the mechanical pans. Capacity of the duplex pan plus trommel is stated to be 4 to 6 yards per hour.

Denver Trommel-Jig Unit is made by Denver Equipment Company, 1400-17th Street, Denver, Colorado. A trommel, jig, gasoline engine, and pump are provided. The trommel contains a scrubber-section with spiral lifting blades to disintegrate clay or cemented gravel. Rated capacities of three different models range from 2 to 6 cubic yards bank run per hour. The jigs are 8 by 12 inches or 12 by 18 inches and engines are 3 hp. or 4 hp. All machines are mounted on steel frames. As a separate unit, an amalgamation barrel is available for amalgamating the gold in the jig-concentrate.

[5] Laizure, C. McK., Elementary placer mining in California, special machines and processes: California Jour. Mines and Geology, vol. 30, pp. 136-227, 1934.

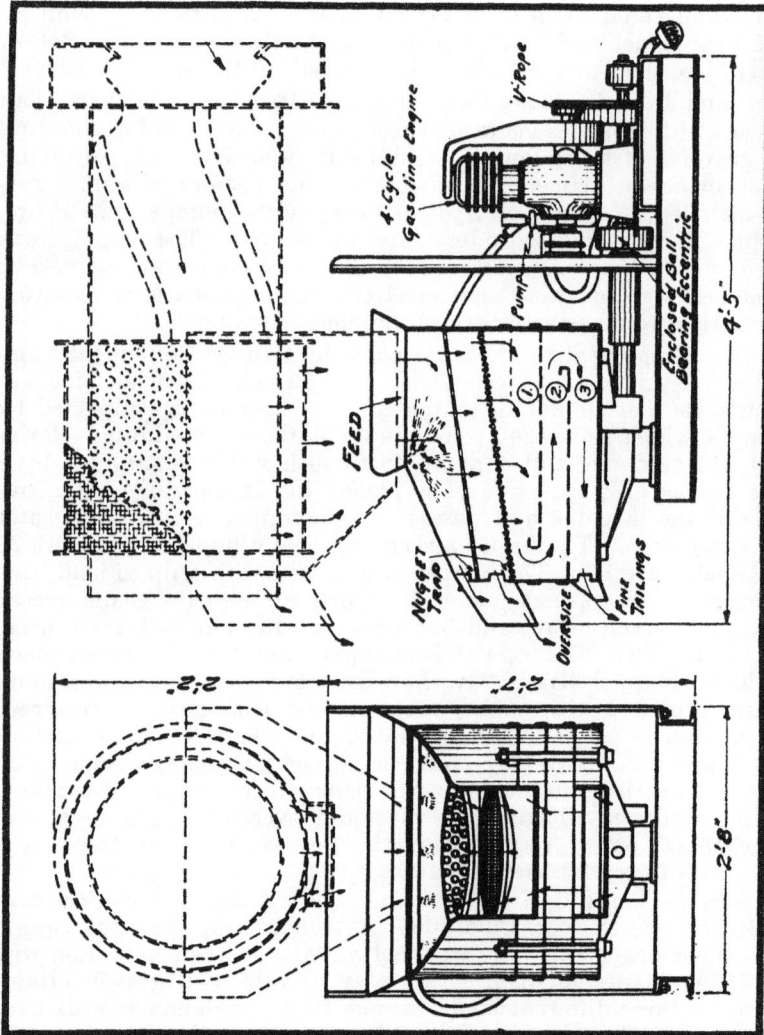

FIG. 8.   Denver Mechanical Gold Pan, single unit with trommel. *Courtesy of Denver Equipment Co., Denver.*

FIG. 9. Denver Mechanical Gold Pan, duplex unit with trommel. *Photo by courtesy of Denver Equipment Company, Denver.*

G-B Portable Placer Machine is furnished by The Mine and Smelter Supply Company, Denver, Colorado. It consists of a hopper, combined scrubber and revolving screen, and molded rubber riffles, which are vibrated at 200 strokes per minute. Tanks are provided for re-use of water, but 60 to 75 gallons of water per cubic yard of gravel are discharged with the tailing, and that amount of make-up water must be provided. Rated capacity on ordinary gravel that contains little clay and is not cemented is about 2 cubic yards per hour. Power is furnished by a 1½-hp. gasoline engine, which drives the scrubber-trommel, pump, and riffles.

Further details about all of the above machines are available from the manufacturers.

DRAGLINE DREDGING

The importance of dragline dredging in California is brought out by table 1, which applies to dragline dredging exclusively. It shows the increasing importance of production from dragline dredges for pre-war years. The sharp decline in 1942 was caused by war conditions. Not only did the War Production Board prohibit gold mining except in special cases, but practically all of the draglines were put on war work both as excavators and as cranes. Manufacturers of dragline excavators must have time to replace these machines before dragline dredging can be resumed on a large scale after the war. The method was in an early stage of development in 1933, but the machinery and methods were rapidly improved, so that gross production rose to a peak of nearly $8,000,000 in 1941.

The name dragline dredge is used in this article to denote a placer mining outfit composed of two separate and distinct units. The digging is done by a standard make of dragline excavator, which travels on the ground by means of caterpillar tracks under its own power. The heavy bucket, which picks up from 1 cubic yard to 3 cubic yards of gravel at one time, is suspended by a steel cable from a structural-steel boom roughly 50 feet in length. Still larger outfits were in use shortly before gold mining was shut down by order of the War Production Board. At the Mocassin mine in Siskiyou County, a dragline excavator of the Monighan type with a 5-cu. yd. bucket was in use, and at Dayton, Nevada, one with a 14-cu. yd. bucket. The one at Dayton did not operate long enough to give a satisfactory demonstration of the performance of an outfit of this size. Washing of the gravel is accomplished on the second unit, which is a barge floating in a pond. For washing out the gold, the barge carries a revolving screen and riffle tables similar to the units used on the bucket-ladder dredges. The dragline excavator digs away at the edge of the pond, which thus advances. To cause the barge to follow, a pull on cables anchored on the shore is all that is needed. The tailing discharged from a belt-conveyor and sand-sluices fills up the pond behind the barge. The dragline dredge has been called a "doodle bug" by many persons, but this is not considered an appropriate name, and it is not used here.

The older type of dredge, on which the digging is done by means of a bucket-elevator comprising a chain of heavy buckets, each of which is connected by a round pin to the next one, will be called here a bucket-ladder dredge. The ladder is the heavy structural steel member that supports the bucket-chain. This type of dredge is described in a later chapter, *Bucket-Line Dredging.*

It is not the purpose of this article to indicate that the dragline dredge is in any way superior to the bucket-ladder dredge. The dragline dredge has opened up a new field to dredging, namely those deposits that are too small to justify the construction of a bucket-ladder dredge. If a deposit is large enough and contains enough gold to amortize the capital investment in a large bucket-ladder dredge, and return a suitable profit, possibly a dragline dredge should not be considered. However, the large sizes of dragline dredges are considered by some operators to be at least as good as the smaller bucket-ladder dredges.

Bucket-ladder dredges have been made portable to a certain extent, and may be used on more than one deposit, but the dragline has the

Table 1. Gold production from dragline dredges, 1933-1943 *

| | Mines producing | Washing plants | Cubic yards washed | Gold recovered | | |
				Fine ounces	Value	Average per cu. yd.
1933	3	3	11,500	75	$1,924	$0.167
1934	4	4	604,000	3,466	121,138	.201
1935	24	23	3,906,000	22,191	776,701	.199
1936	30	26	10,016,000	49,968	1,748,864	.175
1937	51	47	19,364,000	94,142	3,294,970	.170
1938	77	68	24,560,000	118,108	4,133,780	.168
1939	142	109	31,618,000	172,519	6,038,165	.191
1940	198	106	42,747,000	205,181	7,181,335	.168
1941	234	112	45,579,000	225,019	7,875,665	.173
1942	122	79	24,526,000	117,906	4,126,710	.168
1943	3	3	3,180,000	14,196	496,860	.156

* Extracted from Merrill, C. W., and Gaylord, H. M., Gold, silver, copper, lead, and zinc in California: U. S. Bur. Mines, Minerals Yearbook, Review of 1940, p. 219. See also preprints for 1941, 1942, and 1943.

advantage in this regard.   The operating cost per cubic yard is roughly the same on the smallest bucket-ladder dredges as it is on the largest dragline.

The dragline dredge has the following disadvantages:

1. The usual depth to which they have worked in California is roughly 20 feet.   This can be extended somewhat with the largest dragline excavators with very long booms such as the one at the Moccasin mine in Siskiyou County described later in this bulletin.
2. They will not dig gravel that is hard and compact or partly cemented as well as a bucket-ladder dredge.
3. Bedrock must be soft.   No dredge is successful where bedrock is very hard and irregular, but a bucket-ladder dredge will dig harder rock than a dragline.

Subject to favorable conditions regarding depth, ease of digging, and soft bedrock, dragline dredges are successful on deposits too small for bucket-ladder dredges for the following reasons:

1. Less capital is needed to purchase the dragline excavator and washing-plant.
2. The dragline dredges are smaller and float in very shallow ponds because the heavy digging-machinery is not on the barge.
3. The dragline excavator and the tractor with 'bulldozer' blade, which is now practically a standard item of equipment, can quickly throw up small dams so that the barge can be placed on various terraces and in small tributaries higher than the main channel.   If necessary, water for the pond is pumped.
4. When one small deposit has been worked out, the modern outfit with barge of steel-pontoon construction can be quickly moved to another deposit.   Such a move involving dismantling and re-erection has actually been made in a week's time with the regular crew.

Logan[1] and Magee[2] have written articles on dragline dredges. Since those articles were written, washing plants have been much improved, larger units have been put in service, and cost per cubic yard has been much reduced.

A few details of the geology of the dragline field southwest of Redding will be given because conditions are nearly ideal for this type of dredging. Gravels being dredged (1940) are in the channels of present streams and on low terraces adjacent to the present channels. The gravel is seldom more than 10 feet in depth, and most of it is loose enough so that it is not difficult to dig.

Beneath the gravels of the present streams are sediments of Tertiary and Cretaceous age, all of which form soft bedrock that the dragline buckets can dig. Several inches to a foot of it are usually taken up to recover gold lying on bedrock. To the west of the Pacific Highway for a distance of 10 to 15 miles, the Tertiary bedrock is a clay-like volcanic tuff dipping below horizontal at small angles to the east. Gravels of the Pleistocene Red Bluff formation overlie the tuff in large areas, and they should not be confused with gravels of present streams. Apparently no concentration of gold occurs in these widespread Red Bluff gravels. In the vicinity of Gas Point, the bedrock changes from Tertiary on the east to Cretaceous formations toward the west. The Cretaceous dips east at a steeper angle, roughly 20°. It comprises shales, sandstones and conglomerates in general harder than the Tertiary tuff, but a layer near the top is decomposed and is soft enough for easy digging.

The gold has no doubt been carried over these sedimentary formations from an origin in the igneous rocks, schists, and older sediments to the north and west. Clear Creek is one of the principal streams and it passes through the French Gulch[3] district, well known for its rich quartz veins. Erosion of these has unquestionably contributed gold to the placer deposits. In the vicinity of Igo is a deposit of gravel covering many acres to depths reaching 100 feet. It is apparently an old terrace of Clear Creek, now high above the present stream. Part of it has been mined by drifting and hydraulicking. Part of it has not yet been mined. Dry Creek and its tributaries, now (1940) being extensively dredged with draglines, dissect the old Clear Creek terrace, and gold has been carried out by Dry Creek and over Cretaceous and Tertiary bedrock. Hence the placer gold of Dry Creek is derived largely from an older placer deposit.

Some persons have thought of the Cretaceous conglomerates as a possible source of the gold, and it is possible that some of the beds of Cretaceous conglomerate contain gold. However, an examination of the boulders in the placer deposits shows that many are larger than those found in the conglomerates, and that they have apparently been washed in by streams originating in the igneous rocks and schists, and in the older Bragdon conglomerate (Carboniferous). The bulk of the gold must have been washed along with them. Quartz veins in the Bragdon conglomerate are gold-bearing at French Gulch.

[1] Logan, C. A., Placer mining in California with power shovels: California Jour. Mines and Geology, vol. 32, pp. 373-377, 1936.
[2] Magee, J. F., A successful drag-line dredge: Am. Inst. Min. Eng. Tech. Pub. 757, pp. 1-16, 1936.
[3] Averill, C. V., Gold deposits of the Redding and Weaverville quadrangle: California Jour. Mines and Geology, vol. 29, pp. 3-73, map, 1933.
Hinds, N. E. A., Geologic formations of the Redding-Weaverville districts, northern California: California Jour. Mines and Geology, vol. 29, pp. 77-122, map, 1933.

Dragline Excavators

Dragline excavators of such standard makes as Bucyrus-Erie, Lima, Link Belt, Koehring, Marion, Northwest, P. & H., and Thew-Lorain are in use for dragline dredging. Details of various sizes, speeds and horsepower may be obtained from the manufacturers. Thoenen[4] has tabulated some of these data in Information Circular 6798 of the U. S. Bureau of Mines. Fairly high digging and swinging speeds are desirable for this type of work, and hence fairly high horsepower. Most of the draglines in California were equipped with 1¼-cu. yd. and 1½-cu. yd. buckets, but those of 3 cubic yards capacity more recently put in service give a lower operating cost per cubic yard. Some still larger ones have been used but detailed operating costs on them are not available. Probably additional outfits in these large sizes will be developed in post-war years. The 1½-cu. yd. draglines have 50-foot booms, and the 3-cu. yd. draglines have 60-foot booms. Different lengths are obtainable if they are needed to fit different conditions.

Buckets

Both Page and Esco buckets have been used. The Esco with five teeth will dig harder gravel than the Page, but it dumps more slowly. A set of teeth is usually dulled each shift, and must be built up by welding.

Power

The dragline excavator with 1½-cu.yd. buckets for which cost-data are given below are powered by D-13000 Caterpillar diesel engines rated at a maximum of 130 horsepower. The 3-cu.yd. dragline excavator is powered with a 200-hp. electric motor.

Digging Methods

Two general methods of digging are in use. The common method is to move the dragline excavator in the direction of the channel, and reach to each side as far as possible with the boom. Each cut is twice as wide as the horizontal projection of the boom, roughly 60 feet. By utilizing the momentum of the swing, the operator can cast the bucket a little beyond the end of the boom. The other method is to move the dragline excavator across the channel, thus placing the caterpillar tracks at right-angles to the direction of the digging-cable. Wider cuts are possible with this method, and its advocates state that bedrock is cleaned better. This seems reasonable, because in the method mentioned first the arc through which the bucket moves causes a strip of bedrock to remain uncleaned toward the extreme reach of the boom. This can be avoided to a certain extent by overlapping the cuts, but the digging is done under muddy water, and accuracy of this overlap is difficult to attain.

Mats

The dragline excavator can usually travel on the ground in dry weather, but when the ground is muddy or very sandy, mats are needed. These are made by bolting together timbers, about 8 by 10 inches, in sections 4 feet wide and somewhat longer than the width of the tread of the dragline excavator. The boom is used as a crane to pick these up behind the caterpillar tracks and put them down in front.

---

4 Thoenen, J. R., Sand and gravel excavation, Part I: U. S. Bur. Mines Inf. Circ. 6798, pp. 23-39, 1934.

FIG. 10. Dragline dredge under construction in shop, showing trommel and parts of riffle-sluices and stacker-ladder. *Photo by courtesy of Bodinson Manufacturing Company; reprinted from California Journal of Mines and Geology, April 1938, p. 102.*

Tractors

A tractor of the caterpillar type powered by a diesel engine is now practically a standard item of equipment in both dragline dredging and bucket-ladder dredging. It is usually equipped with a scraper or bulldozer blade in front and often has a winch mounted in back. The principal use is for clearing the land of brush and trees. These are either pushed or pulled to one side or piled for burning. Many jobs of handling heavy parts are possible with the tractor, and it is useful in building dams for some locations of the dredge-pond. In dragline dredging, the tractor and bulldozer are particularly useful for smoothing the way ahead of the dragline excavator, so that the latter can be moved ahead with a minimum of lost time. The tractor and Le Tourneau carryall have been used in a few places to remove several feet of soil overburden containing no gold.

Washing-Plants

The washing-plant for a dragline dredge is mounted on a barge, and consists of a hopper into which the gravel is dumped by the dragline, a revolving screen or trommel, and a belt-conveyor to stack the coarse tailing behind the barge. Large streams of water are pumped from the pond into both the hopper and the trommel. The sands that pass through the screen are washed on inclined tables, which are divided by partitions into a number of sluices containing riffles to retain the

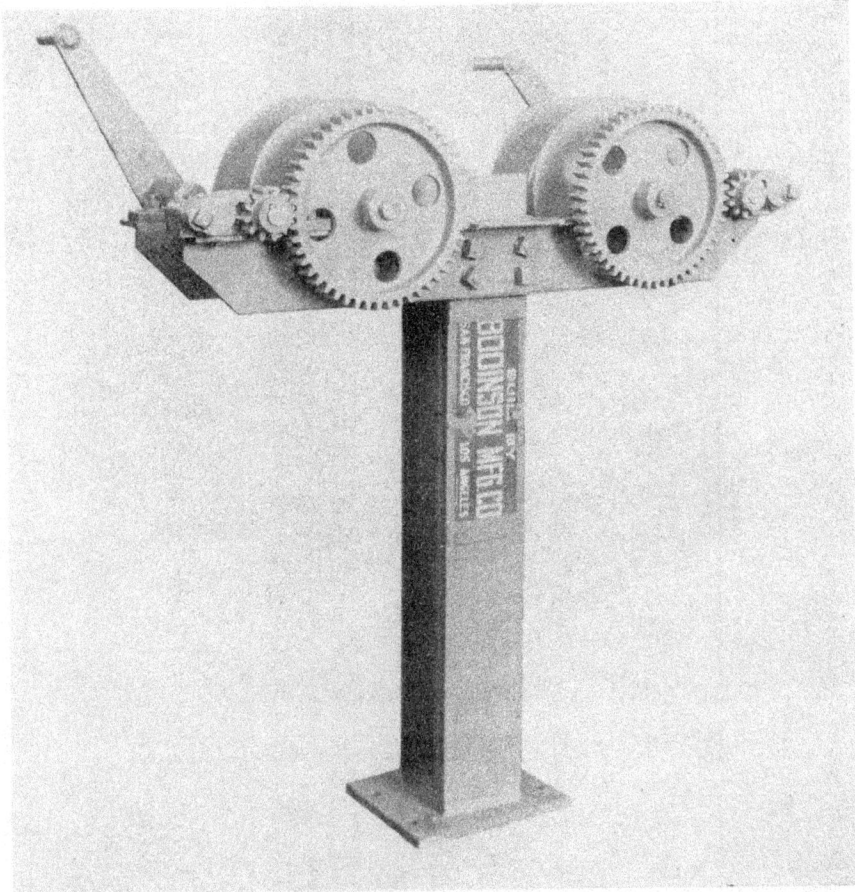

Fig. 11. Hand-winch for dragline dredge. *Photo by courtesy of Bodinson Manufacturing Company; reprinted from California Journal of Mines and Geology, April 1938, p. 103.*

gold. The washed sand flows into the pond behind the barge. The following descriptions of details have been generalized somewhat to cover practice in the state, but they are given with the particular plants in mind for which cost-data are tabulated below. The all-steel plants are made by Bodinson Manufacturing Company, 2401 Bayshore Boulevard, San Francisco. Welded joints are used throughout. Even the corrugated iron housing is tack-welded to the steel frame.

Hulls

The barge for a $1\frac{1}{2}$-cu.yd. outfit is 30 feet by 40 feet and is made of five pontoons, each 8 feet by 30 feet by 42 inches deep. For the 3-cu.yd. outfit, it is 35 feet by 48 feet by 42 inches deep, and comprises six pontoons, each 8 feet by 35 feet. Steel is $\frac{3}{16}$-inch thick, and all seams are electric-welded. Well braced frames for pontoons are made of $2\frac{1}{2}$- by $2\frac{1}{2}$- by $\frac{3}{16}$-inch angles. The earlier barges were made of timber frames covered with 3-inch planks, but these are now considered obsolete and are not used.

FIG. 12.   Power-winch for dragline dredge. *Photo, courtesy of Bodinson Manufacturing Co.; reprinted from California Journal of Mines and Geology, April 1938, p. 104.*

FIG. 13.   Trommel for dragline dredge on truck and trailer. *Reprinted from California Journal of Mines and Geology, April 1938, p. 105.*

Winches

The barge is pulled ahead and swung to distribute tailing by means of cables anchored ashore and attached to winches on the barge. Hand winches are used on the smaller outfits and power-winches on the larger ones. On a plant serving a 3-cu.yd. electric excavator the winch is driven by a 3-hp. electric motor.

Hopper

A heavy hopper usually made of half-inch steel plates welded together receives the gravel dumped from the dragline bucket. A grizzly of 90-pound steel rails spaced at 16-inch centers prevents large boulders from entering the trommel. An effort is made to lay aside with the dragline any boulders that will not pass through this grizzly. On some washing plants the grizzly is inclined downward slightly toward the front of the barge, and boulders are dragged back into the pond with the bucket. Water is discharged from nozzles into the hopper. On the 3-cu.yd. outfit, the hopper is 14 feet by 10 feet and is 13 feet $11\frac{1}{2}$ inches above the deck.

Trommels

Details given here are for the plants on which cost-data are given below. Different sizes of holes and different spacing can be used as required by the particular deposit being worked. For the $1\frac{1}{2}$-cu.yd. outfits, trommels are 24 feet long by 54 inches in diameter. Two end sections of 4 feet each are not perforated. Other sections of 4 feet each are perforated as follows: first, $\frac{3}{8}$-inch holes with $1\frac{1}{4}$ inches of metal between; second, $\frac{3}{8}$-inch holes with three-quarters of an inch of metal between; last two, $\frac{1}{2}$-inch holes and half an inch of metal between. They turn at 14 revolutions per minute. On the 3-cu. yd. outfit, the trommel is 35 feet long by 5 feet in diameter. End sections of 5 feet each are not perforated. The remainder of 25 feet is perforated with $\frac{3}{8}$-inch holes, but the spacing varies in the 5-foot sections as follows: on the first section $1\frac{1}{4}$ inches of metal between holes; second section, $\frac{3}{4}$-inch; last three sections, $\frac{3}{8}$-inch. The different spacing of the holes is to distribute the fine gravel evenly to the riffle-sluices or tables below. The speed of rotation is 12 rpm. A pipe drilled with $\frac{3}{8}$-inch holes extends through the trommel, and water is sprayed from it to wash the gravel.

The intermittent loads dumped into the hopper cause surges of gravel through the screen and sluices. To equalize the flow, some trommels have been equipped with an Archimedean screw of one or more turns in the upper blank or scrubber section. This helps to break up lumps of clay and to feed the gravel more evenly into the perforated sections of the trommel.

On the older outfits, the metal housing around the lower half of the trommel ended a few inches above the riffle sluices, and water and sand dropped directly on the riffles. On the Bodinson washing plants, the trommel housing is carried several inches below the level of the riffles into a narrow, depressed steel box running the full length of the trommel. This is provided with baffles or weirs to regulate the flow to the different sluices. It serves also as an effective trap to retain coarse gold. To recover this at cleanup time, large pipe-plugs in the bottom are unscrewed.

FIG. 14.   Dragline dredge showing trommel, riffle-sluices, pump, and pump-screen.
Reprinted from California Journal of Mines and Geology, April 1938, p. 106.

Riffle-Tables

On the 3-cu.yd. outfit are 10 sluices with riffles on each side of the trommel.   They are all 30 inches wide, twelve are 14 feet long, and eight are 11½ feet long.   They discharge into a pair of sluices of the same width on each side of the barge, running lengthwise of the barge to discharge at the stern.   The lower portions of these sluices are provided with riffles also.   In the upper portions, where the sluices running crosswise of the barge discharge into them, too much turbulence exists for riffles to be effective.   The trommel and all sluices are set at a grade of 1½ inches to a foot.   Some designers use 1¼ inches to a foot.   On the smaller barges for 1½-cu.yd. draglines, the arrangement is the same, except that dimensions are reduced to correspond with those of the barge.

Riffles are of the Hungarian dredge-type of wood, 1¼ inches deep, ¾-inch wide, spaced at 1¼ inches.   They are made up in sections of a length equal to the width of the sluice, and about a foot along the direction of flow.   These small sections are easily handled during the cleanup. The top of the wood is beveled off for an eighth to a quarter of an inch, so that the top is nearly level when the riffle is in the sluice.   It is shod on top with strap iron, 1 inch by ¼-inch, held in place with countersunk wood screws.   On the Roaring River dredge rubber is substituted for iron.   But the rubber and the iron are wider than the wood beneath, and overlap the wood a little on both edges.

Most operators use expanded metal lath of 1-inch mesh over burlap, coconut matting or English corduroy in the upper half of the sluices running crosswise of the barge, that is just beneath the trommel.   The metal lath is raised with tongue-and-groove flooring so that the top is even with top of the riffles in the lower part of the sluice.   Quicksilver is sprinkled on this at the start of each shift, and the metal lath holds the quicksilver close to the under side of the flow of sand and water, where it is more effective in amalgamating the gold than in the deeper riffles.

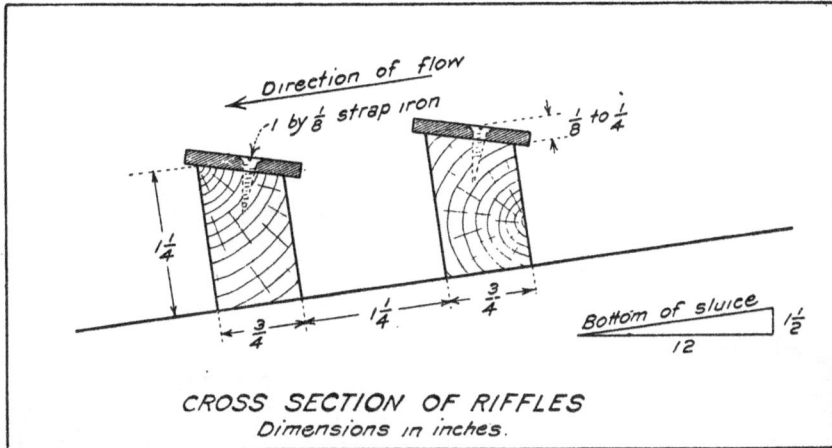

FIG. 15. Cross-section of dredge riffles. *Reprinted from California Journal of Mines and Geology, April 1938, p. 106.*

Stackers

For stacking the coarse tailing behind the barge, the 1½-cu.yd. plant is equipped with a belt-conveyor system, 50 feet long between pulleys, with a 30-inch belt. The stacker for the 3-cu. yd. plant is 50 feet long with a 36-inch belt. Some of the boats with diesel power have an electric generator and motor so that the stacker can be driven by the upper pulley. One plant has the upper pulley driven by a shaft running the full length of the stacker.

Power

On one of the 1½-cu.yd. outfits for which cost-data are given below, power is furnished by a D-7700 Caterpillar diesel engine rated at 50 hp. on continuous sustained loads or 63 hp. maximum; on the other by a D-8800 Caterpillar diesel engine rated at 64 hp. and 80 hp. respectively. Electric lights are furnished by 2000-watt Koehler plants.

Power on the 3-cu.yd. outfit is furnished by the following electric motors: 50 hp. on pump, 30 hp. on trommel, 10 hp. on stacker, 3 hp. on winch and 5 hp. on auxiliary pump.

Water

During the first half of the year, for which cost figures are given below, practically all water needed was obtained from the natural flow of the streams. During the dry season, impounded water bought from a company which furnishes water primarily for irrigation may cost $500 per month total for all three outfits.

Water for washing the gravel on the barges is pumped from the pond on the 1½-cu.yd. outfits with a 7-inch centrifugal pump. The 3-cu.yd. plant has a 10-inch pump discharging into the hopper and trommel; also a 4-inch auxiliary pump to supply additional water to the sluices. The 4-inch pump is used to furnish water for cleaning up the riffles. The proportion of water and sand is variable according to the character of the ground being mined. The mixture of sand and water discharged at the stern of the barge is roughly 10 to 15 percent solids.

FIG. 16.   Dragline dredge under construction in field.   *Reprinted from California Journal of Mines and Geology, April 1938, p. 107.*

FIG. 17.   Dragline dredge in an early stage of construction.   *Reprinted from California Journal of Mines and Geology, April 1938, p. 113.*

Delays

The plants are kept in operation 24 hours per day. "Operating hours" listed below include only that time in which the dragline was digging gravel and delivering it to the hopper on the barge. All other time is counted as delays. These include time for moving, for lubricating and servicing the dragline and other machinery, and for repairs and cleanups.

For a move to a new location involving dismantling of equipment, 7 days are required for the 1½-cu.yd. outfit, and 8 days for the 3-cu.yd. outfit. The regular crew of roughly 14 men is used in either case. As extensive replacements of worn parts are usually made at this time, an accurate estimate of the cost of such a move is not available. Parts and cost of installing them should be charged to maintenance and not to moving. In such a move the dragline is used as a crane to pick up a pontoon or other heavy part and load it on a truck and trailer. It is interesting to note that $1,000 should be ample to cover the cost of dismantling and re-erection when the length of the truck-haul is moderate.

Merrill [5] in Peele's Mining Engineers' Handbook gives a table of actual costs of six such moves ranging from $1,470 to $1.656. Transportation, partly by truck and partly by rail, was the largest item in each case, and ranged in cost from $900 to $1,200. Distances ranged from 50 to 300 miles.

Cleanups

Cleanups are probably made on the average of about once a week. Some operators could no doubt improve their recovery by watching the condition of the riffles more closely and cleaning up when the riffles are loaded instead of at regular intervals. One operator who uses expanded metal lath over burlap near the trommel cleans up the lath after every 80 hours of running time, and makes 80 to 90 percent of his total recovery in this way. The metal lath is taken up, then the concentrate on the burlap is hosed off into a tub. To clean it thoroughly, it is finally held in a vertical position over the tub and hosed again. When the Hungarian riffles are cleaned up, the sections about a foot in length are taken up one at a time, and lighter sands are washed overboard with a hose. Amalgam and several tubs full of the heavier sands are saved for further treatment. This treatment varies with different operators, and long toms, tables of the Wilfley type, and amalgamation-barrels are all in use. Amalgam is squeezed and retorted, and the resulting sponge-gold is ready for the mint.

One operator who recovers platinum makes the final concentration by panning, dries the concentrate, and blows away the last of the sand. The metal is then treated with nitric acid, washed and dried, and sold to platinum-buyers.

Crew

The crew employed on the three outfits for which cost-data are given below comprises the following: 18 men on barges, 9 on draglines, 3 oilers, 3 tractor-drivers, 3 mechanic-welders, 3 extras used as truckdrivers, etc.,

[5] Merrill, Charles White, Dragline dredging, in Peele, Robert (editor), Mining engineers' handbook, Vol. I, sec. 10, pp. 600-606, New York, John Wiley & Sons, Inc., 1941.

FIG. 18.   Dragline dredge to accommodate 3-cu. yd. excavator. *Photo by courtesy of Bodinson Manufacturing Company; reprinted from California Journal of Mines and Geology, April 1938, p. 115.*

one cleanup man, and one superintendent.   This crew of 41 men operates the three plants for 24 hours per day.

Capital Investment

The following figures are intended to give a rough idea of the cost of the principal items of equipment of high quality and bought new about 1935.

	1½-cu.yd. diesel	3-cu.yd. electric
Dragline excavator	$22,000	$30,000
Barge and washing plant	20,000	28,000
RD7 Caterpillar tractor, diesel	6,500	
RD8 Caterpillar tractor, diesel		8,000
Attachments for tractor (bulldozer, winch)		3,300
Miscellaneous welding, etc.	1,500	1,700
	$50,000	$71,000

In addition to these main items, the following may or may not be needed depending on the location and other variable conditions: truck, shop, camp, stock of spare parts, electric power line and transformers, storage for diesel fuel.   A shop of some kind is usually provided.   It may contain a part or all of the following: welding equipment, machine tools, retort for amalgam, machinery for cleaning sands, and possibly for recovery of platinum.

Operating Costs

The following figures on operating costs, covering the first six months of 1937, were furnished by the auditor of an experienced operator who had been in the business for some time.   All of the equipment was bought new for the purpose of dragline dredging and had been in operation for an average of about a year before January 1, 1937, when the period covered below starts.   Depreciation of $1,000 per month on each outfit is charged by the operator with the idea in mind that the machinery runs continuously, as near 24 hours per day as possible, not intermittently like equipment used by a contractor. The figures are believed to be accurate, but with less accuracy in the figure for cubic yards than in the others. Yardage was calculated on the basis of an actual survey for area, but depths were estimated. Emphasis is again placed on the fact that this dredging was done under conditions practically ideal for a dragline. Depth of gravel was less than 20 feet, it was recent gravel of present streams, loose and easy to dig; bedrock was soft, and a foot of it was easily dug by the dragline; the outfits were operated for periods of a year and more on the same deposits, and no time was lost in dismantling for moves; freezing of water during winter was so slight that it caused no trouble. The only condition not ideal was the presence of growing trees and brush on much of the land. Cost of removing this is included in the same account as repairs. Wages were $1.00 per hour for dragline operators and $0.625 for other classes of labor during the period.

The lower cost per cubic yard for the 3-cu.yd. electric is due chiefly to the fact that a much larger yardage is handled by the same size of crew as is used on the smaller outfits. The cost per cubic yard for power is a fraction of a cent lower on the electric.

	1½-cu.yd. diesel	1½-cu.yd. diesel	3-cu.yd. electric
Gravel handled, cubic yards	394,050	330,900	696,000

Cost of Operation

Dragline, payroll	$4,659.10	$4,003.38	$4,269.20
Fuel oil, lubricating oil, gasoline	1,500.00	1,471.38	57.03
Maintenance	1,405.58	1,264.15	889.96
Cable	592.73	777.03	1,394.97
Direct expense	8,157.41	7,515.94	6,611.16
Washing-plant, payroll	5,283.11	4,489.94	6,307.50
Fuel oil, gasoline	1,255.00	1,278.68	160.39
Maintenance	877.22	825.73	949.48
Direct expense	7,415.33	6,594.35	7,417.37
General operation			
Power	--------	--------	3,599.13
Water	--------	--------	159.00
Repair, labor and materials, including clearing of land with tractors	8,282.11	7,658.11	8,301.89
Compensation insurance	586.70	593.26	585.62
	8,868.81	8,251.37	12,645.64
Office, taxes, general	1,001.00	875.39	948.80
Depreciation	6,000.00	6,000.00	6,000.00
Total operating expense, 6 months	$31,442.55	$29,237.05	$33,622.97
Cost per cubic yard	0.08	0.088	0.048
Operating hours	2,175	2,489	2,686

No land-costs and no royalties are included in these figures.

Gardner and Allsman [6] give a table of costs at 21 placer mines with floating washing plants, but few details of methods of accounting used to arrive at the figures are given. The range for most of the plants is 9 cents to 12 cents per cubic yard. Depreciation is included in most cases, but not royalties. They state that when all costs except royalties are included, 12 cents would probably be about average. Royalties usually are 10 to 15 percent of the recovered gold. However, some operators bought the land and reduced this cost. Most of the draglines covered by this table had buckets of $1\frac{1}{4}$ or $1\frac{1}{2}$ cubic yards capacity.

---

[6] Gardner, E. D., and Allsman, P. T., Power-shovel and dragline placer mining: U. S. Bur. Mines Inf. Circ. 7013, pp. 64-65, 1938.

During the early 1930's a number of the so-called 'dry-land' dredges were built in northern California. Most of them were so poorly designed and constructed that they had no chance to succeed, and were used for very short periods. Even the best of them were built on timber skids to be pulled forward by the power shovel. Gravel accumulated under the skids, irregular bedrock interfered, and much time was lost in moving. Another common fault was tailing-sluices on trestles, which needed rebuilding every time a move was made. Lack of head-room often resulted in the tailing backing up against the rear end of the washing plant. Most of these outfits were built of second-hand material including second-hand gasoline engines. Contrast these with the latest dragline dredges, which were built of new material of excellent quality, and which were powered by diesel engines or electric motors. Gasoline engines may be considered obsolete for such service. Diesel engines soon pay for themselves in fuel-savings.

A few outfits of later design were operated with some degree of success, such as the one used by Pantle Bros., in the Lincoln district, Placer County. The outfit included a movable land plant, not self propelled, consisting of a hopper, trommel, centrifugal bowls and stacker. It was mounted on a steel frame supported at the rear on caterpillar treads 5 feet long, and at the front on 8-inch steel wheels. The gage was 12 feet and the distance from front to rear axle was 14 feet. The gravel was charged to a 3-cu.yd. hopper and was hand-fed to a 4 by 10-feet trommel. The trommel consisted of two screens, one inside the other. The inner trommel was perforated with 1-inch holes for a length of $8\frac{1}{2}$ feet. The outer screen was 6 feet in diameter and was perforated for a length of 4 feet with 1- by $\frac{1}{8}$-inch slots, and for a length of 2 feet with 1- by $\frac{1}{4}$-inch slots. The undersize from the trommel was distributed to four 36-inch Ainlay centrifugal gold savers running at 100 rpm. The 48-foot stacker, with an 18-inch belt, which could be swung laterally by hand or raised and lowered with block and tackle, was provided for discharge of the coarse tailing. Power was provided by a 35-hp. electric motor, and a 2-hp. motor on the stacker. The entire outfit weighed 15 to 16 tons. It was fed by a 1-cu. yd. dragline excavator.

Humphreys Gold Corporation built some very large machines of this type at Clear Creek, Colorado, and Virginia City, Montana, and operated them successfully. The one at Clear Creek, Colorado, was self-propelled by a 90-hp. gasoline engine and was mounted on the chassis of a crawler crane. The one at Virginia City, Montana, was a huge machine weighing 500 tons. It was mounted on two tractor chassis, and a third set of caterpillar treads. It was self-propelled by 2 motors, each driving one of the tractor chassis. The gravel was dug with two $2\frac{1}{2}$-cu.yd. 100-hp. electric draglines, and an auxiliary $1\frac{3}{4}$-cu.-yd. power shovel was used to clean bedrock.

Costs, exclusive of depreciation and royalty, of the outfit at Clear Creek, Colorado are given by Gardner and Allsman [1] as $0.25 per cubic yard; at Virginia City, Montana, as $0.1183 per cubic yard.

---

[1] Gardner, E. D., and Allsman, P. T., Power-shovel and dragline placer mining: U. S. Bur. Mines, Inf. Circ. 7013, p. 65, 1938.

4—

FIG. 19.   Dry-land dredge.   *Reprinted from California Journal of Mines and Geology, January 1941, p. 39.*

The same corporation had one of these large outfits in California, but during their last gold mining in California this was idle and they were using the floating washing plants such as are described under the heading of dragline dredging in the preceding chapter.

These washing plants that are designed to operate on dry ground have not yet been standardized as well as the floating washing plant. They are more subject to mechanical trouble and lose more time on account of bogging down, which results in a higher cost per cubic yard.

Another somewhat similar method of handling placer gravel is hauling it with trucks to a stationary washing plant, consisting of a hopper, trommel, and riffle-sluices for recovering the gold. The oversize from the trommel is usually discharged into a bin, from which it must be hauled away with trucks.

Wm. Von der Hellen had a successful operation of this kind on McConnell Bar on the Klamath River, Siskiyou County, 6 miles west of U. S. Highway 99. The washing plant and trucks handled 1200 cubic yards per day, at an estimated cost of $0.35 per cubic yard, exclusive of depreciation and royalties. Gravel was excavated from a pit 40 feet deep below river level by means of a $1\frac{1}{2}$-cu.yd. gasoline shovel of the dipper-stick type. The large amount of handling in trucks is chiefly responsible for the high cost.

This method can be used if the gravel is too tight to dig with a dragline or if other conditions are unfavorable to the use of a dragline, but the cost is so high that the gold content of the gravel must be much higher than that needed for dragline dredging; otherwise the method will not pay.

BUCKET-LINE DREDGING

By Charles M. Romanowitz* and Herbert A. Sawin**

Successful bucket-line dredging in California, as known today, is an industry of nearly 50 years standing. During that long period, it has, on the whole, demonstrated soundness and resourcefulness. Principles today are no different from those applying when dredges of 1898 to 1906 were being designed and built. Placer gravel containing gold, platinum, or other mineral products, recoverable by dredging, must (1) be dug, (2) be screened, (3) be washed and the metal saved, (4) be disposed of to rock and sand tailings. Thus are set up the primary problems; how they have been met is a story of constant improvement in materials, methods and operating 'know how.' What once was considered impossible often has been accomplished. What may look to be insurmountable today, in dredging practice, probably will be done in the future. Early dredges weighed a hundred tons or less. Today's great dredges weigh as much as 3,750 tons. The digging ladder and bucket line on a large dredge might weigh 1000 tons. The investment for a dredge alone, without property or royalty costs, can run from $100,000.00 to over $1,000,000.00.

Successful dredging is not entirely mechanical; it involves good judgment by owners and operators. A dredge might be operated profitably in one area, but if moved to another, without redesigning or rebuilding to meet new conditions, be a complete failure. Many inexperienced investors hesitate to spend a comparatively small amount of money for prospecting. One feature of placer dredging not common to many forms of mining, is the relative simplicity of proving property value, extent, and characteristics. Long experience with drilling and shafting has developed methods for logging and mapping a gold placer property which make it possible to design a dredge best suited to that property. The presence of boulders, cemented gravel, unusually hard bedrock or other serious conditions can be discovered before a dredge is ordered. Depth of deposit and variations in bedrock elevations dictate the digging ladder length which will work to best advantage. These are physical characteristics to be considered; there are other obstacles, which must be understood, such as local ordinances requiring resoiling or leveling or prohibiting dredging under zoning restrictions. Stream pollution laws will influence the type of tailings disposal equipment needed. Pond water sometimes must be lowered or changed by pumping and plans made accordingly. These remarks only briefly touch upon the many problems in dredging: the thought behind them to suggest careful consideration by anyone planning to dredge placer gravel and full investigation by a competent engineer with a background of dredging experience.

Gold dredging in California first was attempted as early as September 1850 when the small river boat 'Phenix' was fitted out as a dredge and attempts made to mine placer gold from river gravel about 9 miles above Marysville in the Yuba River. According to newspaper accounts of the day, the dredging principle was similar to that now in use. An endless chain of scoops brought the mud from the river bottom up to

* Director of Sales, Yuba Manufacturing Company, San Francisco, California.
** Sales Engineer, Yuba Manufacturing Company, San Francisco, California.

a rocker-washer which was propelled by the same power operating the scoops. Screens separated the fine material from the coarse and the 'Phenix' was equipped with a 'Bogardus Patent Amalgamator' which caught free gold by the use of quicksilver.

J. Wesley Jones, in an article entitled *Jones' Pantoscope of California*,[1] described the 'Phenix' as follows:

"The 'Phenix' dredging machine is seen in the Yuba River, a cumbrous arrangement, by which it was designed to drag up sand from the bed of the river, and obtain gold in large quantities. It was soon found, however, that this machine dredged more money from the pockets of the owners than it did gold from the bed of the Yuba, and this kind of dredging was very soon abandoned."

Today, technically minded travelers in the dredging areas of California learn that a modern dredge is a marvel of mechanical efficiency, and while according to Mr. Jones, it was a method of mining "soon abandoned," he, of course, had no way of visualizing 20th Century dredging which began in the Oroville area in 1898 and has continued successfully through the years to become a principal source of California's gold production.

Today, two mammoth dredges in California excavate gravel 112 feet and 124 feet respectively below water level against banks 50 feet or more high. In other words, gold-bearing gravels 180 feet below ground level, laid down centuries, perhaps ages, ago by ancient rivers, are being dredged and today contribute to the welfare of the community, state and nation. Dredges like these represent an investment well over $1,000,-000.00 each. Thousands of men are employed, both directly in operation and in maintenance and manufacture. Dredges are great consumers of capital goods, steel products especially. Electric power is a large item of operating expense which indirectly supports many men and their families. Lubricating greases and oils are used by the carloads. Rubber belts, electric motors, cables and supplies of all kinds find constant use in dredging. Quicksilver, for saving gold in riffles, is used over and over again but eventually must be replaced and dredges provide a constant market for quicksilver producers.

It is rather startling to come across a dredge working in a field. From a distance, it is not always possible to see the pond that it digs and which moves along with the dredge. Constantly excavating, digesting the gravel, saving gold, and stacking tailings astern; these mechanical 'gold-diggers' operate 24 hours a day. Ordinarily, an operating crew consists of three or four men per shift with three shifts per day. A dredgemaster is in full charge and plans his work to comply with the owner's instructions. Average actual running time is better than 22½ hours for three shifts. It is only by steady and constant operation that gold dredging can be made to pay. Only by handling a tremendous yardage and by operating constantly can returns justify dredging. Since dredging started in California, the average value of ground dredged is less than 12 cents per cubic yard or in other words, about 7½ cents per ton. The greater portion of California land dredged is of no value for other purposes except possibly for a few weeks grazing in the spring. Agriculturally-minded people raise the cry periodically that good land is being destroyed but they overlook the fact that when land is of more

---
[1] Jones, J. Wesley, Jones' pantoscope of California: a 'lecture' together with pencil sketches depicting the journey across the plains to California, section third: California Hist. Quart., vol. 6, p. 244, 1927.

value for mining than it is for agriculture, mining must be given preference. This is true not only in gold mining but in other branches of the industry. One fact often overlooked is that, in California, there are approximately 4,000,000 acres of so-called arable land, untilled, and potential dredging land is only a small fraction of 1 percent of this large acreage. Dredging operations return to the owners of the land dredged, in royalty payments, far more than the land could earn if planted. The owner's royalty properly invested will pay dividends greater than the land would earn if cropped.

Dredging lands are found adjacent to mountain ranges and have been formed in past ages through the action of streams or glacial ice depositing gravel bearing reconcentrated gold values. The California placers were formed by streams cutting through beds of ancient streams, several of which flowed at right angles to the present stream courses. In this manner, original deposits of placer gold were reconcentrated and deposited in sufficient quantities to make placer dredging profitable. The value per yard, as stated above, is not high but occasionally properties are found where placer gold deposits range from $0.20 per cubic yard upward. Much of the gold is quite fine and some gold has been recovered which would pass through a screen of 300-mesh size. Closer to the foothills, the gold is more coarse and naturally, as one gets into the hills, nuggets are found.

The principle of dredging is quite simple. Good California practice is to dig with the maximum depth reached while the ladder is at 45° with the water level. Digging is started at the top of the bank, and as the bucket line moves upward, the dredge swings to the left and right, pivoting about the spud which is at the stern with its point imbedded in rock tailings. The spud takes the thrust of digging, distributing the load to the fore-and-aft trusses. Spring-mounted spud keepers help in absorbing shocks and distributing the load evenly. The side swinging is accomplished by port and starboard bow lines which are carried from the under water end of the digging ladder to shore blocks and back to the bow fairleads on the forward deck, thence to the swing-winch, usually mounted inside the deck house on the starboard side. As one drum takes up the line, say, on the port side, the other pays out a slack line to starboard. As the swing to one side is completed, the operation is reversed.

Material, after it is dug, is elevated in the bucket line to the main hopper and is classified in a revolving screen which discharges oversize tailings to a rubber stacker belt. These large tailings are stacked in a pile and form the rock tailings which can be seen in parts of the West. These rocks often are used for road building and other purposes after being crushed and graded in separate plants built for the purpose.

Fines (usually minus $\frac{1}{2}$-inch) are discharged through the screen to gold-saving tables equipped with Hungarian riffles with mercury trap riffles usually used in the ratio of about 4:1. Free gold readily amalgamates with quicksilver and is cleaned up weekly and retorted ashore. There is endless discussion concerning gold losses which occur with the discharge of fine tailings overboard from tail sluices. On a well-constructed dredge, mining clean placer gold which amalgamates freely, it is possible that losses are less than the cost of additional equipment and labor to prevent them. However, in recent years, jigs of one type or another have been installed on several dredges, used either as a complete recovery system or in conjunction with tables and riffles, either

Fig. 20. Carrville Gold Company dredge. This Yuba dredge built for Carrville Gold Company in 1939, has 12-cu. ft. buckets and is capable of digging 50 feet below water level. It is operated in Trinity County, under the hardest digging conditions known to exist in California. Its machinery and power units are equivalent to those of an 18-cu.-ft. dredge.

ahead or behind jigs. Jigs are old in mining, but new developments give them a place in gold dredging. They have been used for tin in the Orient, but were a long time in finding favor among gold men. Amalgamators and other mechanical devices are needed with the jigs and extra men are required to operate this department. Yuba Consolidated Gold Fields' dredges in California, using riffles for gold saving for many years, are now being equipped with modern Yuba jigs of new design and high efficiency. Long tests indicate that improved recovery will justify the added investment.

Most California dredges are electrically operated with power conducted from connections on shore through submarine type cables. Transformers on board step down power to usable voltages, usually 440. Several dredges in California, notably the newer ones owned by Natomas Company, convert a-c power to d-c power on board. Bucket lines and other units have d-c motors for a wide range of speed. In barren ground the buckets dump at speeds as high as 40 per minute; in hard digging the speed can be reduced to a few per minute. The new Natomas dredges were designed and built by Natomas Company engineers. Bow swinging lines and hoisting lines are synchronized with the bucket speeds. The company claims many advantages for this type of control. Natomas Company also has been 'jig-minded' for many years. Its new dredges make use of Pan-American and Bendelari jigs as primary gold savers and riffles for secondary gold saving.

Easy digging gravel once was almost a general condition for California dredges but most of it was soon worked out. As dredges moved away from river and bench gravels and started to work on old channels, much harder digging became the rule. Dredges must dig into bedrock to save rich material lying on it or in crevices. Today's dredges are built to cut and dig into hard bedrock or cemented gravel. One Yuba dredge in California, owned and operated by Carrville Gold Company near Carrville, Trinity County, digs at a depth of 50 feet below water level using 12-cu.ft. buckets. Its power and digging ladder construction, however, are equivalent to those of a dredge with 18-cu.ft. buckets. As the buckets dig the bedrock they are forced into the hard bottom by pressure of the digging ladder resting upon the backs of the buckets. This dredge has the hardest digging of any known successful dredging operation in California. It was built upon the site of an earlier dredge of much larger bucket capacity which was abandoned because it lacked power and strength to operate successfully. Manganese steel chips from the bucket lips are found on the gold saving tables; evidence of the tremendous power used in cutting into the bedrock.

A recent Yuba development in main drive arrangement provides for the use of two a-c motors mounted just aft of the upper tumbler. The motors are of equal power and sufficiently synchronized to operate without trouble. Each motor is connected through V-belts to the pulley shaft and the double drive is typical of dredging practice; pulley shaft pinions to intermediate gears to bull pinions to main-drive bull gears. The intermediates are best if of herringbone type. This type of main drive is used on dredges such as Yuba No. 20 at Hammonton and Capital No. 4 near Folsom, each an 18-cu.ft. dredge using two 300-hp. 440-V motors.

No great change has been made in the revolving screens used on dredges in California. In some foreign fields two or more screens are used sometimes but it is California practice to use but one. Material

Fig. 21.   Repairing bucket-line, Yuba No. 20, showing latest bucket-design

for screen and liner plates has been greatly improved. Early screen plates were of low carbon steel; later of high carbon steel (.40-.50C) with drilled tapered holes. Cast manganese·steel screen plates with cored tapered holes were introduced and improved greatly the life of screen plates. One disadvantage was the cored holes; many operators preferred that holes be drilled because a closer spacing could be secured. To meet this demand, about 15 years ago experiments were made with U. S. Steel abrasion-resisting steel plates. Today many dredge screens have ARS screen plates which are taper drilled. As the name implies they resist abrasion caused by sand in the screen; also, they are hard because of chemical content. Cast manganese-steel plates, where subject to pound-ing by rocks in the screen, 'work harden' and are preferred by some operators. Where no pounding action takes place the life of such plates is not longer than ARS plates which resist abrasive action. Screen plates have tapered holes so as to pass all small gravel entering. Hole spacing and diameter must be determined by experience to secure best results. Plates in different courses of the screen usually differ in hole diameter and spacing to secure better distribution through the screen to the tables or jigs.

Gold saving table arrangement has undergone much thought and change. Older dredges distributed material, which came through the screen, to athwartship sluices which dumped into fore-and-aft sluices; two or more of the former emptying into one of the latter causing greatly increased volume and greater velocity. Because only a small percentage of the cleanup was found in the fore-and-aft sluices, it was long felt that practically all of the gold was caught in the athwartship sluices. Even-tually it was found that such gold as reached the fore-and-aft sluices probably was washed overboard because of the increased volume and resulting greater velocity in those sluices. The old theory that greater table area contributed to larger savings was upset by the new findings. A new practice based on total width of all sluices making up the tables and carrying a controlled volume was substituted. Yuba No. 20 at Hammonton, when designed and built, had tables and sluices based on the new ideas. It and later dredges are arranged so that material from the screen can be conveyed from the upper end to any lower position before reaching the tables. All gold saving is done in athwartship sluices which are fed from head gates in two distributor troughs under the screen and extending aft beyond the lower end of the screen. This new arrange-ment secures a more even accumulation of amalgam and avoids the older system of collecting most gold in a triangular section of forward tables near the upper end of the screen. It is found that even with less table area a much higher efficiency results. Double-banked tables are often used; the upper about 6 inches above the lower. The top tables are provided with counter-weights to aid in lifting them on hinged supports to make easy access to the lower tables.

An outstanding improvement in dredging equipment concerns the bucket design and method of attaching lips. Over a long period man-ganese steel foundries, making dredge buckets, have worked with dredge designers and operators in improving the shape of buckets to secure clean, fast dumping. In the past 10 years the lips have been improved to secure a firm-locking fit with the buckets and several types of bolted lips have been devised and used. The bolted lip, most widely accepted by the operators, makes use of two vertical bolts to hold the lip in place on the

FIG. 22.   Perry Idler.

bucket with the tongues of the lip snugly fitted into recesses in the inner walls of the bucket.   At one time it was customary to take five or six buckets from the line at each weekly cleanup and send them ashore for relipping (riveted).   Now a long bucket line can be relipped on the dredge without removal from the line during a cleanup period of 4 to 8 hours.

As long ago as 1912, alloy steels were used on California dredges. As dredge machinery became larger, stronger bucket pins and shafts were required but the diameters could not always be increased.   The first full heat of alloy steel sold by a mill to one customer came to Yuba at its old Marysville plant in 1912.   It was made to a specification developed by the company's metallurgist and from then on bucket pins have been of alloy steels.   Today's pins are usually nickel-chromium or nickel-chromium-molybdenum steels carefully forged, machined, and heat treated to develop full strength and wearing qualities.   Wearing plates for upper tumblers and idlers are usually of heat-treated forged alloy steels or cast manganese steel.   Bucket bushings, which are removable, are of manganese steel and it is customary to have bushings of varying thickness to adjust the bucket line assembly as pins, bushings, and bucket back eyes wear.   The length of a bucket line must be kept about constant and as the line wears, three or more buckets may be removed to shorten the line to its proper length.

Lower tumblers are of cast manganese steel pressed onto alloy steel shafts.   Bearing seals of many types have been tried, all with the intent of reducing wear in the journals by keeping out grit.   Ladder rollers and their bearings are of many types and those under water especially are sealed against intrusion of muddy water.   Upper tumblers usually are of the one-piece type in California having the body and shaft cast integral. Wearing plates are bolted in place and are replaceable to maintain a fairly constant pitch diameter to match the bucket pitch.

Dredge hulls have received much attention by designers; in early days they were largely box-like structures of wood and later steel and made of a size sufficient to support the dredging machinery. As dredges became larger and heavier it became necessary to give a lot of thought to hulls, insuring a workable freeboard and stability for the dredge. Accidents to hulls sometimes resulted in flooding and capsizing, causing great damage and occasional loss of life. Modern dredge hulls are designed with water-tight compartments using every precaution to comply with safety requirements. A more modern development led Yuba Manufacturing Company in 1933 to experiment with portable pontoons for hulls. Many Yuba dredges and perhaps others today are equipped with fully portable pontoon hulls which form a series of water-tight compartments, bolted together and having strength equal, at least, to any so-called standard construction. Such hulls are especially adapted to dredges which may be moved from one property to another, thereby making a high salvage value. The pontoon system of constructing hulls was put to extensive use by the U. S. Navy during World War II as reported and pictured in the press and trade journals.

Deep gravels in California fields and in other parts of the world are being dredged successfully because of two developments of recent years. The Perry Patented Bucket Idler and the Yuba Mud Pumping System are jointly responsible for the success of such dredging. The Perry Idler is a cylindrical device mounted under the digging ladder in a structural steel frame. It supports the bucket line about midway in its return trip to the lower tumbler. This support balances and divides the catenary into two parts. For dredges digging 100 feet below water level, a Perry Idler is positively necessary to insure success. It was developed by the late Colonel O. B. Perry, a well-known engineer, who had much to do with dredging in California and elsewhere over a long period of years. From his experience he recognized the need for such an idler to make deep dredging with a bucket line a practicable operation.

Coupled in use with a Perry Idler, deep-digging dredges in California use a Yuba Mud Pumping System. Designed to remove silt and mud from pond bottoms where accumulations reach a depth of 30 feet or more, the system was first used on Yuba No. 17 at Hammonton. A more detailed description of the Perry Idler and the Yuba Mud Pumping System will be found in another chapter where an article concerning the operation of Yuba Dredge No. 20 is reprinted.

Space does not permit a long discussion of many points of dredge design which need careful attention. The bow gantry supports the digging ladder. The stern gantry supports the stacker. The loads and suspension points must be carefully calculated. The spud used for digging takes the full thrust and its design must be such as to withstand tremendous loads without failure. Small dredges today ordinarily use one spud and move ahead and about the pond by shore lines at the bow and stern. Large dredges usually make use of two spuds; either can be used for digging and both used alternately for stepping ahead. The stacker belts of rubber are typical of the California-type dredge. Long study by rubber manufacturers has resulted in special belts for dredging service; especially non-skidding types, such as American Rubber Manufacturing Company's "Lightning Ribbed," which minimize the tendency of wet rocks to roll backward causing unnecessary wear on belts, idlers, and other equipment. Rubber also is used in dredging for riffles and sluice liners. Molded solid-rubber riffles are used on dredges having a long operating life and despite the high initial cost are economical because there is practically no wear caused by the abrasive material constantly flowing over them.

Dredges in California use and wear out literally hundreds of miles of wire rope. Bow lines and ladder suspension lines in particular are bought with length of service in mind. These ropes take severe punishment and only the best quality is advisable. Special constructions, by wire rope manufacturers, have lengthened the life of such ropes and the experience gained in dredge work has been of great value in recommending ropes for other services.

In summarizing it may not be amiss to speculate upon the future of dredge construction. The history of placer dredging is one of continuous improvement. Accidents have occurred in the past and in almost every case have pointed to an improvement which might avoid a repetition of similar accidents. The human element is always present and lapses of memory or attention on the part of a winchman or other members of a dredge crew can result in disaster. Dredges of the future will become more automatic in operation than ever. Trends in dredge design today are toward greater dependence on electric controls. These will relieve the winchman, to a large degree, of responsibility for maximum yardage, high running time, and accidents due to human failings. Control devices will provide, automatically, efficient operating speeds for maximum yardage under all ground conditions, full use of motor capacities under varying loads, and thereby maintain highest possible running time. Some electrical devices now safeguard machinery against human failure. For instance, a "Lilly control" on a ladder hoist will prevent a winchman from dropping a ladder too fast, endangering the motor windings and also from raising the ladder too high or dropping it too far. Improvements, such as are contemplated above, add to the initial cost but will more than pay for themselves in the long run.

Dredge operators in California have one great advantage over operators elsewhere in the world. Our dredging areas are not far from main highways and transportation is not a problem. Spare parts can be secured in California on short notice, since the dredge-building industry centers in the San Francisco Bay area. New ideas can be tried in the field and builders of dredges can easily get back and forth from dredges while watching new developments. It is not necessary to experiment at the customer's expense in a far-off country and wait months for a report. The success or failure of newly designed equipment is soon known to the dredge builder. This fact has helped in making California the center of the placer dredge building industry.

Dredging in California for gold is in line with the State's heritage. California got its fast start toward statehood because of gold mining and all through its history, gold has been a mainstay. There is a theory that basic industries are those which can operate successfully and ship their products from the point of production profitably; thus considered, gold dredging is a basic California industry and can continue to be one in the future.

BECKER-HOPKINS SINGLE-BUCKET DREDGE

By H. A. Sawin *

A new type placer dredge, known as the Becker-Hopkins Single-Bucket Dredge, was introduced prior to the war. Its inventors are G. E. Becker and H. H. Hopkins of San Francisco. Yuba Manufacturing Company acquired manufacturing and sales rights under the Becker-Hopkins patent. Becker-Hopkins dredges are 'single-bucket' excavators; each dredge being a self-contained floating unit, designed particularly for operation on shallow properties, in limited areas, or in narrow canyons, where it is impracticable to operate bucket-line dredges or other types of equipment to advantage.

The digging unit of a Becker-Hopkins dredge consists of a bucket, built integral with a sluice-type boom, which conveys dredged material from the bucket to the screen. The dredge operates from a fixed position on the pond surface, being moored by bow and stern lines. The bucket is dropped vertically at the rear end of the well and a cut made horizontally by pulling the bucket forward into and through the material being dredged. The telescoping boom extends in length automatically, permitting a horizontal bottom cut. When the bucket reaches a point under the bow of the dredge, a latch is released and the bucket is elevated radially to a point where the dredged material slides down the sluice-type boom and is evenly distributed to the screen.

Boulders can be successfully dislodged and in many cases put through the screen and disposed of over the stacker. Boulders too large for the bucket can be brought to the surface and cast aside by use of a tractor, usually available on dredging properties. The ready control, which the operator has of the sluice-type boom, makes it possible for him to slide a boulder into the screen gently; this avoids wear which might result from dropping heavy boulders at high speeds. Control of the slope of the boom also prevents heavy intermittent overloading of the screen, even distribution being assured because the movement of dredged material down the boom can be accelerated or retarded.

* Sales Engineer, Yuba Manufacturing Company, San Francisco, California.

Fig. 23. Becker-Hopkins single-bucket dredge.

FIG. 24.   Bucket detail, Becker-Hopkins single-bucket dredge.

Each cut follows the previous cut, until the desired depth has been reached.   The horizontal cutting action is controlled, which makes it possible to clean bedrock thoroughly.   The dredge is moved to new digging positions on the pond by use of the sidelines, and from its new position, the digging cycle starts again.

This is a brief description of the digging operations of Becker-Hopkins Single-Bucket Dredge.   The other functions of dredging are similar to those on a bucket-line dredge.   Power can be either electric or Diesel engine to suit conditions, and all units are made for easy dismantling, shipping, and re-erection.

The original designers built and operated a small unit in California in cooperation with A. R. McGuire of Fresno.   Later, Mr. McGuire was interested in the operation of two such dredges (a 1-cu. yd. and a 2-cu. yd.) in Alaska prior to the war.   Just before the war, Yuba built its first one for use on Butte Creek in Butte County, California.   Experiments with it demonstrated worth-while qualities but also pointed to several 'bugs' which resulted in design changes to improve its operation. War conditions closed down the work but as this is written (January, 1946) the dredge is being rebuilt on a property in Yuba County and it is planned to operate it experimentally for several months to test the new design.

JIGGING APPLIED TO GOLD DREDGING [*]

By P. Malozemoff [**]

Jigging is one of the oldest processes used by man in separating the heavy minerals from the lighter gangue. At the turn of the present century it was extensively used for the concentration of base metal ores and for the washing of coal. Comparatively recently it has been widely applied for treating placer gravels in recovering tin, tungsten, and gem stones.

Its extensive application for the recovery of placer gold is of even more recent origin. One of the earliest large scale jig tests on board a gold dredge was made by J. W. Neill in 1914 on the Yosemite dredge in California.[1] Subsequently, tests were made by the Natomas Consolidated Company, also in California. This company used Neill jigs in fore-and-aft sluices in an effort to effect a saving of some of the gold that was being lost in the tailings.

However, these were isolated examples for some twenty years, until, in 1932, the Bulolo Gold Dredging Company initiated a series of thoroughgoing tests, as the result of which one of the company's dredges in New Guinea was completely equipped with an installation of Bendelari jigs. This installation proved the practicability of treating the entire output of the dredge (screen undersize) with jigs, and the company took steps to prepare for extensive application of jigs on their boats. The company's engineers soon designed a new machine, now known as the Pan-American Placer Jig, in the effort to make a jig particularly adapted for use on board the dredges. Subsequently, several more boats in New Guinea and in Colombia were equipped with jigs of new design. As the result of the success of these installations, interest in jigging on board the dredges was again aroused in the United States, and in 1936-37, installations were made by the Yuba Consolidated Goldfields, Ltd. in California, and by Fisher and Baumhoff in Idaho. Of recent months several other companies are reported to have made jig installations on their dredges.

These efforts may well mark a turn in the history of gold dredging industry, spelling, as they do, the directing of attention towards improving the gold recovery by the elaboration and refinement of the gold saving practices on board the dredges. Not that efforts in this direction were lacking in the past, for great improvements in the design of riffles and their adjustment were made; but riffles obviously have their metallurgical limitations, especially under the conditions imposed by dredge practice. Jigging, on the other hand, is free of some of these limitations, and, if properly employed, capable of meeting even the more exacting dredge conditions successfully. Jigging on board the gold dredges is a modern development in dredge practice that no operators can afford to ignore, and one, it seems safe to predict in the light of the interest already aroused, which is destined to gain almost universal application in the very near future.

Now that the groundwork has been laid, and jigs on dredges, when properly installed, have been proven beyond any reasonable doubt to be

* Written for and published as pamphlet by Pan-American Engineering Company, 820 Parker Street, Berkeley, California. Printed in condensed form in Engineering and Mining Journal, vol. 138, no. 9, September 1937. Reprinted by permission of Pan-American Engineering Company.
** Formerly metallurgical engineer for Pan-American Engineering Company.

1 Neill, J. W., Application of jigs to gold dredging: Min. and Sci. Press, vol. 109, p. 839, November 28, 1914.

both practicable and profitable, the time seems propitious to rehearse for the benefit of the technical fraternity the advantages that jigging on board the dredges offer, and the problems that will confront the operator who contemplates an installation of jigs.

The subject is so vast and its ramifications are so numerous that it would be impossible to deal with it in detail within the scope of a single magazine article. The main purpose of this paper, then, is to define the problems involved in general terms so as to present an introduction to a subject that is both timely and little known.

The use of jigs on dredges will be considered only because of their ability to effect a saving of gold in addition to that made by the riffles. That losses exist when riffles only are employed is recognized by every intelligent dredge operator; and despite exhaustive efforts towards improving riffles and other conditions on a dredge, appreciable losses are ever present in the majority of cases, greater in some than in others, yet usually of sufficient magnitude to warrant serious consideration of some other means of reducing these losses. The gold lost by the riffles is predominantly fine, although sometimes it might be comparatively coarse if flat and difficult to amalgamate. The presence of "rusty" gold in the ground is usually indicative of appreciable losses, for it is a general contention among placer operators that the gold which does not amalgamate or amalgamates only with difficulty, such as "rusty gold," is difficult to save by the riffles.

Limitations of the riffle as a gold saving device are inherent in it: its action is such that the gold must settle and be entrapped by the riffle in a swift current of water, the velocity of which must be great enough to transport the material, both coarse and fine, across the riffles. The less the velocity of the current of water, the greater is the tendency of the gold to settle and be saved by the riffles, yet at the same time, the less the carrying power of the water stream; and consequently the yardage capacity of a given area of riffles is reduced. Since space on the dredges is limited, only a certain limited riffle area is available for a given yardage to be handled by a dredge. In the effort to increase the output of a dredge, the yardage is often boosted beyond the optimum capacity of the available sluices, and the amount of water and the slope of the riffles is so regulated as to induce sufficient velocity of water to transport all the material; the recovery of the gold will naturally suffer under such conditions. Still other factors, such as packing of the riffles, especially with heavy black sand, and sudden surges in feed which tend to dislodge and wash out the gold that has already been entrapped by the riffles, are all inimical to the optimum recovery of the gold, and can not always be fully corrected.

The action of the jig, on the other hand, disposes of several factors that cause loss of gold in the riffles. The jigs operate continuously, and the bed can be adjusted so as to permit settling and, consequently, trapping of the gold at all times. Once trapped, it is removed from the stream and there is no more danger of losing it as there is in riffles that pack, or from which gold is dislodged and washed out because of occasional surges. The presence of a large amount of black sand in the ground, which causes especially severe packing of the riffles and consequent loss of gold, will not cause loss of gold by jigs. Furthermore, dilution of the feed to the jigs need not be so great as to the riffles, since in the case of jigs the transporting of the material is aided by the alternate pulsations, whereas the carrying power of water is the sole transporting agency in the case of riffles.   For this reason the conditions on top of the jig bed can be made far more quiescent than for riffles, thus affording a better opportunity for the gold to settle.

Naturally enough, the jig has its limitations also.   It is only a gravity machine, therefore will recover only that gold which will settle by gravity under the conditions that obtain on the jig bed.   Some of the finest gold will be lost by the jigs as well, but at the present stage of development of the arts of the recovery of gold, this loss is in almost all cases below the economic limits of the known methods.

Flotation, for example, can probably be applied in some form or other for the recovery of the fine gold that will be lost even by jigs, but the cost of doing so, it appears at present, will come very close to, or exceed the amount recovered.   In 1933 an installation of six full size flotation machines, which treated 300 tons per day, was made on one of the dredges operating on the American River in California.[2]   Three months' operation showed that the recovery of gold by flotation from the sand wheel overflow, which should contain the bulk of the finest gold in the dredge tailings, was only 2 to 5 cents per ton on heads of 3.5 to 9 cents per ton, as calculated from the concentrate and tailing assays.   An average of 120 assays made on the flotation heads directly gave a value of only 9.3 cents per ton; each assay was made on six to eight assay ton charges. These values are based on $35.00 gold.

Besides, it must be remembered that all placer deposits were formed as the result of the material settling by gravity, and in most of the placer deposits now being exploited the finest gold had already been eliminated by natural agencies during the process of deposition, except in rare cases in which the conditions of deposition were such that even the finest material settled and formed the deposit.   For this reason there will be very little gold that cannot be recovered by gravity, provided the method of recovery is sufficiently refined to reproduce the settling conditions that obtained during the formation of the deposit, when material settled by gravity under natural agencies.

There are, of course, some exceptional deposits in which the gold has been liberated by chemical agencies from the gold-bearing sulfides, or those in which the gold is still locked up within the sulfides or oxides that have been deposited with the sand and gravel comprising the deposit. Such occurrences of gold present individual problems that involve methods of milling in one form or another for their solution and will not be considered in the present paper.

The use of jigs for the treatment of placer gravels presents problems some of which are entirely different from those encountered in the use of jigs in milling in the concentration of base metal ores.   In the base metal ores the ratio of the specific gravities of the valuable mineral to that of the gangue is relatively low; for this reason, and in order that the jigs make proper recoveries, it has been general practice to classify the feed as to size rather closely, and to provide a jigging surface that is much longer than it is wide.

In placer gravels, on the other hand, the ratio of the specific gravities of the gold to that of the sand and gravel to be discarded is very much higher than for base metal ores.   Close classification of the feed, though

---

[2] A test made by the Pan-American Engineering Corporation, Ltd.

5

desirable, is not so essential, nor does the ratio of length to width of jigging surface need be so great. Accordingly, the load per square foot of jigging surface can be materially increased without appreciably affecting the results that can be obtained.

This is fortunate, for were it not so, jigging as practiced in metal mining industry could hardly have been applied to the gold dredges. Close classification of the screen undersize on the dredge would not be practicable; and floor space is always at a premium, thus requiring the gold saving machinery on board the dredge to have large capacity per unit of floor space.

When attention turned to jigs to be used for dredges a great many different designs of jigs used in washing of coal and in concentration of base metal ores were available, but none of them were designed to conform to the altogether peculiar conditions imposed by the gold saving dredge. For the most part, the jigs were heavy and cumbersome, occupying considerably more space than is required by the effective jigging area. One of the first jigs tried out on a gold dredge, the Neill Jig, conformed to the demand for economy of floor space and total room occupied was confined to the screen area of the jigging surface.

The first modern jig applied to gold dredges was the Bendelari Jig, which actuated the water through the screen and bed from below by means of a plunger sealed with a rubber diaphragm. This permitted the floor space to be defined by the screen area of the jig.

Somewhat later a new jig was designed by the Placer Development Company's engineers and used with success in New Guinea at the Bulolo Goldfields. In this jig, now known as the Pan-American Placer Jig, a part of the hutch in the form of an inverted cone is moved by means of an eccentric to transmit the pulsations through the bed. Free discharge of concentrates and uniform distribution of pulsation throughout the area are thus aided. The weight of the machine was cut down to the minimum to meet the demand of the dredges, especially the smaller ones, for least weight of the gold saving equipment.

In principle, these modern jigs are no different from the old type machines, such as the Harz Jig. To secure activation of the bed the water pulsation is transmitted mechanically by means of an eccentric. In the Harz this was done by a plunger working in a separate water compartment outside the hutch; in the Placer Jig this is done by a moveable cone-shaped hutch, and in the Bendelari by a diaphragm inside the hutch.

Jigs for gold placer operations are designed so as to produce the concentrate continuously as a hutch product. A certain amount of shot bedding is usually used to reduce the amount of concentrate thus obtained. It is quite important to have delicate control over the suction so as to be able to secure maximum recoveries with the maximum ratio of concentration. This requirement was answered by a hutch water connection closely controlled by a plug cock. A screen area of 42 by 42 inches has become largely standard for the jigs used on dredges.

Testing of Dredge Tailing Losses

Eventually, perhaps, it may be found that all gold dredges that employ riffles should install jigs to improve the recovery of gold. But because jigging on dredges is a comparatively recent development, and data on wide variety of dredge operations are still lacking, such a generalization is premature. Hence, whenever a jig installation is contemplated, it should rightly be preceded by suitable testing, which can definitely prove that the jigs are capable of effecting a recovery in addition to that obtained by the riffles.

The existence of dredge tailing loss may be qualitatively detected with relative ease, yet to arrive at a definite, reliable figure in cents per yard is an extremely difficult task. The difficulty of this problem will readily become apparent if the factors that complicate such a determination are considered.

The values recoverable by jigging from dredge tailings are usually very low, in the majority of cases less than 5c per yard; the gold is free and not uniformly distributed; so that even if a very large sample of several tons be taken the sample would hardly be representative of the dredge operation as it is conducted from day to day. Anyone acquainted with the dredging operations is familiar with the fact that the gold usually occurs in the ground in "pay streaks" of comparatively small thickness, and before the "pay streak" is dug a large proportion of the total digging time is consumed in moving material that is barren or nearly barren. Thus, even the most elaborate intermittent sampling will almost always be open to doubt.

However, a quantitative determination of the sluice tailing losses can be made accurately by determining the gold content of a small, continuous cut of the total flow of the tailings; such for example, as continuous jigging of the total flow of one tailing sluice, or of a stream representing a small part of the total dredge tailing flow.

Numerous methods can be applied for this purpose; a fairly comprehensive discussion of them and of the interpretation of the results is beyond the scope of the present paper. The subject is sufficiently broad to merit its presentation in a separate, full length paper.

Suffice it here to say, that a reliable determination of the dredge tailing losses is possible, and should be made for the border line cases, in which it appears that the existing loss is small and may not be sufficient to justify the use of jigs.

Besides the determination of the amount of gold that can be saved by the jigs, testing may at times be required for other purposes. For example, a special method of treating the rougher concentrates may be found necessary, especially when the ground to be dredged contains a large amount of heavy mineral constituents, such as black sand and pyrite; or, again, when a part of the gold in the ground is present in the form of included values within the heavy mineral constituents. In these cases large scale testing on the dredge may be necessary in order to be able to devise a suitable method of treating these concentrates for the final recovery of the gold.

Factors Affecting Jig Installation and Jig Recovery

The number of jigs necessary, the manner in which they are installed, and the method of treating the rougher concentrates will depend on a large number of local conditions that will vary widely from one operation to another. A few of the more important ones are: total yardage dug; proportion of screen undersize to oversize; ease with which the ground is disintegrated and washed, which in turn may determine the dilution of the feed to the jigs; the nature of that feed; and the nature of the gold, i.e., whether coarse or fine, flat or granular.

Although the mechanical capacity of a 42 by 42-inch jig has been established at about 30 cubic yards per hour, the effect of some of the above factors may necessitate a radical revision of this figure. It needs be reduced for sandy ground, when excessive dilution obtains, or when fine, flat gold is present. At times it may be possible to rate the capacity of the jigs to fit the conditions that obtain when the dredge is digging pay gravel, even though they will be overloaded when it is digging the top ground, which may contain a greater proportion of sand than does the pay gravel. As long as this material contains little or no gold no harm will be done.

Excessive dilution of the screen undersize will usually mean excessive top water velocity over the jig bed, and will cause the loss of fine gold that does not have a chance to settle in the swift current. When this excessive dilution cannot be reduced because the ground requires a large amount of water for proper disintegration and washing of the material in the screen, the jigs should be rated at a lower capacity than normal and some means of controlling the velocity of the top water across the jigs should be provided; boiling boxes or retarding baffles are the usual remedy. Another method to obviate this difficulty is to dewater the entire jig feed, or some part of it. Although this may not always be possible, it will always be desirable. Dewatering elevators, dewatering tanks or sumps may seem a revolutionary, complex innovation on a dredge, but it is one that can certainly be justified when jig recovery can thereby be appreciably increased.

After the jigs are properly installed, there is yet their adjustment to make, which is a problem that is at times quite complex. The material passing through the dredge screen from hour to hour is generally variable to the extreme, both as to quantity and character. Since the amount of water added to the screen will usually remain constant, with the variation in the quantity of solids delivered to it, the dilution of the screen undersize will vary in inverse proportion. Because of the variation as to quantity and dilution, the distribution of the load fore and aft will change constantly. Moreover, the list of the boat from side to side, and the pitch fore and aft will affect the distribution of load accordingly. These difficulties, more serious on the small dredges than on the large, may be overcome to some extent by a carefully considered installation, and by an adjustment of jig controls that would permit effective jigging even at the worst of conditions. This, however, would only be a compromise and would never be altogether satisfactory, since under other conditions that would obtain such adjustments may not be the optimum. The only effective remedy for aggravated cases of this sort is the installation of a central distributing system, which would consist of a central, preferably mechanical, distributor, to which the entire product from the screen is delivered, and from which the load is equally distributed among the several jigs installed on the boat.

Jigging Practice

Jigging practice consists of several different operations. The first is the roughing treatment by one row of jigs on each side of the dredge screen. This is usually followed by a second treatment over the scavenger jigs. The scavenger jigs are necessary to assure the more complete recovery of the fine gold. Only in special cases will it be possible to limit the operation to a single treatment by one row of jigs: such, for example, in which only a small proportion of the total gold in the ground is fine; but when the gold to be recovered is predominantly fine, the number of jigs installed has to be increased so that each is required to treat considerably less than its normal rated capacity.

Besides the roughing and scavenging some method of treatment of the rougher and scavenger concentrates must be provided for. Generally, the feed to the jigs, depending on the size of the dredge, will be from 100 to 400 cubic yards per hour. The ratio of concentration that can be achieved by rougher and scavenger jigs will on the average be 100:1. This means that from 1 to 4 cubic yards of concentrate per hour must be treated for the recovery of gold. Obviously, this cannot readily be done by batch treatment but must be done continuously.

There are two methods that have been developed for this treatment:

(1) The concentrates may be treated directly for the recovery of gold by grinding in the ball mills in the presence of mercury and then passing the resulting product over amalgamation plates. Grinding need not be very intensive, as the polishing of gold and its ready amalgamation is the main purpose of such treatment.

(2) Another method, which is the simpler and therefore the preferable one, is to pass the rougher concentrates over suitable cleaner jigs in order to reduce the amount of the final concentrate to be treated for the recovery of gold. Of recent months a hydraulic jig known as the Pan-American Pulsator Jig, has been used for this purpose to an advantage. It allows the production of a cleaner concentrate that bears a ratio of 1:30 to 1:100 to the primary concentrates. The final concentrate is thus reduced to a small bulk that can be readily amalgamated in batches, as in an amalgamating barrel, and the amalgamated product either rejigged or streamed down for the recovery of mercury and the contained amalgam. With this method, the loss of gold in the intermediate products that are discarded is negligible.

With different combinations of the two methods a large number of variants can be had. For instance: the primary concentrate may first be rejigged, then the resulting cleaner concentrate subjected to continuous or intermittent grinding in the presence of mercury, and finally passed over a trap and an amalgamation plate for the recovery of the mercury and amalgam. The obvious advantage of such a procedure over the direct grinding and amalgamation of the primary concentrates is that the size of the grinding installation can be only one-thirtieth to one one-hundredth of that necessary for direct treatment.

Installation

Various methods of applying jigs to existing or proposed dredges may be employed. They can conveniently be grouped into two classes: (1) installing jigs as auxiliary recovery equipment, conforming to the existing sluice layouts; (2) installing jigs as essential recovery equipment which may or may not entail the elimination of the existing sluices.

When jigs are installed as auxiliary recovery equipment they may be placed either in the fore-and-aft sluices or at the end of the lateral sluices as is shown by the accompanying figures 25a and 25b, respectively. When in fore-and-aft sluices, as in figure 25a, the jigs will probably be required to take more than their share of the load because the amount of material that these sluices usually handle is greater per inch of width than it is on the lateral tables. Also, because the riffles require more

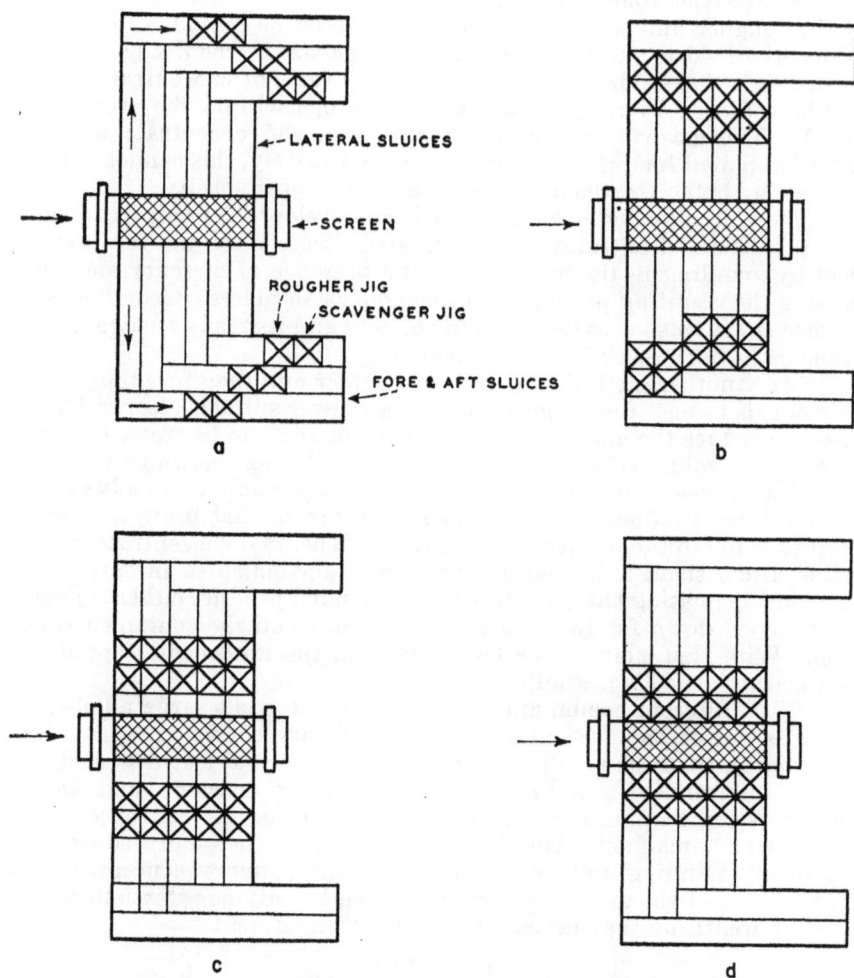

FIG. 25.   Jig arrangements: four methods of applying jigs on gold dredges. *a*, Jigs installed in fore-and-aft sluices; *b*, jigs installed at end of lateral sluices; *c*, jigs installed in lateral sluices, preceded by short sections of sluice; *d*, jigs installed in lateral sluices, next to the screen.

water than is best for jigging there will always be excessive dilution impossible to reduce, and consequently, an excessive velocity of top water that will be difficult to control. Besides, headroom will be limited at this point.

When installed at the end of the lateral sluices, as in figure 25b, the load over the jigs will usually not be excessive, as the width of the total jig installation will be equal to the width of the total table area. But here again, excessive dilution and velocity of top water obtain, and headroom is not always available.

Either one of these methods requires a double clean up. The sluices ahead of the jigs will recover the major proportion of the gold and must be cleaned up every so often; the jigs will yield an additional amount that is cleaned up continuously and therefore separately from the major riffle clean-up.

It would therefore seem preferable to install the jigs as close to the screen as is possible, so that they may constitute the essential recovery equipment. In this location dilution of the feed can be more closely controlled, the headroom is almost always available and only a single clean-up is necessary. If riffled sluices are contemplated below the jigs they will have a much better chance of doing good work because the jigs will remove the heavy mineral constituents which are such a source of trouble in packing the riffles. Figures 25c and 25d show a possible method of installing jigs as essential recovery equipment. The jigs may or may not be preceded by a short section of sluices, the main purpose of which would be to distribute the feed uniformly across the width of the jigs. These short sections of sluice may also be riffled and made to trap out coarse gold and tramp iron. The installation of jigs in this position will not necessarily involve the scrapping of existing sluices, as the jigs may be cut into them.

Although at present the installation of jigs on a dredge will be contemplated only if the jigs can be shown to produce greater recovery of gold than the existing riffles, the time may not be far distant when it may become recognized that the jigs offer sufficient other advantages to be installed even when they do not yield increased recoveries of gold. As mechanical improvements of dredge machinery are made, the percent lost time owing to repairs will decrease, and the periodical clean-ups of the riffles may necessitate loss of operating time in addition to that due to dredge repairs. This obtains even now in a great number of dredge operations. With proper jigging installation it may not be necessary to have any riffles on board the dredges at all, consequently the entire clean-up can be effected continuously, involving no loss of time. There is no reason for the jig repair to be responsible for any appreciable loss of time, since, if properly cared for, they can be made to operate almost without interruptions. In milling, which employs similar machinery, it is not unusual to have 98 to 99 percent operating time year in and year out.

This paper is the outgrowth of some of the work done by the Pan-American Engineering Company at the instance of the Placer Development Company, Ltd., Yuba Consolidated Gold Fields, Ltd., and Fisher and Baumhoff, to which companies credit is due for initiating the installation of jigs on dredges recently. Their cooperation in that work, which made the writing of this paper possible, is gratefully acknowledged. Also, the writer wishes to acknowledge with thanks the helpful guidance given by V. E. Bramming and F. W. Collins.

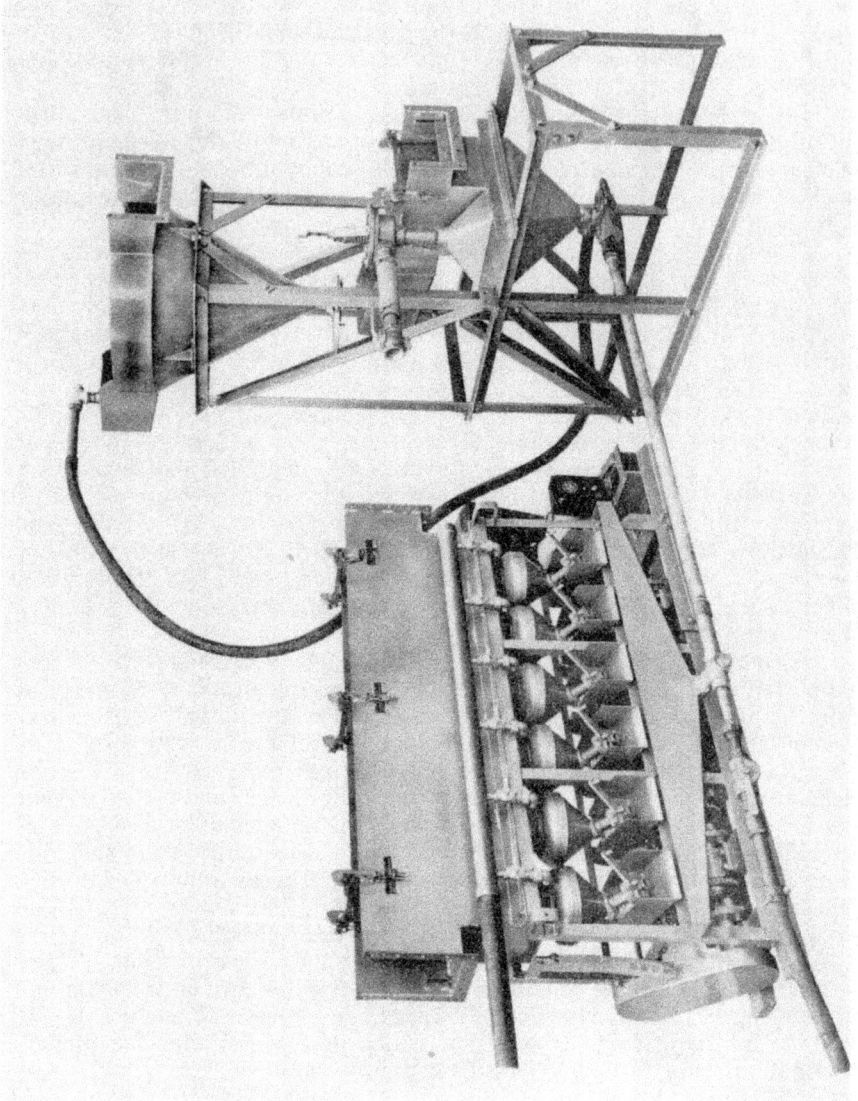

FIG. 26.   Jig testing set.

NOTES ON JIGS FOR GOLD DREDGES

BY F. W. COLLINS*

The article written by P. Malozemoff entitled *Jigging Applied to Gold Dredging* and published in condensed form in Engineering and Mining Journal, vol. 138, no. 9, 1937, is a very well considered paper, and it is reprinted in full above. This paper of Malozemoff's shows the picture as it was late in 1937, and since then a great deal of detailed information has been developed that was not then available. High points of these developments are briefly outlined below.

Methods and equipment have been developed for testing tailing from existing dredges to determine the amount of gold present that can be recovered by jigs. Figure 26 shows one of these testing sets. The method is to cut a sample of about 2 cubic yards per hour from one or more tail sluices and this is fed continuously to the 10 by 60-inch rougher-jig. The tailing is measured to determine volume treated and the rough concentrate is jet-pumped through a rubber hose to the dewatering cone ahead of the cleaner-jig. The overflow from the cone goes to the pond and the underflow to a 12-inch Pulsator jig. The photo shows a 12-inch single-cell jig, but a two-cell jig is usually used. The cleaner-jig concentrate is amalgamated. A number of dredges have been remodeled to use jigs after being tested with this equipment and results to date indicate that the test results are reliable.

From 1932 to 1936 the trend was to install jigs as scavengers following conventional riffled sluices but the error of this is so apparent that now the jigs are being installed as close as possible to the screen and are depended on entirely to save the gold. However, it is desirable to have each rougher-jig followed by a sluice 3 or 4 feet long and as wide as the jig. This sluice should carry burlap under expanded metal and its purpose other than carrying the tailing to waste is to act as an indicator of the work being done by the jig. If any gold shows up in the burlap then the dredge master should check up on the jig-operators.

The distribution of the feed to the rougher jigs is of prime importance and this has been worked out in a very satisfactory manner along lines first developed by Natomas Company. The distributor immediately below the screen is made as a modified Jones riffle and automatically makes a 50-50 split of the screen undersize to the two sides of the boat. It may then go directly to the jig opposite its point of exit from the splitter or be discharged into a fore-and-aft distributor sluice. There is one such distributor sluice on each side of the boat and any material in these sluices may be fed to any jig aft of the point where it enters the sluice. As the major portion of the fine material tends to come through the screen near the forward end, this system permits the fine material to bypass the forward jigs and reach those farther aft that would be 'starved' if their feed carried only that part of the fines that comes through the lower part of the screen.

The attached flow sheet is for a 9-cu. ft. (nominal size) dredge that normally digs about 11,000 cubic yards per day and has dug 13,000 cubic yards in 24 hours on several occasions. The screen openings are half an inch in diameter, and about 50 percent of the bank run is under-

* Mechanical Engineer, Pan-American Engineering Company, 820 Parker Street, Berkeley 2, California.

FIG. 27. Rougher jigs, four-cell block.

size. The pumping of the rougher-jig concentrate is necessary on this particular boat but can be avoided in some cases. A mechanical dewatering device for the rougher-jig concentrate is more desirable than the settling tank shown on the flow sheet.

Placer-type jigs or similar machines are always used as roughers and in such fields as Hammonton and Natomas where the 'black sand load' is comparatively light they may also be used as cleaners but when the deposit to be dredged carries a considerable quantity of heavy material the Pulsator jig should be used for this purpose.

Two systems of recovering the gold from the rougher-jig concentrate have been developed and are in use on a number of dredges. The first is to pass the rougher-jig concentrate directly to 'auger hole' riffles, which amalgamate the clean gold. Tails from these riffles are dewatered and fed to a cleaner-jig, which makes a tailing and a concentrate. The concentrate goes to a small grinding mill, and it is preferable to dewater ahead of this mill. The mill-discharge goes to another set of 'auger hole' riffles and then over a scavenger jig and to the pond. The scavenger-jig concentrate may be returned to the grinding mill for a short time but the circulating load of 'tramp' iron soon builds up to a point where it becomes necessary to remove this concentrate and clean it up by hand.

The second system is as shown on the flow sheet of the 9-cu. ft. dredge above mentioned and involves the use of a Titan Amalgamator. Of the two systems the one with the Titan Amalgamator seems to be the better and will probably eventually supersede the other system entirely.

On two dredges where the gold is very badly tarnished the arrangement has been modified as follows: The cleaner-jig concentrate goes

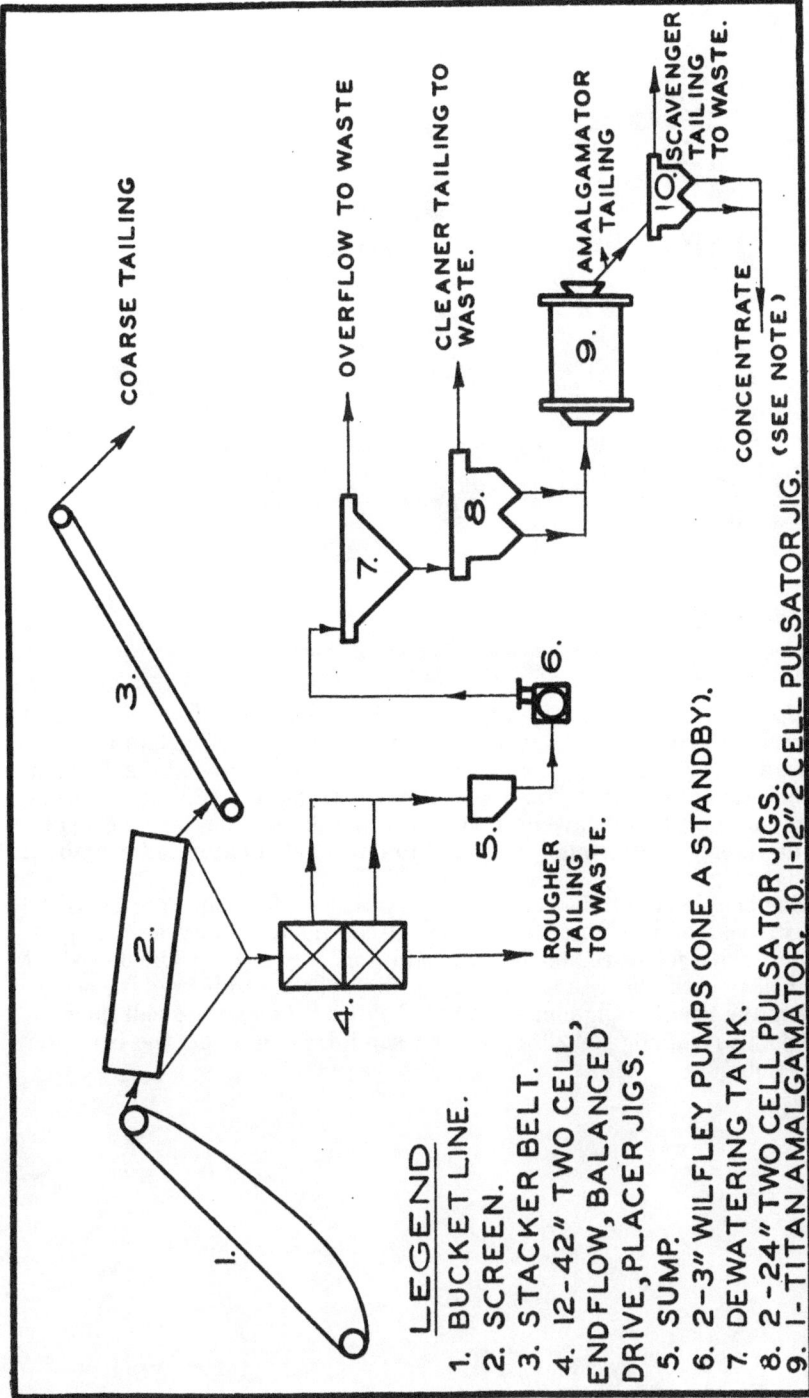

COARSE TAILING

OVERFLOW TO WASTE

CLEANER TAILING TO WASTE.

AMALGAMATOR TAILING

SCAVENGER TAILING TO WASTE.

CONCENTRATE

ROUGHER TAILING TO WASTE.

LEGEND

1. BUCKET LINE.
2. SCREEN.
3. STACKER BELT.
4. 12-42" TWO CELL, ENDFLOW, BALANCED DRIVE, PLACER JIGS.
5. SUMP.
6. 2-3" WILFLEY PUMPS (ONE A STANDBY).
7. DEWATERING TANK.
8. 2-24" TWO CELL PULSATOR JIGS.
9. I-TITAN AMALGAMATOR, 10.1-12" 2 CELL PULSATOR JIG. (SEE NOTE)

FIG. 28. Flow sheet for use of jigs on 9-cu. ft. dredge. *Note:* Amalgam is removed from amalgamator at 10-day intervals. The scavenger-jig concentrate which is small in quantity is returned intermittently to the amalgamator feed. At intervals it is cleaned up by hand and a tailing discarded in order to purge the circuit of metallic iron.

FIG. 29.   Sand-drag, Sumpter Valley Dredging Company, Sumpter, Oregon.

directly to a Titan Amalgamator which discharges to a mechanical dewatering device which feeds a 2- by 4-foot ball mill.   The ball mill discharges to a second Titan Amalgamator which is followed by a 12-inch, two-cell scavenger jig.   This has worked extremely well, but the same results could probably have been obtained had the cleaner-jig concentrate been dewatered and ground, then fed to one amalgamator followed by a scavenger jig.

The last dredge to be remodeled of which we have any record is the 9-cu. ft. boat of the Sumpter Valley Dredging Company at Sumpter, Oregon.   This job as remodeled started up in July 1945.   The flow sheet is identical with the one shown excepting that a sand-drag is used to dewater the rougher jig concentrate.   Figure 27 shows a 4-cell block of rougher jigs and figure 29 shows the sand-drag used on the Sumpter Valley job.

Black sand accumulates as a concentrate in the riffles or jigs used at placer mines to catch the gold. It occurs also as a natural concentrate on many of the ocean beaches of California. Under the sundry civil act approved March 3, 1905, Congress directed the U. S. Geological Survey to investigate the useful minerals contained in the black sands of the Pacific slope, and this investigation was subsequently enlarged to embrace the United States. Most of the samples collected were concentrates from the sluices of placer mines, but samples from the beaches near Crescent City, Upper and Lower Gold Bluff and Humboldt County were included; also from the beaches from San Francisco south to San Luis Obispo, from the elevated beach at Aptos on Monterey Bay, and from the beach at Ocean Park near Los Angeles. A report on this investigation was published by Day and Richards[1]. The investigation included methods of separating gold and platinum, and the point is brought out that separation is easily accomplished on sized sand with shaking tables of the Wilfley or similar makes. The sizing is done in a hydraulic classifier or with screens. Magnetic separation was investigated also.

Following is a summary of results from about 200 samples collected in California in pounds per ton. Each figure represents the largest amount in any sample.

	Maximum found in lbs. per ton
Magnetite	1856
Chromite	1800
Ilmenite	1500
Garnet	1874
Hematite	1120
Olivine	836
Monazite	56
Limonite	552
Zircon	928
Quartz	2000
Unclassified (in one case largely pyrite)	2000

"Unclassified" includes grains containing more than one mineral.

Thus we see that a black sand may be entirely quartz or it may be nearly pure magnetite, chromite, or garnet, or it may be 75 percent ilmenite. Various combinations of the minerals mentioned above and others (rutile) are possible. Gold and platinum are often present. The sands from various localities have a wide range in composition.

Many persons have expressed an interest in separating the minerals mentioned in the above table and placing them on the market as individual minerals. As no commercial process has yet been devised by means of which this can be accomplished at a profit, attention will be given here only to separating gold and platinum from the sand. The amount of gold and platinum in the sand is determined by standard methods of assaying and chemical analysis. Claims of secret processes either for assaying the sand or for recovering the gold and platinum should be regarded with suspicion.

The platinum is usually so small in amount that elaborate machinery for its recovery is not justified because of excessive cost. The expendi-

---

[1] Day, D. T., and Richards, R. H., Useful minerals in the black sands of the Pacific slope, U. S. Geol. Survey, Mineral Resources U. S., 1905, pp. 1175-1258, 1906.

FIG. 30. Beach-sand being worked with dip-box, northern Humboldt County.
Reprinted from California Journal of Mines and Geology, October 1941, p. 504.

ture of $5000 for machinery that would ultimately recover only $1000 worth of platinum is unwise to say the least. Many persons who have attempted to recover platinum from the beaches in Del Norte and Humboldt Counties have failed to realize this. At times the action of the waves concentrates gold and platinum in small areas of the beaches, and men are able to make good wages searching for these small patches of concentrate and working them by small-scale methods such as long toms. Many attempts have been made to work them with machinery but all have failed because the deposits are too small to pay for the machinery.

A similar condition applies to black sand from placer mining. Only the large dredging companies have enough of it to justify the expense of much equipment. A dragline dredge is not likely to recover enough black sand, but several of them together will produce enough to justify a little equipment for treatment. Several profitable businesses were operated in 1940 by men who collected black sand from a number of dragline dredges and hauled the sand to small plants equipped with amalgamation barrel, small shaking table of the Wilfley or similar type, and a melting furnace. The concentrate often contains lead shot and bullets, which amalgamate. Enough gold sticks to the amalgam and lead to make refining in the furnace worth while. A little cyanide or lye is often used in the amalgamation barrel to clean tarnished or coated gold.

Recent practice of two large dredging companies in handling black sand is described below. Earlier methods are described in a report by C. McK. Laizure.[2]

One dredging company reduces the bulk of the sand removed from the riffles on the dredge-tables during clean-up to a volume of 90 cubic feet with a long tom on the dredge. This amount is sacked and transported to a clean-up room ashore. It is ground in an amalgamation barrel in which several flat weights are placed to polish the gold, then discharged to a sluice and run over a pool of quicksilver. A baffle is placed above this pool of quicksilver so that particles of gold will be forced downward against the quicksilver. The mixture of sand and water then flows over a copper plate treated with quicksilver to amalgamate gold. Riffles in a sluice 1 foot wide below the plate recover platinum in about 100 pounds of sand. This is panned down by hand for final recovery of platinum.

Another dredging company accumulates larger quantities of black sand and has a more elaborate plant ashore. Black sand is dumped into a bin from which a bucket-elevator raises it to a small pulsator jig.[3] One of the functions of this jig is to remove tramp iron and other scrap metal. About two buckets are filled per day from the hutch and this is worked down by hand for tarnished or coated gold and platinum. Overflow from the jig goes to a Straub[4] rib-cone ball-mill then to a shaking copper plate treated with quicksilver for amalgamation, then to a hydraulic cone. Some quicksilver and a little amalgam are recovered in the underflow from this cone, but practically nothing else. This underflow is hand-

---

[2] Laizure, C. McK., Elementary placer mining in California and notes on the milling of gold ores: California Jour. Mines and Geology, vol. 30, pp. 228-233, 1934.
[3] Pan-American Engineering Company, 820 Parker Street, Berkeley, California, makes such a jig.
[4] Made by Straub Manufacturing Company, 507 Chestnut Street, Oakland, California.

worked.   The overflow goes by pipe-line to a Wilfley table 4½ by 8 feet. Concentrate from this table goes to a Wilfley table of laboratory size, 1½ by 3 feet.   Concentrate from the small table is hand-panned; middling is returned to the same table.   Tailing from the small table and middling from the large table go back to the ball-mill.   Tailing from the large table goes to a hydraulic cone, from which the overflow goes to waste.   Underflow goes back through the entire plant.

If the three products that are hand-worked contain much rusty gold, they are treated in a second small ball-mill in batches for polishing. Amalgam is sometimes treated in a pebble-mill made of a 3-gallon stoneware crock containing flint pebbles.   This tends to free much of the platinum that may be entrained in the amalgam.

DRIFT MINING
GENERAL DESCRIPTION *

Drift mining in the United States has been applied chiefly to the exploitation of buried Tertiary river channels in the foothills of the Sierra Nevada in California. It has also been applied extensively, although on a smaller scale, to the mining of rich streaks on or near bedrock in more recent gravels where pay dirt is covered with a thick mantle of unproductive material. Ground may also be drifted where there is insufficient grade or water for hydraulicking or where conditions are unsatisfactory for dredging. Bedrock under rivers has also been drifted where it was impracticable to divert the stream; however, loose gravel containing a large quantity of water cannot be mined successfully by drifting. Usually the method is one of last resort and can be applied only to rich gravel. Even under favorable conditions 6 feet of gravel on bedrock generally must average at least $2.50 per ton to be mined profitably by drifting. Ground that has been drifted by the oldtimers with limited capital has been worked by other methods later; in these instances the overburden carried enough gold to pay for mining on a large scale.

In the latter part of the nineteenth century many large and productive drift mines were operated in California; according to Hill,[1] 11 million dollars in gold was produced in California by this method from 1900 to 1928, inclusive. In the summer of 1932, however, there were no large-scale operations in the United States, and the production of gold by this method was relatively unimportant. Two well-equipped properties, Vallecito Western and Calaveras Central, were doing development work but no regular breasting.[2] The washing plants were used when enough gravel had accumulated to run the plant most of a shift. A few men were employed at a number of old properties in an endeavor to find new deposits of gravel. At a few other old mines lessees were taking out a very limited tonnage from around old workings. Throughout the western placer districts small operations were under way, but relatively little systematic breasting was being done.

Most of the present drift mines are operated through shafts, although in the past some large and productive mines were worked by adits. In many districts large quantities of water must be pumped.

In mining, the gravel is either drilled and blasted or picked by hand to break it down, then it is shoveled into cars and trammed to the surface or to the hoisting shaft. At the surface the gravel is sluiced or put through a washing plant to recover the gold. The gravel from most drift mines requires mechanical methods of washing to disintegrate it and free the gold.

Milling practices bear no direct relation to mining methods at drift mines and are treated separately in this paper.

---

* The following general description of drift mining consists of extracts from *Placer mining in the western United States, Part III, Dredging and other forms of mechanical handling of gravel, and drift mining*, U. S. Bur. Mines Inf. Circ. 6788, 81 pp., 1935, by E. D. Gardner and C. H. Johnson.
Additional details on methods of drift mining and methods of washing the gravel for recovery of gold are contained in this bulletin in Section IV, in which individual mines are listed by county and described. See Calaveras Central, Calaveras County; Ruby mine, Sierra County; and Vallecito-Western, Calaveras County.
[1] Hill, J. M., Historical summary of gold, silver, copper, lead, and zinc produced in California, 1848-1926: U. S. Bur. Mines Econ. Paper 3, 22 pp., 1929.
[2] "Breasting" is the term used in drift mining to designate the mining of the gravel; it corresponds to "stoping" as used in lode mining.

(81)

6

Development

General Development

The general development plan of a drift mine usually resembles that of a lode mine where similar flat-lying deposits are exploited. Lateral development and the blocking out of the pay gravel are modified to fit local conditions.

Bench deposits or old channels exposed by later erosion or covered by only moderate depths of overburden may be opened and mined through adits. Ventilation shafts, however, may be required in extensive workings.

Deeply buried deposits must, of course, be mined through shafts. This form of entry also is used for mining relatively shallow deposits where adits are not practicable. Occasionally long drain tunnels will be run and the gravel mined through a series of shafts sunk along the course of the pay gravel. Moreover, shafts may prove more economical for mining shallow deposits where their use obviates long underground trams. Conversely, adits may be run for drainage and to work gravels which have been developed through shafts.

Some of the ancient channels are buried as much as 500 feet deep by later gravel and lava flows or beds of volcanic ash. The gravel is hoisted through a central shaft; one or more auxiliary shafts usually are required for ventilation. A buried gravel deposit generally is prospected by a drift along the course of the channel and crosscuts from the drift to either rim. Raises also are occasionally put up to prospect for possible rich strata above. As stated elsewhere, the buried Tertiary channels of the Sierra Nevada are not related to the present stream system; competent geological advice is needed to plot their probable course and aid in their development.

Adits should be run at such a horizon or shafts sunk deep enough to insure drainage in the workings. Drifts generally are run upstream to allow drainage to the shaft or out of the entrance adit. Where water is not a serious item drifts may be run both ways from a crosscut or a shaft; any water from the downstream branch is pumped into the drainage system. The breasting is done upgrade by retreating toward the shaft or crosscut. At drift mines in the frozen gravels of Alaska the common practice is to drift in both directions from a shaft.

Ideal conditions, of course, would be an even bedrock and a grade sufficient to allow drainage but not too steep for easy tramming; such conditions, however, seldom exist. A prospecting drift may be run partly in bedrock to avoid swinging it from trough to rim and back again so as to keep a practical grade for tramming. With a rapid rise of bedrock, however, as where a waterfall or rapids existed in the original stream, the drift has to be run entirely in bedrock with raises put up to prospect the gravel above or the drift continued on a higher level with a transfer point at the break. This, of course, increases the cost of handling the material. If the size of the deposit as shown by the development work justifies the initial expense, tramming drifts may be run on an even grade in bedrock and the gravel from breasting operations above dropped into raises from which it can be drawn into cars. Then the development drifts and crosscuts are used for extracting the gravel.

Sometimes drifts at different levels are run from the shafts to mine deposits at these horizons. More than one channel may be worked from the same shaft.

In shallow deposits little or no mechanical equipment may be used except for hoisting; in small-scale work hoisting also may be done by hand. The development and mining of deeply buried channels require expensive installations and usually must be done on a moderately large scale. Hoisting and pumping equipment and air compressors such as those used for lode mining are required for mining this type of deposit, as well as air drills and mechanical haulage equipment.

Shafts. Shafts seldom have over three compartments; in small-scale work one compartment usually suffices. Untimbered shallow shafts may be as small as 2 by 5 feet, the minimum section in which a man can dig.

Sinking practices are similar to those at lode mines except that blasting is seldom done; the gravel is loosened by picking or moiling. The shaft lining usually consists of lagging back of standard framed-timber sets.[3]

Considerable water may have to be handled in sinking deep shafts in gravel, in which case ample pumping capacity is needed. Ordinary sinking pumps usually are employed. Steffa[4] has described the sinking of a 2-compartment shaft at Vallecito, California, in which a novel method of handling the water was used; other sinking practices at this mine, however, conformed to the general practice. He states:

"The shaft of the Vallecito Western was located at a point 50 feet north of the actual channel in order that the shaft station, at a depth of 153 feet below the collar, might be in the solid slate bedrock. At the point selected the shaft passed through 143 feet of volcanic cobble, ash, and sand and gravel before reaching the slate. It was sunk a total depth of 167 feet, providing a 14-foot sump below the station.

"The shaft is 4 feet by $7\frac{1}{2}$ feet in the clear and has one 4- by $4\frac{1}{2}$-foot skip compartment and a $2\frac{1}{2}$- by 4-foot manway. It is timbered with 8- by 8-inch Douglas fir, excepting that 6- by 8-inch material was used for dividers, and is lined with 1- by 12-inch boards.

"The shaft was sunk to bedrock without blasting, picks and gads being sufficient to loosen the material for shoveling. The 24 feet through rock was sunk by hand drilling, using 10 to 12 holes per round, light charges of powder, and electric delay detonators.

"A 12-inch churn-drill hole was sunk first at one end of the shaft to handle the flow of water which was struck at a depth of 8 feet and amounted to about 35 gallons per minute throughout the work. The hole was sunk to a depth of 187 feet and cased with perforated 7-inch inside diameter stove-pipe casing. A deep-well type of turbine pump was installed which was powered with a 20-hp. vertical electric motor, the motor resting on staging about 4 feet above the shaft collar. Three-foot lengths of pump column were used, and as the shaft deepened from day to day enough blocking was removed from under the motor support to keep the pump intake at the level of the bottom of the shaft. When blasting, during the latter part of the work, the casing and pump column, exposed in one end of the shaft, were protected from damage by a heavy plank hung from the bottom end plate directly in front of the drill hole.

"Numerous strata of sand and volcanic ash were encountered, one such bed at a depth of 70 feet being 7 feet thick. A large part of this fine material was carried to the surface by the pump. A test showed that at one time the pump discharge was one-third sand by volume. The pump impellers wore rapidly, three sets being used. Moreover, the drill hole rapidly filled with sand to the level of the pump, after which the pump could not be lowered farther. Twice the pump was removed and the hole cleaned with a sand pump. Finally, at a depth of 75 feet, this difficulty was remedied by cutting a slot in the casing of the hole, wide enough to insert a hand to clean out the sand. As the shaft deepened the slot was likewise cut down. To secure suction with a shallow sump, such as could be dug out easily by hand in this manner, a 4-inch strainer was substituted for the original 3-foot one. The pump was run continuously

[3] Gardner, E. D., and Johnson, J. F., Shaft-sinking practices and costs: U. S. Bur. Mines Bull. 357, pp. 48-60, 1932.
[4] Steffa, Don, Gold mining and milling methods and costs at the Vallecito Western drift mine, Angels Camp, Calif.: U. S. Bur. Mines Inf. Circ. 6612, p. 7, 1932.

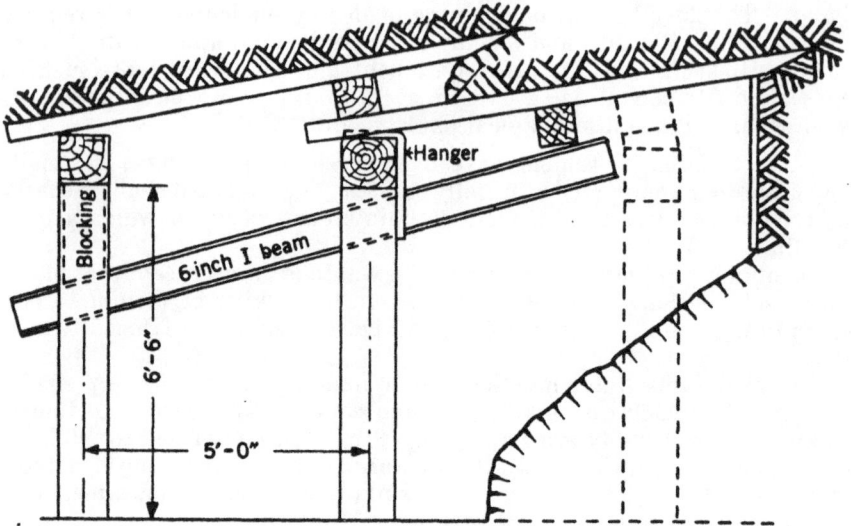

VERTICAL SECTION

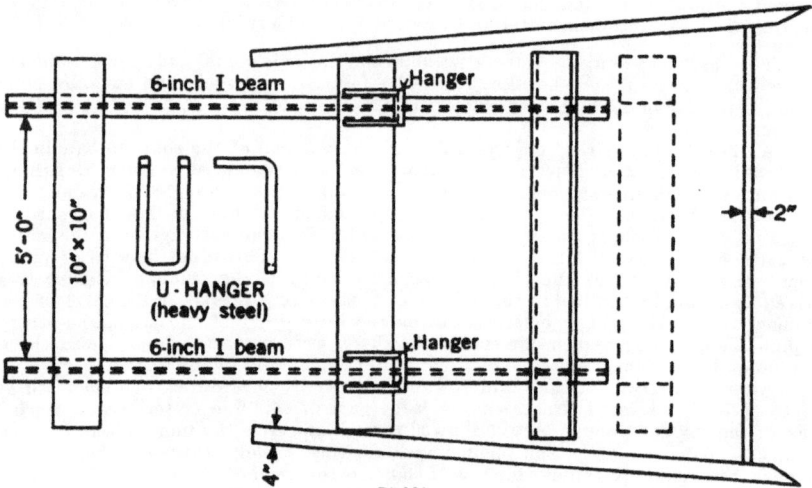

PLAN

FIG. 31. Method of spiling in loose ground.

and regulated by the gate valve on the discharge pipe to the exact amount of water flowing into the small sump.

"It required 90 days to complete the shaft. The average progress in sinking, including timbering, was slightly less than 1 foot per shift, working two 8-hour shifts per day. The cost was $39.50 per foot. Shaftmen and the foreman received $6 per day and engineers $5. Timber and lumber laid down at the shaft cost $42 per thousand board-feet."

Drifts and Crosscuts. As used in this paper, a "drift" designates a development working parallel to the major axis of the deposit; a "crosscut" is a transverse working. This distinction is not observed strictly in the terminology of the mining districts.

Drifts may be run as small as $3\frac{1}{2}$ by $5\frac{1}{2}$ feet in section where the handling of a minimum of material is desirable. In pay dirt they may be run up to 7 by 9 feet in size or as large as they can safely be held. The size of crosscuts depends upon the service required of them.

The gravel in the ancient channels generally is compact enough to stand without timbering; blasting usually is required. The number of holes required to the round depends upon the compactness of the gravel. A simple toe-cut round—that is, one with the cut holes pointing downward—usually suffices for breaking the ground. It is desirable when blasting pay dirt in both development work and breasting to pulverize it as much as possible to facilitate washing operations. Heavy blasting, however, should be avoided so as not to scatter the gold-bearing gravel. In loose gravels the main difficulty in driving may be to prevent caves until the timbering is in place; the gravel is excavated by picks and shovels.

Wheelbarrows may be used in short drifts or buckets on trucks in small-scale work where the broken material is hoisted. In more elaborate workings, however, cars running on rails are employed.

For drifting in pay dirt, a wide drift may be run and the boulders piled at the side to form dry walls. Where timber is brought from a distance regular drift sets of square timber generally are used for supporting the drifts, but if round timber is available locally sets usually are made of it. The posts of the sets generally are stood with a batter so that the drift may be given a section more nearly approaching an arch.

In loose or running ground spiling or forepoling must be used. The first step in spiling is to place bridging over the foremost standing set. Bridging usually consists of a 4- by 8- or 4- by 10-inch lagging laid parallel to the cap on top of 6-inch blocks at either end. This lagging is blocked solidly to the ground above, leaving a space 6 inches high above the cap through which the spiling is driven. If side spiling is necessary bridging is placed on the outside of the posts. Spiling usually consists of 2- to 5-inch timber 4 to 10 inches wide and as much as $9\frac{1}{2}$ feet long, depending upon the weight to be borne and ease of driving; one end of the spiling is sawed on a sharp bevel. The top spiling is driven at an upward angle into the caved or loose ground. In mines having compressed air a drilling machine with a special tool may be used for driving the spiling. The spiling extends over the cap far enough to provide room for placing a complete set. The upward angle is sufficient to allow bridging to be placed over the new set. The first spiling usually is driven at one side of the bridging close to the bridging block at such an angle that the forward end when in place will be 6 or 8 inches beyond and above the cap and close to the wall. The remaining ones are driven at such angles that they "fan" and form a complete covering for the set of timber to be put in place. As each spiling is driven ahead some of the gravel is cleared away from underneath it so that if any large boulders are encountered ahead of the spiling they can be barred out of the way or taken down. After the top spiling is in place side spiling, if necessary, is driven in the same manner, beginning at the top. Two 6-inch I-beams, or heavy timbers, are then hooked on the last cap by heavy steel hangers. The ends of these beams are extended forward to just back of where the cap of the next set will be when in position. A crosspiece is then placed across their forward ends and brought up snugly against the spiling; the back ends of the beams are blocked down under the second cap back. When the gravel is removed the next set is put in. The beams support the top spiling while the set is placed. Posts and caps of ordinary drift sets are used.

The same method of top spiling is used for breasting in running ground. The I-beams or timbers with overhead lagging may be used in firmer ground to protect men working ahead of the last set in position from falling material, both in drifting and breasting operations.

Steffa [5] gives the drifting practice at the Vallecito drift mine as follows:

"Both gangways and crosscuts are generally 7 by 7 feet in section. The usual drill round consists of six holes drilled 5 or 6 feet deep and breaking an average of 4 feet per round. The gravel drills easily, $2\frac{1}{2}$ hours generally being sufficient to drill the round. Drill steel is of $\frac{7}{8}$-inch hollow-hexagonal material, sharpened with cross bits. Slightly more than 9 pounds of 25 percent strength powder is used per round, with four sticks in each of three lifters, three sticks each in the two cut holes, and two in the single back hole. Caps are treated with a standard waterproofing compound.

"The broken gravel is shoveled by hand into 18-cu. ft. end-dump, roller-bearing cars holding 1 ton each. Track consists of 16-pound rails laid to 18-inch gage. The grade of the channel has proved uniform over considerable distances and averages 75 feet to the mile. Track has been laid therefore on a grade of $1\frac{1}{2}$ percent upstream. It has seldom been necessary to take up bedrock to maintain the grade; wherever a dip in the floor has been found the track has been kept on grade, and bedrock has always been found at the expected elevation when reaching the opposite side of the dip.

"In the opening of new areas by drifts or crosscuts, samples are taken from the skip at the collar of the shaft, a sample consisting of one full pan or about 20 pounds of gravel. Samples taken at this point have the advantage, as compared with samples taken from the solid face, of being representative of a larger volume of ground and of being mixed thoroughly by the blasting and by the handling of the gravel from muck pile to car and to skip. Thus an experienced panner is able to make fairly accurate estimates of the value of the gravel developed.

"Drifts and crosscuts are driven by crews of three or sometimes four men, making an average advance of 4 feet per shift. The cost of driving main headings averages $16 to $17 per foot. In a pay area 65 feet wide, where gravel can be breasted 10 feet high, each foot of heading developed 45 tons of gravel. (It is estimated that the gravel expands one quarter on being broken, and a ton of broken gravel has a volume of 18 cubic feet.)"

The cost of running a drift under average conditions at a small-scale mine where no other work was being done was shown by the Golden Belt Gold Mining Company which was developing a drift mine on Magpie Gulch near Helena, Montana, in the summer of 1932. An 80-foot shaft had been sunk, and a drift was being run up the channel; the drift was 160 feet long and 5 by 6 feet in section and was timbered with 8-inch round timber sets placed 4 feet apart. The top and sides of the drift were lined with split lagging. The timber was cut and sawed on the ground. The gravel was picked by hand and trammed in a 6-cu. ft. car.

⁵ Steffa, Don, Gold mining and milling methods and costs at the Vallecito Western drift mine, Angels Camp, Calif.: U. S. Bur. Mines Inf. Circ. 6612, pp. 8-9, 1932.

It was hoisted in the body of the car, which at the surface was placed on a truck and trammed to the washing plant. The cost of running the drift was $6 per foot, excluding supervision. The surface equipment at the shaft consisted of a headframe and a hoist run by a 15-hp. electric motor. Power cost 1.07 cents per kilowatt-hour.

An example of the cost of running a drift under adverse conditions in small-scale operations was illustrated at the Lucky Charles Mining & Milling Company small drift mine on North Clear Creek, Blackhawk, Colorado, which in July 1932 was being developed through a 40-foot 2-compartment shaft; 50 feet of drift had been run but no breasting done. The property was well equipped with an electric hoist, a deep-well pump, a substantial headframe, and an ore bin. About 20 gallons of water was being pumped per minute. A 10-hp. motor operated both the pump and a hoist which had an 18-inch drum. The gravel was hoisted in a 7-cu. ft. bucket attached to a $\frac{1}{2}$-inch cable.

The gravel was 3 to 5 feet thick and was overlain with 5 feet of quicksand which required both top and side spiling. The drift was 6 feet high, 5 feet wide at the top, and 6 feet wide at the bottom. Sets of 6-inch round timber were placed 2 feet apart. Top spiling was 3 by 6 inches by $5\frac{1}{2}$ feet; side spiling was 1-inch boards.

An advance of 1 foot per day was being made by 2 men underground and 1 man on the surface. The cost per foot of drifting was as follows:

Labor (3 men at $4)	$12.00
Power (hoisting)	1.00
Timber	1.80
Other supplies	1.00
Total	$15.80

Breasting

A number of different methods of breasting are employed at drift mines, depending mainly upon the nature of the deposit. Drift-mining methods were evolved in the early Californian diggings; present methods do not differ materially from those of the early days.

In narrow channels the gravel may be mined on either side of the drift as it is advanced or the drift advanced the full width of the pay streak. In wider deposits the drift may first be run to the limit of the deposit and then the gravel mined, retreating toward the shaft. In extensive deposits the gravel usually is divided into blocks preparatory to mining. The blocks generally are mined by retreating. Pillars usually are employed only to protect haulageways. A modified room-and-pillar system, however, has been used at some mines in which the pillars, if in rich gravel, were later removed.

Breasting may be done from crosscuts or from drifts run parallel to the haulageway. Breasting from the crosscuts may parallel the haulage drift on the retreat toward it. When working from drifts the line of retreat usually is parallel to the drift although sometimes toward it. The spacing of crosscuts or drifts at different places ranges from 40 to 200 feet, depending mainly on the system of breasting. Crosscuts generally are turned off at such an angle as to give the proper gradient for tramming.

Cuts or slices range from $2\frac{1}{2}$ to 8 feet wide. If cars are used in long faces the tracks are shifted after each cut. Usually all of the gravel rich enough to mine and enough of the overlying gravel to provide headroom is taken out. Rooms generally are 6 or 7 feet high; the minimum height in large operations is 5 feet. At the Vallecito mine, described later, the thickness of the pay dirt varied up to 14 feet, although at most mines it was less than 6 feet. The rooms may be broken to a strong strata of ground where such strata occur. In some California drift mines volcanic ash makes a strong roof.

In compact or cemented ground the breasts are broken by blasting drill round; holes may be $2\frac{1}{2}$ to 6 feet apart. At most places, however, breasting is done with picks. At many places the gravel is undercut, usually in the upper and softer part of the bedrock; the remaining gravel in the face is then broken to the undercut. Usually 1 or 2 feet of bedrock is taken up. Often, bedrock with deep crevices containing gold can be picked. Hard bedrock is cleaned carefully by hand, as in surface mining. Boulders and gravel too low in grade to take out are piled back of the working face.

Low-built cars usually are preferred for the sake of easier shoveling and tramming in the low workings. Scrapers in drift mines have not proved successful, but with the recent improvements in equipment and technique this method of moving gravel offers possibilities.

Some timbering usually is required in breasting, if only an occasional stull which may be recovered later. Regular timbering consisting of stulls with headboards is used at most mines. If the bedrock is soft, footboards also are used; in soft ground lagging is required overhead. Heavy ground generally is supported by lines of sets. In narrow channels tunnel sets with long caps may be used.

A SYNOPTIC PRESUMPTION
REGARDING CALIFORNIA'S DRIFT MINES

By L. L. Huelsdonk*

In years past, including but not considering the war and closing order L-208 of the War Production Board, there have been many theories advanced, from the rapid extinction of the old-time gravel miner and the inability of the present generation to absorb his art to the exhaustion of the ancient river channels, for the sick condition of the California-sired and once booming drift mines. The actual reason is without doubt economic, and depletion of the easily accessible channels is probably the chief contributing factor. However nearly all of the successful drift mines operated at a time prior to the epoch of laws, rules, regulations, restrictions and taxes governing compensation, unemployment, social security, sales, income, corporation, labor, forest, water, transportation, tailings, and many other seemingly unimportant riders regulated under some admixture of the ABC's which directly or indirectly affect present-day operations.

In the so-called 'good old days' a drift miner often wore knee-pads, worked long hours under a goose-cooker, back-filled boulders and waste and loaded only select pay dirt into his breast buggy. This was transferred to cars and trammed to the outside washing bins by Chinese labor where even in the larger mines, the superintendent had the time and did in most cases wash the gravel and make the clean-up. He fed the white miners beans and the Chinese rice, and if after meeting the payroll and bills an ounce of gold remained the mine was a profitable one. He needed no accountant, income tax expert, attorney or engineer to estimate ore reserves, values, percentages of depletion or other incomprehensible guesses to determine if the profit was actual or merely one on paper.

The foregoing is not an advocation for the return of the 'good old days' or is it without exaggeration or exceptions. Its main intent is to express in brief generalities, for comparable purposes and the sake of argument, conditions that existed during California's early-day drift mining history. As a conservative estimate these mines had an over-all cost of possibly $3.50 to $4.00 per man-day as compared to an immediate pre-war average of $10.00 and a probable $15.00 post-war outlook. This means simply that on the basis of a 50-man crew the early day operation had a monthly cost of approximately $5600.00 as against $15,000 in the late 1930's and $22,500 for crystal ball operations. So in order to keep the unit cost of gravel washed on an equal basis it will without much doubt be necessary for the post-war operator to wash four times the number of units per man as did the old timer and one and one-half times as many as those washed just prior to the war. In comparing these (excluding tax and other ABC nuisances which have been included in the man-day costs) many other factors must be considered, such as the highly publicized $35.00 gold and the possibility of a future increase, the amount of waste mined with pay, the efficiency of up-to-date equipment and

---

* Superintendent, Ruby mine, Downieville, California. Mr. Huelsdonk explains in a letter that these remarks apply to drift mining in general and that exceptions exist, such as the Ruby, where rich beds of coarse gold are found; but that these have little bearing on the industry as a whole.

machinery, working hours, modern explosives, high-grade control, gold recovery, management and engineering facilities. Since most of these are incidental we will consider only the first group, that is $35.00 gold, its possible increase, the efficiency of modern equipment and the ratio of waste to pay mined. The old timer mined very little waste, he worked by hand, the gravel was prospected, he skipped the poor, mined the good, and at times took only a few inches of bedrock gravel by back-filling waste and leaving just enough room to work. Even if the post-war man would submit to this type of mining the over-all cost per miner would without doubt be prohibitive and therefore any post-war drift mine plans must include the use of modern equipment if success is to be reasonably expected. However, this is not an over-all answer or is it without drawbacks. What we might consider modern drift mining equipment are mucking machines and slusher scrapers supported by various accessories such as power augers, jack legs, air bars, drifters, electric locomotives, etc. To begin with, even the smallest mucking machine requires a seven-foot high face to work in and if in a timbered drift, eight feet. It requires a track, and grade must be maintained. Its operation cannot be held up while boulders are sorted out and back-filled, as its prime purpose is a muck mover and any delays must be proportionately charged to costs. Also since the bulk of the gold lies on the bedrock or a few inches above it in most drift mines, breasting with a mucking machine usually causes undue dilution. Its use, however, is indispensable around the modern drift mine for running bedrock tunnels and where applicable for opening up ground for slushing.

By breasting with slushers a fairly moderate roof height can be maintained. This must, however, be at least five feet in order to make working room and conform substantially to modern working conditions. By using blasting boards to keep from scattering the pay over the worked-out areas and at the same time utilizing them for a scraper way during the mucking cycle a good condition is created for back-filling boulders in the open breast. Heavy rocks can be pulled over the boards by use of the tugger and a chain net sling. A scraper also has the advantage of following over irregular bedrock (when the gravel is reasonably dry) without grade or drainage. It is also a good bedrock cleaner when properly applied. In other words, the slusher, under ideal conditions is a hard combination to beat for breasting purposes. It is not however, without faults and disadvantages. The scrapers have a tendancy to cut and follow troughs in the broken gravel making it hard to crowd the face and mechanically clean up the breast for drilling. In wet ground they stir and mud the gravel making a sticky mass which is very difficult to handle in the chutes, cars and ore bins. Under these conditions they sometimes bury themselves and cut deep into the softer bedrock spots forming puddle holes which catch more water and tend to further wet the muck. Also selective breasting cannot be carried on successfully as the main theme must be the spirit of high production and low costs and therefore the drilling and mucking cycles cannot be interrupted. In other words with a hundred-foot breast face with fifty feet averaging $6.00 and fifty feet averaging 10 cents the entire length must necessarily be taken rather than rearrange set-ups to take select sections. This further tends to dilute the pay when comparing it to the old-timer's work. Also tugger stations contribute waste and their setups absorb man power.

In summing these groups we might say that modern breasting, no matter how closely guarded, will add an equal amount of waste to the pay gravel mined by the old-timer, so therefore, although the present price of gold is $35.00 per ounce it has a modern drift mine value of only $17.50 as compared with $20.67 for the old-timer. Consider with this the four to one mining ratio anticipated for post-war operations and the gold for this operator will have a value of less than $5.00 per ounce on a comparable basis. In other words, to balance the two periods, the post-war drift mine operator should get $41.34 per ounce for his gold to compensate for dilution necessary with modern methods, and since he will be required to wash four times the amount of diluted pay dirt to obtain an equal unit cost he should receive a total of $165.36 per ounce for his gold to reach the boom basis of the drift miner's heyday.

These figures, although subject to considerable variation, will serve in a general way to explain the sick condition of California's drift mines. Providing that gold remains at its present value or enjoys an increase with a compensating sur-tax the post-war drift mine operator, in order to effect a cure, must be a strong-minded, hard-headed doctor willing to suffer public criticism by experimenting with and practicing ultra-modern methods such as the rapid back-filling and tamping of waste into the 'worked-out' areas by specially designed machines that will insure safe working conditions and eliminate to a great degree one of the industry's main bugaboos, the expense of timbering. He will have to work out a ratio between the expansion of his broken ground and the tightness of his back-fill whereby he will be able to haul out and wash only his richer bedrock gravel and thus minimize transportation and milling costs. His development program must be carefully laid out and his plans must include the mining of the entire bedrock area as there can be no applied rule for following the pay streak in this type of mining and as some gold usually spreads over the bedrock aside from the run of gold the effort would in most cases be compensated for by low mining costs.

The small amount of gold accompanying the upper gravels which would go into the back-fill would no doubt be cheap pay for the fill material.

In conclusion, it might be generally said that if gold remains on a par with its present value and if the post-war drift miner is to enjoy the higher (or any) income tax brackets, he must develop and adopt a more streamlined mining system rather than knock his brains out against Davy Jones' locker with the present day conventional methods.

FIG. 32.   Flume for hydraulic mine under construction.   *Photo by C. V. Averill.*

HYDRAULIC MINING *

Application

In hydraulic mining a jet of water issuing under high pressure from a nozzle excavates and washes the gravel. The gold is recovered partly by cleaning bedrock after the gravel has been stripped away but chiefly by riffles in the sluice box through which the washed gravels and water flow to the tailing dump.

Almost all types of placer deposits can be worked by hydraulicking if water is available but certain physical characteristics have an important bearing on the cost of the operation. If the gravel is clayey, the washing is more difficult but more important. If the gravel is cemented, it can be cut only by high-pressure water. If the grade of bedrock is flat, the duty (cubic yards per miner's inch or other unit) of the water is relatively low, and where gravity disposal of water and tailings is impossible or impracticable elevators must be used to raise them from the pit, further decreasing the capacity of the installation.

Apart from the deposit itself, the water supply is the most important factor in determining the application of hydraulicking and the scale of operation. Under any given conditions, the daily yardage is roughly proportional to the quantity of water used. The quantity excavated likewise is proportional to the head used on the giants, but the higher pressure is of less value in driving and washing and of none at all in sluicing the gravel through the boxes to the dump. As the cutting and sweeping capacity of the giants usually exceeds the carrying capacity of water a stream of flowing water, known as "by-wash", or "bank water", is directed through the pit and into the sluices. If run over the bank, as in ground sluicing, it aids materially in cutting the gravel. The proper relative quantities of high pressure and bank water can be determined only by trial. Frequently the by-wash is supplied by the natural flow of the stream at the mine, the giant water being brought from a considerable distance up the stream or from another source. When an excess of bank water is available it may be used for ground-sluicing, thus increasing the capacity of the plant.

The preparatory or development work necessary to start hydraulicking usually is greater than that for any other form of placer mining except dredging or drift mining. A deposit preferably is opened at the lower end to permit gravity drainage and progressive mining of the entire deposit in an orderly fashion. If the gravel is thick or the grade of bedrock flat, a very long cut may be necessary to reach bedrock at the desired point. This may involve the mining of large quantities of barren or at least unprofitable gravel. A more important element of preparatory cost is the water supply. As heads of 50 to 300 or 400 feet are desired, a mile or more of ditch or flume is almost always necessary to bring water onto the property by gravity flow. A single mine may have many miles of

* The following information on the general subject of hydraulicking consists largely of extracts from *Placer mining in the western United States, Part II, Hydraulicking, treatment of placer concentrates, and marketing of gold,* U. S. Bur. Mines Inf. Circ. 6787, pp. 3-108, 1934, by E. D. Gardner and C. H. Johnson. Many of the tables contained in that publication have been omitted. Tables showing flow of water in pipe-lines and ditches and over weirs will be found in books on hydraulics such as *Hydraulic tables,* by G. S. Williams and Allen Hazen, 115 pp., John Wiley & Sons, Inc., New York. Gardner and Johnson published tables of prices of pipe and other equipment, but these have since changed and are likely to change further from time to time. Hence such prices should be obtained from manufacturers.

ditch, costing perhaps $2,500 per mile, as well as dams and reservoirs and thousands of feet of flumes, tunnels, or inverted siphons. The mechanical equipment of a hydraulic mine ordinarily consists of a few hundred to a few thousand feet of 10-to 30-inch, or larger, iron pipe, one or more monitors, and a varying number of sluice boxes; the cost of equipment ordinarily is small compared to the expenditures necessary for ditches and tailraces.

Although it is obvious that the recoverable gold content of the gravel must pay a profit over operating costs, which usually range from 5 to 20 cents per yard, a surprising number of ventures in hydraulicking have failed because the promoters have not allowed for all the preparatory expenses noted above. Each yard of gravel mined must carry its share of this cost, therefore the size of the deposit is of utmost importance in considering a hydraulic mining venture.

Hydraulicking under suitable conditions is a low-cost method as it yields a larger production per man-shift than any other method except dredging. The initial investment required is less than that for dredging; hence, hydraulicking in small or medium-size deposits may be more economical even though dredging would result in a lower operating cost. When the operations are on a very large scale hydraulicking costs are lower than dredging costs on a comparable basis. Very clayey or bouldery gravels should be hydraulicked as dredging usually is unsatisfactory in such ground.

There is enough similarity in all hydraulic operations that no natural classifications of the method can be made. The methods of attacking the gravel vary too little to make any general distinctions. Factors such as conditions of the gravel, percentages of boulders and clay, grade of bedrock, and quantity and head of the hydraulic water affect the costs, but no general grouping is possible in accordance with any of these heads.

Ditches

Open ditches are used commonly to bring water close to, yet high enough above, the mine to furnish a satisfactory pressure for the giants. At several hydraulic mines in the Western States and Alaska ditches 30 to 40 miles long have been built, and even relatively small operations usually have 5 to 10 miles of ditch line.

Hydraulicking is feasible with heads as low as 40 or 50 feet if the gravel is not tight; however, heads of 80 to 200 feet usually are desired, and if the gravel is cemented it is not uncommon to employ high-pressure equipment and heads ranging from 300 to 400 feet. This consideration fixes tentatively the location of the lower end of the ditch. Its final location may be a matter of compromise, as the head usually can be increased only at the cost of a lengthened ditch or a decrease in the grade. The latter reduces the quantity of water that can be carried in a ditch of given size.

The grades of most hydraulic-mine ditches lie between 4 and 8 feet per mile, or ¾ to 1½ feet per 1,000 feet. Early Californian ditches were run on much steeper grades, but the consequent high velocities caused erosion of the banks and serious breaks were common. Small ditches may be run at grades of 6 to 12 feet per mile without excessive velocities.

Practical velocities range between limits of which the minimum is determined by silting and the maximum by erosion. If the entering water contains sediment it may be deposited in the ditch. This should be guarded against by installing a sand trap near the intake and by designing for a velocity of not less than 1 foot per second. On the other hand, a velocity of more than 3 feet per second is apt to erode the channel and cause breaks.

Table 1. Recommended maximum mean velocities for ditches in various materials

Material	Mean velocity	
	Feet per second	Miles per hour
Loose sand	1	0.7
Sandy soil	2	1.4
Loam	3	2.0
Stiff clay, gravel	4	2.7
Coarse gravel, cobbles	5	3.4
Conglomerate, cemented gravel, soft rock	6	4.0
Hard rock	10	7.0

The figures in table 1 represent mean velocities, the corresponding bottom velocities being 20 or 30 percent lower and the corresponding surface velocities as measured by floating objects possibly being 25 to 35 percent higher.

The velocity, hence the capacity of a ditch, depends upon its slope, the nature of the walls, the size and shape of the water section, and the straightness and regularity of the channel. All these factors, except straightness and regularity of cross-section, are involved in the well-known Kutter formula:

$$V = \frac{\dfrac{1.811}{n} + 41.65 + \dfrac{0.00281}{S}}{1 + \dfrac{n}{\sqrt{R}}\left(41.65 + \dfrac{0.00281}{S}\right)} \times \sqrt{RS}$$

in which

V = mean velocity (in feet per second).

n = roughness coefficient.

S = sine of slope (fall divided by length).

R = hydraulic radius (area of water section divided by wetted perimeter of channel) in feet.

The proper values to use for the coefficient n are a matter of judgment. The values of n recommended by modern designers are shown in table 2.

Earth canals for irrigation usually are designed with $n = 0.025$ or even 0.0225; however, the usual hydraulic-mine ditch is not straight, uniform, nor smooth, and probably the coefficient 0.030 or 0.035 should be applied. Any increase in the assumed value of n results in an approximately equal percentage decrease in the calculated velocity, or a doubled percentage increase in the required slope.

Although the shape of the ditch has a bearing on its capacity, in practice the section is influenced more by the method of excavation. However, for a given area, the section should be so shaped as to have the largest hydraulic radius consistent with economical construction. The usual earth or gravel ditch for hydraulic mines has a trapezoidal section,

Table 2. Values of roughness coefficient n *

Surface	Best	Good	Fair	Bad
Coated cast-iron pipe	0.011	[1]0.012	[1]0.013	
Commercial wrought-iron pipe:				
Black	.012	.013	.014	0.015
Galvanized	.013	.014	.015	.017
Smooth brass and glass pipe	.009	.010	.011	.013
Smooth lock-bar and welded "OD" pipe	.010	[1].011	[1].013	
Riveted and spiral steel pipe	.013	[1].015	[1].017	
Vitrified sewer pipe	.010–11	[1].013	.015	.017
Common clay drainage tile	.011	[1].012	[1].014	.017
Concrete pipe	.012	.013	[1].015	.016
Wood-stave pipe	.010	.011	.012	.013
Plank flumes:				
Planed	.010	[1].012	.013	.014
Unplaned	.011	[1].013	.014	.015
With battens	.012	[1].015	.016	
Concrete-lined channels	.012	[1].014	[1].016	.018
Cement-rubble surface	.017	.020	.025	.030
Dry rubble surface	.013	.014	.015	.017
Semicircular metal flumes:				
Smooth	.011	.012	.013	.015
Corrugated	.0225	.025	.0275	.030
Canals and ditches:				
Earth, straight and uniform	.017	.020	[1].0225	.025
Rock cuts, smooth and uniform	.025	.030	[1].033	.035
Rock cuts, jagged and irregular	.035	.040	.045	
Winding sluggish canals	.0225	[1].025	.0275	.030
Dredged earth channels	.025	[1].0275	.030	.033
Canals with rough, stony beds; weeds on earth banks	.025	.030	[1].035	.040
Earth bottom, rubble sides	.028	[1].030	[1].033	.035

[1] Values most used.
* Part of a more complete list by Horton, R. E., in Eng. News, vol. 75, p. 373, 1916; quoted by Metcalf. L., and Eddy, H. P., in Sewerage and sewage disposal, 2d ed., p. 130, McGraw-Hill Book Company, 1930.

with a flat bottom 2 to 10 feet wide, sides sloping about 45°, and a water depth of one-third to three-quarters the bottom width. The sides should be excavated at a slope that will be stable in use, otherwise caving will result in irregularity of section and consequent loss of capacity. The side slopes recommended for ditches in various materials are given in table 3.

Wimmler,[1] who tabulates data on 35 Alaskan ditches, states that side slopes of 45 to 65° are common but that the higher slopes cut down quickly.

On steep hillsides relatively steeper sides and deeper sections may be cut if the soil is firm to avoid excessive excavation on the uphill side of the ditch. In rock the sides may be vertical; the width should be twice the water depth, as in rectangular channels this results in the least excavation for a given capacity and slope. Likewise, in rock the size may be decreased and the grade increased, thus reducing the yardage of rock excavation. Ditches should be designed to run not more than three-fourths full, allowing 1 to 3 feet of freeboard.

[1] Wimmler, N. L., Placer-mining methods and costs in Alaska: U. S. Bur. Mines Bull. 259, pp. 40-56, 1927.

Table 3. Side slopes recommended for ditches

Material	Side slopes	
	Horizontal to vertical	Degrees
Firm soil, coarse firm gravel	1 : 1	45
Ordinary soil, loose or fine gravel	1½ : 1	35
Loose, sandy soil	2 : 1	25

In porous soil considerable water is lost by seepage. Peele [2] quotes Etcheverry as stating that seepage losses range from as little as 0.25 cubic foot per square foot of wetted surface per 24 hours in impervious clay loam to 1.0 cubic foot in sandy loam and 2 to 6 cubic feet in gravelly soils. It is easily computed that a medium-size ditch, 5 miles long, carrying 1,000 or 2,000 miner's inches, may lose 5 or 10 percent of the intake water by seepage, even in good soil, and in porous soil, as much as 20 percent.   Remedies where the loss is serious are to decrease the size of ditch and increase the velocity; to reduce the velocity to a point at which the silt will deposit and tend to seal the ground; to line the channel with sod, canvas, or concrete; or to substitute flumes for ditches.   According to Wimmler, sod lining often is used in frozen muck in Alaska, sometimes with entire success.

Very few ditches have been built in recent years, and no modern costs are available.   Many methods are available for such work, ranging from hand-shovel and pick work to excavation by power shovel or mechanical ditchers.   A common method is to plow the surface and excavate as near to grade and correct section as possible with teams and scrapers, then finish by hand.   Some instances have been noted where hydraulic giants were used for ditch excavation.   This, of course, is possible only when water is available from a higher ditch line.   Incidentally the hydraulic miner uses high-pressure water for excavating wherever practicable.

The alinement of ditches should be such that excavation to grade will provide just enough bank material to form a channel of the desired size.   Wherever the water level is to be above the original ground surface it is well to plow the surface before excavation starts to form an impervious joint between the bank and ground.   If the material is not such as to form tight banks it may be advisable to excavate the entire water section below the original surface.   The grade must be maintained exactly and the desired section adhered to as closely as possible, as all irregularities have a retarding effect on the flow.   Curves should be made smooth and regular for the same reason.

If there is danger of water from floods or other sources filling the ditch beyond capacity, spillways must be provided at intervals to prevent breaks in the line which would stop operation and be costly to repair.

---

[2] Peele, Robert, Mining engineers' handbook, 2d ed., p. 2147, John Wiley & Sons, 1927.

Measuring Weirs. The simplest method of accurately measuring a flow of water in a stream or ditch is by means of a weir. Numerous types of weirs are used, and there are many formulas for calculating the flow over weirs.

The width of the weir notch should be at least six times the depth of the water flowing over the crest. The bottom of the notch should be level and the sides vertical. The weir notch is beveled on the downstream side so as to leave a sharp edge on the upstream side. The weir should be installed so that the water in the pond above is comparatively still. It must also be high enough so that there is free access of air to the underside of the overflow sheet of water. A stake is driven in the pond 5 or 6 feet above the weir with the top of the stake level with the notch of the weir. The depth of flow over the weir is measured with a rule or square placed on top of the stake. The Francis formula is commonly used for calculating the flow.

$$Q = 3.33\ wd^{3/2},$$

where

Q = quantity of water in cubic feet per second,
w = width of notch in weir,
d = depth of water going over weir.

Flumes

As already stated, most hydraulic-mine ditch lines contain some flume sections. Flumes may be necessary where the line passes around cliffs or over ravines or desirable over porous or shattered ground where a ditch would lose much water or tend to cause slides. On steep hillsides or where ditching would require much costly rock excavation a flume may prove economical; finally, the cost of the line may be lessened and considerable saving made in the total fall by building a flume or trestles across valleys instead of ditching the greater distance around the head.

The same conditions should be considered in designing a flume as in designing a ditch, and the Kutter formula applies equally to both. The formula is used most conveniently in the form of tables or charts.

The low friction coefficient of board flumes may be used to advantage either by building a flume of smaller section or by decreasing the grade below that of the ditch line. If the latter is done a saving in head may be made at the mine. Usually, however, smaller sections and higher velocities are used than for the ditch line. The width of the flume should be twice the water depth and a freeboard of 1 to 2 feet allowed. According to Egleston[3] the usual water velocity is 3 to 6 miles per hour (4 to 9 feet per second). The same author gives the range in grade of 28 prominent California flumes as 9 to 18⅔ feet per mile. The extreme range of 86 well-known flumes in the Western States was 5 to 53 feet per mile, the usual range 10 to 18, and the average slightly under 14. Bowie[4] states that grades of 25 to 35 feet per mile are used where practicable. Such steep grades would permit the use of a relatively small flume section, but the authors believe that usually they would involve inconveniently high velocities; moreover, a longer flume would be required to give the same head.

---

[3] Egleston, Thomas, The metallurgy of silver, gold, and mercury in the United States, p. 152, John Wiley & Sons, 1890.
[4] Bowie, A. J., A practical treatise on hydraulic mining in California, p. 143, D. Van Nostrand Company, New York, 1889.

The construction of wooden flumes has changed little since the early days of placer mining in California. The flume was built in 12- or 16-foot sections of 1½- to 2-inch lumber, 12 to 24 inches wide. The longitudinal joints were made tight by nailing over each a batten ½ inch thick and 3 or 4 inches wide. A flume 34 by 24 inches built about 1930 for water power carries about 600 miner's inches on the flat grade of one-fifth foot per 1,000 feet and would serve excellently for a small hydraulic water-supply line. It differs in construction from the other type chiefly in having splines between all the boards of the boxes and lacking framing in the sills and caps. It was built over 6,800 feet of rugged country at a total cost of $2.50 per foot.

Where the flume is on grade the box units should be set on stringers laid on a carefully cleared and graded surface or on a bench cut in the hillside. Trees or branches that might fall and wreck the flume should be removed. In cold climates the flume may be covered and heaped with earth to prevent freezing. Where the flume is on trestles a walk must be provided; usually a line of plank is nailed over the caps or on alternate sills extended a couple of feet to one side of the box. The grade must be uniform, and at curves the outer edge of the flume should be raised sufficiently for the smoothest possible flow of water, the elevation being determined by trial. Three-foot iron placer pipe cut in two lengthwise has been used successfully for flumes at placer mines in British Columbia.

Diversion Dams and Reservoirs

Diversion dams for hydraulic ditch lines usually are of earth-filled timber cribs or rock-filled cribs faced with boards. Small streams often are dammed by throwing logs across and facing the upstream side with boards. Diversion dams usually are only a few feet high but should be built where possible on solid rock or hardpan, sufficiently wide to be stable and provided with suitable spillways to prevent erosion and scouring out of the foundation.

At mines where the water supply is insufficient for 24-hour operation or where the stream flow is less than is needed to operate at the desired capacity for one shift, reservoirs often are used. If it is impracticable to have the resevoir in the stream itself above the diversion dam, it is usually located at the lower end of the ditch, just above the intake to the pipe lines. Reservoirs may be built by damming a canyon, by excavating a basin on level ground, or merely by enlarging a section of the lower end of the ditch. As a reservoir break might be disastrous to a mine lying directly below it, the work should be done carefully, all leakage checked, suitable gates and spillways provided, and regular inspection maintained.

As both diversion dams and reservoirs tend to act as settling basins it may be convenient to provide gates close to the bottom through which sediment may be flushed as often as necessary.

Mining Equipment

The chief items of equipment used in most hydraulic mines are pipe lines to carry the water under pressure to the places where it is used; giants or monitors for cutting, washing, and driving the gravel; derricks, winches, or other machinery for handling boulders; and sluice boxes for saving the gold and disposing of the tailings. Picks, shovels, and forks are the common hand tools used at placer mines. Power drills run by compressed air or steam may be used if the gravel contains an excessive

FIG. 33.   Pipe installation, 42- and 54-inch.   *Photo by courtesy of Swanson Mining Corporation; reprinted from California Journal of Mines and Geology, January 1941, p. 56.*

quantity of large boulders.   However, hand drills are used at most mines to drill boulders and sometimes to drill cemented gravel or hard-clay strata.   Churn drills are employed occasionally for drilling cemented gravel ahead of hydraulicking.

Pipe Lines

Pipe.   As described previously, ditch lines are used to bring the necessary water to a convenient point above the mine.   From that point a pipe line is laid down the hillside to the pit.   Occasionally, where the grade of a creek is steep, the water will be diverted from the stream directly into a pipe line.   Although wooden stave pipe is used at a few properties, steel pipe is preferred at nearly all hydraulic mines.

Pipe may be made from steel sheets in the mine shops or bought from pipe manufacturers.   Unless a large quantity of pipe is to be used or transportation is difficult, it usually is more economical to buy the pipe already made up.   Various types of steel pipe are used, but light-weight riveted pipe with slip or stove-pipe joints generally is preferred in the Western States as it is cheaper, lighter, and more easily transported and installed than other steel pipe.

Spiral riveted pipe will stand greater pressures and harder usage than the straight riveted pipe, but it is more expensive.   Moreover, flange joints, which are an added expense, generally are used with the spiral pipe.   Ordinary riveted pipe of 10 to 16 United States standard gage material 7 to 46 inches in diameter was being used in western mines in 1932; the diameters used most were 36, 32, 24, 22, 18, 15, 11, and 9 inches.   Large pipes are easily damaged if made of material thinner than 14 gage.   Usually two or more diameters and gages of pipe are used

Table 4.　Maximum quantity of slip-joint water pipe that can be loaded on flat car 8 feet 6 inches wide by 40 feet long, using side stakes 10 feet high

Diameter, inches	Maximum number	
	Sections	Feet
4	1,152	22,320
5	760	14,725
6	544	10,540
7	420	8,137.50
8	310	6,006.25
9	264	5,115
10	220	4,262.50
11	180	3,487.50
12	144	2,790
13	128	2,480
14	112	2,170
15	98	1,898.75
16	84	1,627.50
18	60	1,162.50
20	60	1,162.50
22	40	775
24	40	775
26	40	775
28	30	581.25
30	30	581.25
32	30	581.25
34	30	581.25
36	30	581.25
48	10	193.75

in the same line, mainly as a matter of convenience since this permits nesting in transit. A saving may be made in ocean freight and occasionally in truck hauls by nesting the pipe.

Slip-joint pipe is made in standard lengths of 19 feet 7½ inches each. The sections may be made longer or shorter, however, as required by transportation purposes, provided they are in multiples of 4 feet. The extra pipe required for a slip joint is about 3 inches per section. The standard length of sections of spiral riveted pipe is 20 feet. Placer pipe usually is coated inside and out with an asphalt paint.

A pipe of smaller diameter will withstand a greater pressure than a larger pipe of the same wall thickness; therefore, it is common practice to use smaller diameters as the pressure increases. Reducing the diameter increases the friction in the pipe, and a balance must be struck between loss of effective head in the pipe line and first cost of the line. Branch lines usually have a smaller diameter than the main supply lines.

The minimum carload weight is 20,000 and the maximum 80,000 pounds in California. Carload shipments take fifth class rate. Less than carload shipments of pipe up to 12 inches in diameter take third class and over this diameter one and one-half times the first-class rate.

As used pipe is available in nearly all placer districts, very little new pipe is purchased except for installations of some magnitude. There are no established prices for used pipe.

The Y's and T's needed for branch lines usually are purchased from the pipe manufacturers. A header box may be used when more than two branches are taken out at one place.

Joints and Valves. In making the pipe with slip joints the diameter of one end is slightly contracted. This joint in straight pipe lines will withstand most pressures encountered at placer mines. Slip joints, however, become battered from frequent laying, and other types are desirable where the pipe is moved often.

Flanged or bolted joints are used in some pits for siphons and in very high-pressure lines. The Taylor flanged joint is of forged steel and is welded to the pipe; the price includes bolts and gaskets. The 8-inch size is good for a working pressure of 200 pounds; 8- to 12-inch size, 125 pounds; 12- to 20-inch size, 110 pounds; and above 20-inch size, 75 pounds. The American flange also is complete and attached to the pipe. Most sizes are good for a working pressure of 300 pounds per square inch.

Sometimes a lead joint is used; this consists of a sleeve three-fourths of an inch larger than the pipe, placed around the two ends to be connected. The space between the ring and pipe is filled with molten lead.

Riveted elbows furnished by the pipe manufacturers generally are used for making turns in pipe lines. Taper joints are used where reductions are made in lines. Sudden reductions in size are to be avoided because of the loss of head and strain on the line.

Standard valves are used for diverting water or closing off flow in pipe lines. Valves should be closed slowly and with great care in high-pressure lines; the pressure exerted by the sudden stoppage of flow in the water column may burst the pipe.

Air vents are needed at all crests in hydraulic pipes to prevent a vacuum being formed and subsequent crushing of the pipe. Venting also is necessary to prevent air pockets in the line. The device consists of a leather-faced flap on a hinge bolted on the inside of the pipe. A bail attached to the flap goes through an oblong hole 1¾ by 3 inches in size, cut in the pipe. As the water fills the pipe the flap fits tightly against the inside; as the water falls the flap drops, making a vent.

Pressure Boxes. To give the water entering a pipe line an initial velocity pressure boxes or penstocks are used. A head of 4 to 6 feet usually is provided. A length of large-diameter pipe may be used at the top of the line instead of a penstock. A screen usually is placed at the head of the line to keep out trash. In some installations settling boxes are provided where solid matter may settle out before the water goes into the pipe, as such material may cause rapid wear of the nozzles of the giants.

Laying Pipe Lines. Pipe lines are laid by beginning at the bottom and working upward. Sharp curves are avoided wherever possible, and where used the pipe must be anchored securely to prevent the thrust of the water pressure from pulling the joints apart. Curves in a vertical plane are especially undesirable as they may cause air pockets in the pipe. The pipe should be filled gradually for the same reason. In crossing small ravines a trestle should be built first and the pipe laid on plank for the complete distance.

In laying new pipe with slip joints the outside pipe is started over the end of the other, then heated with a blow torch, which expands the outer pipe and melts the tar previously placed on the end of the lower pipe. As the heating is completed the upper pipe is driven home by hammering on a block of wood placed at the upper end. The tar makes a water-tight connection. Where the pipe has been battered from previous handling, burlap or sacking may be wrapped around the joint before driving. If leaks develop they may be stopped by driving in wooden plugs; sometimes an outside band is required.

In placing pipes with flanged joints they are laid end to end and the bolts put through and tightened up. The flanges usually are attached to the pipe at the factory. This prevents nesting of the pipe in shipping but permits a better joint to be made.

When pressures are very high or when the pipe has vertical or lateral curves, lugs should be riveted on the ends of the pipe with slip joints and the two pipes wired together after the connection is made to prevent the joint pulling out. Similar lugs can be used for anchoring the line to stumps or posts.

In straight pipe lines expansion joints should be placed at intervals of 100 to 2,000 feet, depending upon the conditions to be met. Where pipe lines have lateral curves expansion joints are not needed, as the expansion or contraction of the pipe is taken up in the curved sections. A long, empty pipe line may contract several feet between a warm day and a cold night, and unless provision is made for this contraction the pipe will pull apart. When the pipes are kept full of water this contraction does not occur. Pipe lines are buried in some locations but seldom at western placer mines.

The cost of laying pipe lines depends upon the size of the pipe and the topography and cover of the country. Ten men working 90 days laid 5,000 feet of 36- to 16-inch pipe at the Browning mine, Leland, Oregon, in open country in the spring of 1932.

Flow of Water Through Pipes. The quantity of water that will flow through a pipe line at a placer mine depends mainly upon the diameter of the pipe, the effective hydraulic head, and the size of the nozzle used on the giant at the end of the pipe. Generally the nozzle used is of such a size that the pipe will carry the available water. As the water supply is reduced smaller nozzles are used on the giants.

The effective head on a pipe is the static head minus the loss of head caused by friction. The loss of head depends upon (1) the velocity of the water, (2) the roughness of the interior of the pipe, (3) the diameter of the pipe, and (4) the length. The pressure available and the amount of flow at the end of a long pipe depends mainly upon the last three items. The pressure of the water in the pipe has no effect, by itself, on the loss of head. Formulas have been derived for calculating the loss of head in which coefficients of roughness are used. These coefficients have been derived by experiment for different types of pipes; specifically, however, consideration must be given to the service conditions encountered. No standard of roughness exists, and the degree of roughness of the interior of a pipe does not remain constant. Usually a pipe is chosen about 20 percent larger than would be indicated if there was no loss due to friction. Flow through an unobstructed pipe line of uniform diameter can be calculated from a number of formulas. The Kutter modification of the

FIG. 34.   Cutting with giant.   *Photo by C. V. Averill.*

FIG. 35.   Small giant in operation.   Extra water (by-water) for sluicing in background.   *Photo by C. V. Averill.*

Chezy formula appears to be preferred by hydraulic engineers. The Chezy formula may be stated as:

$$V = C \sqrt{RS}$$

The Kutter modification of the Chezy formula is:

$$V = \frac{\dfrac{1.811}{n} + \dfrac{0.00281}{S} + 41.66}{1 + \dfrac{n}{\sqrt{R}}\left(41.65 + \dfrac{0.00281}{S}\right)} \sqrt{RS}$$

where

V = mean velocity of flow, feet per second;
C = "coefficient of retardation," so-called;
R = mean hydraulic radius of the pipe, that is, $\frac{1}{4}$ the diameter in feet;
S = hydraulic grade or slope, in feet per foot of length of a pipe of uniform size;
n = "coefficient of roughness," so-called.

The value of n for riveted lap-joint pipe up to and including $\frac{3}{8}$ inch thick can be taken as 0.015.

Giants

A giant or monitor is a device with a nozzle for directing and controlling a stream of water under a hydraulic head. The giant can swing horizontally through a full circle and from 11° below to 55° above the horizontal. A box of stones is used to counterbalance the weight of the spout. A giant generally is set up in a pit by being bolted to a log or to timbers securely anchored in bedrock. Nozzles of different diameters can be used up to the diameter of the outlet of the giant to make allowance for variation in the quantity of water used. The giant and nozzle are constructed so that a rotary motion of the jet is prevented, and the water is discharged in a solid column. Giants are made for a wide range of service in 10 sizes, numbered 0 to 9, inclusive.

With heads of 100 feet or more deflectors are used for pointing the larger giants. A common type of deflector consists of a short section of pipe that projects over the nozzle. It turns on a gimbal joint and is controlled by a lever. As the deflector is turned against the jet the force of the stream turns the giant in the opposite direction.

Table 5 shows the sizes, weights, and prices of giants and deflectors made by one manufacturer.[5] Other companies make similar equipment at competitive prices.

Discharge Through Nozzles. Table 6 gives the discharge through different sizes of nozzles under heads from 100 to 400 feet. In this table 40 miner's inches is considered as 1 cubic foot per second. The theoretical flow of water through nozzles exceeds the figures in table 6 by about 10 percent; allowances have been made for friction losses. The flow through nozzles not shown in the table or for different heads can be calculated from the equation:

$$Q = 8CA \sqrt{h}$$

where

Q = cubic feet per second,
A = area of nozzle (square feet),
h = effective head at nozzle (feet),
C = coefficient of discharge ranging from 0.8 to 0.94 (usually taken as 0.9, which makes allowance for friction).

To convert cubic feet to gallons multiply by 7.48.

---

[5] Joshua Hendy Iron Works, San Francisco, California.

Table 5. Sizes, weights, and prices of double-jointed, ballbearing giants and deflectors

Size no.	Giants						Deflectors	
	Diameter of pipe inlets, inches	Diameter of butts with nozzle detached, inches	Shipping weight, pounds	Weight of heaviest part, pounds	List price[1]		Weight, pounds	List price[1]
					Flanged inlet	Slip-joint inlet		
0	5	3	350	105	--------	--------	(2)	-------
1	7	4	390	120	$180	$165	30	$29.00
2	9	5	520	150	225	210	40	32.00
3	11	6	890	210	320	295	45	37.00
4	11	7	1,075	225	365	330	55	40.00
5	13	8	1,475	335	485	460	70	56.00
6	15	9	1,850	520	620	580	75	62.00
7	15	10	2,100	520	725	685	80	67.00
8	18	10	2,300	600	850	805	80	67.00
9	18	11	2,450	690	925	865	90	71.00

[1] Subject to discount because of fluctuations in prices of iron and steel.
[2] None required.

Elevators

Ruble Elevators

The Ruble elevator is named for the Ruble mine in Josephine County, Ore. It consists essentially of an inclined grizzly on a pitch of about 17°, up which the gravel is driven by the stream from a giant. The oversize goes over the grizzly to a rock pile, and the undersize runs down a chute under the grizzly and thence into sluice boxes, usually set at right angles to the elevator. The spacing between bars of elevators in use in 1932 ranged from ¾ to 2½ inches. A 10- or 12-foot apron is used in front of the grizzly. The gravel generally is swept to the foot of the elevator by one giant and through the Ruble by another. The gravel must be washed thoroughly before it is elevated, and the stream of the elevator giant must be used with caution; otherwise, considerable gold may be driven over the top.

Under favorable conditions one giant can handle as much material through the Ruble as another can cut and sweep to it. Under other conditions less than half of the material can be put through the Ruble that one giant working steadily can get to it.

Hydraulic Elevators

Hydraulic elevators are used to raise gravel, sand, and water out of placer pits into sluice boxes. An elevator consists of a pipe with a constricted port or throat and a jet which provides a high-velocity ascending column of water. The relative diameter of pipe, throat, and jet must be proportioned according to the conditions under which the elevator is used. The elevator may also be used as a water lifter.

Table 6. *Flow of water through giants* [1]

Giant no.	Diameter of nozzle, inches	Effective head, feet							
		100		200		300		400	
		Cubic feet per second	Miner's inches	Cubic feet per second	Miner's inches	Cubic feet per second	Miner's inches	Cubic feet per second	Miner's inches
0	1⅛	0.6	22	0.8	31				
0	1⅜	.8	33	1.2	47				
1	2	1.6	63	2.2	89	2.7	109	3.1	125
1	3	3.0	120	4.3	173	5.3	213	6.4	257
2	3	3.3	133	4.7	187	5.7	227	6.6	267
2	4	5.6	227	8.3	333	10.3	410	11.9	477
3	3	3.7	148	5.0	200	6.5	245	7.1	283
3	4	6.0	240	8.6	343	10.6	423	12.2	488
4	4	6.3	253	8.9	357	10.9	437	12.6	504
4	6	13.3	535	19.2	770	23.7	950	27.5	1,100
5	5	9.8	395	13.9	560	16.7	670	19.7	790
5	6	13.5	540	19.3	770	23.8	950	27.7	1,110
6	6	13.8	550	19.6	780	24.1	960	27.9	1,120
6	7	18.7	750	26.7	1,070	33.2	1,330	37.7	1,510
7	6	14.2	570	20.0	800	24.5	980	28.3	1,130
7	7	19.0	760	26.9	1,080	33.3	1,330	38.0	1,520
8	7	19.2	770	27.2	1,090	33.8	1,350	38.3	1,530
8	8	15.2	1,010	35.3	1,410	43.7	1,750	48.7	1,950
9	9	32.0	1,280	45.0	1,800	55.3	2,210	63.7	2,550
9	10	39.3	1,570	55.3	2,210	68.2	2,730	78.7	3,140

[1] Adapted from table in catalog of Joshua Hendy Iron Works, San Francisco, California.

The height to which gravel can be lifted is one-tenth to one-fourth of the effective head of the pressure water at the nozzle of the elevator. Usually the lift will be about one-fifth the head.

The volume of gravel that can be handled by an elevator depends primarily upon the head and volume of pressure water available and to a lesser extent upon the quantity of other water that has to be raised by the elevator. The solids in the water usually are 1.7 to 2.5 percent.

Where little drainage water has to be handled and other conditions are favorable the proportion of the water delivered to the elevator and the giant, respectively, should be about equal, provided the pressure is the same in both. Usually, however, about twice as much water or a correspondingly higher head is required for the elevator. The discharge of the elevator should be high enough to provide dumping ground, otherwise a giant may be needed to stack the tailings. Where plenty of water is available a compound or step-lift elevator may be installed in which one-third of the pressure water is used in the first lift and two-thirds in the second, with a correspondingly larger area of upraise pipe. Thus, the height of the lift may be nearly doubled. Double lifts sometimes are used; that is, the discharge of one elevator goes to the intake of another.

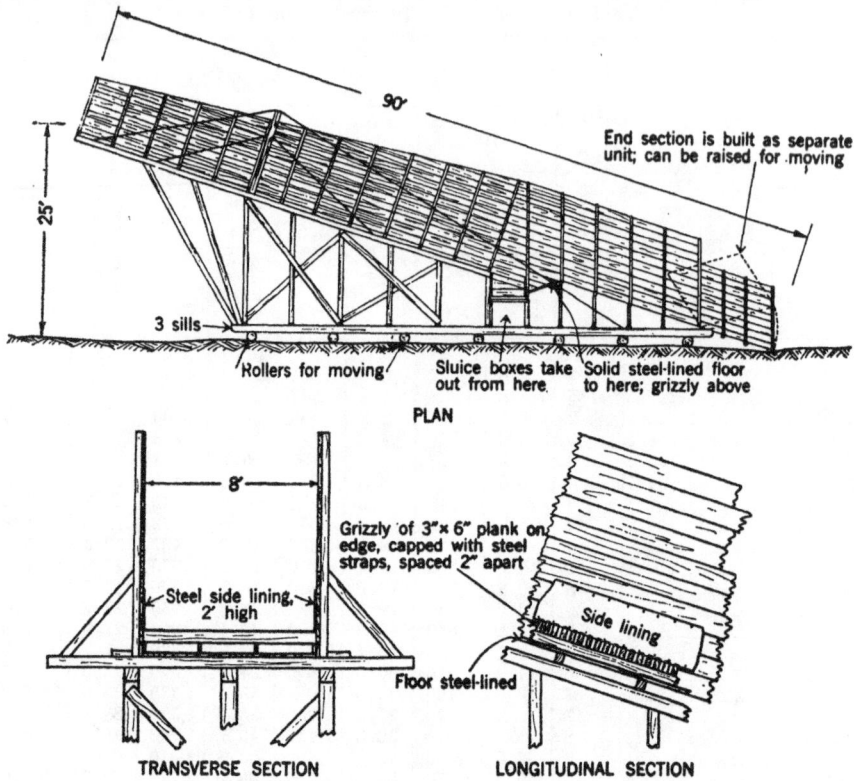

90'

End section is built as separate
unit; can be raised for moving

25'

3 sills

Rollers for moving   Sluice boxes take   Solid steel-lined floor
                          out from here.   to here; grizzly above

PLAN

8'

Grizzly of 3"x 6" plank on
edge, capped with steel
straps, spaced 2" apart

Steel side lining,
2' high

Side lining

Floor steel-lined

TRANSVERSE SECTION   LONGITUDINAL SECTION

FIG. 36.   Ruble elevator used at Redding Creek mine, Douglas City, California.

The elevator discharges upon a cover plate to take the wear in the head of a sluice. Boxes may or may not be used in the pit. The size of the gravel handled is limited by the size of the throat of the elevator. Grizzlies generally are used at the intake. Coarse material reduces the capacity of the elevator; sometimes a Ruble elevator is used in the pit, and only the under-size is sent to the hydraulic elevator.

In clayey ground a hydraulic elevator tends to break up the clay as it goes through the elevator, thus permitting a higher extraction of the gold.

Gravel pumps have been used successfully in alluvial tin mines and in at least one placer mine in British Columbia.[6] As far as known, they have not been used successfully in placer mining in the Western States.

Hydraulic Mining Practices

Conditions varied widely at the hydraulic mines operated in the Western States in 1932. The practices at these mines illustrate the different phases of hydraulic mining and are discussed in this paper. In earlier days, however, when the large hydraulic mines of the West were being worked, more elaborate equipment and larger installations were used than at present. Higher banks were worked, and very large daily yardages were washed, with correspondingly lower costs.

---

[6] Operations of B. Boe on Cedar Creek, Quesnel District: Ann. Report of the Minister of Mines of British Columbia, 1932, p. A112.

Gravels

The gravels being worked at hydraulic placer mines in the summer of 1932 ranged in average depth from 5 to 100 feet; at Relief Hill, where an old mine was being reopened, the depth was 200 feet. The condition of the gravel ranged from soft, easily washed material to gravels that had to be loosened by blasting. The percentage of boulders over 1 foot in diameter ranged from less than 1 to 20. Usually, 5 to 15 percent of all material handled consisted of boulders. Boulders up to 20 inches in diameter were put through the sluices. Clay constituted zero to 15 percent of the total material. At one mine, the Elephant, $2\frac{1}{2}$ feet of gravel was overlain with 40 feet of volcanic ash.

Bedrock at nearly all mines was soft, and the top could be piped off in cleaning up. The slope of the bedrock ranged from $\frac{1}{10}$ inch to 2 inches per foot.

Water Supply

Very few hydraulic mines can operate the entire year. Advantage generally is taken of high-water periods for working the mine. In California the season may begin in November or December, when the winter rains commence, and continue into the dry season of June or July. At most California mines the winter temperature is not low enough to interfere seriously with placer operations. Elsewhere in the West, however, hydraulic placer mining must cease with the advent of cold weather in October, November, or December. At such places, work can not begin until spring when the snow melts and the ground thaws. In many localities placer mining can be carried on only while the snow is melting on the mountains above during the spring months. The length of the 1932 season at the mines visited by the authors ranged from 25 to 225 days. The precipitation during the winter of 1931-32 was normal or above normal in nearly all districts; immediately preceding years, however, were dry, and the number of days operated at the majority of places was much less than in 1932. In exceedingly dry years some mines do not have enough water to operate at all.

Reservoirs are used at most mines. As the flush supply gives out the water may be stored and used periodically for mining. Usually cutting operations cease when water is not available for piping at least $1\frac{1}{2}$ or 2 hours per day. The dwindling supply then will be used for cleaning bedrock and cleaning up the boxes.

Water rights in most of the older placer districts have been adjudicated. The rights of some old placer companies are still intact, and the water can be used without hindrance for operating these mines. However, other water rights in streams have been obtained by power or irrigation companies, and water for placer mining must be acquired from those controlling the rights. In some instances, however, water can be appropriated for placer mining.

As stated before, water under a relatively low pressure may be used for undercutting a bank to assist ground sluicing. Generally, however, a head of at least 40 feet must be available for hydraulicking sand and loam and the easiest cutting gravel. An 80- or 90-foot head usually is required to cut average gravel banks. When the gravel is tight or contains boulders a head of at least 125 feet should be available for hydraulicking. For very tight or cemented gravel, heads over 200 feet should be available. Higher heads give greater cutting and driving power to the giants and thus increase production. High pressures are necessary

FIG. 37. Ruble elevator in use at Redding Creek mine, Douglas City, California. *Reprinted from Division of Mines Bulletin 110, p. 134.*

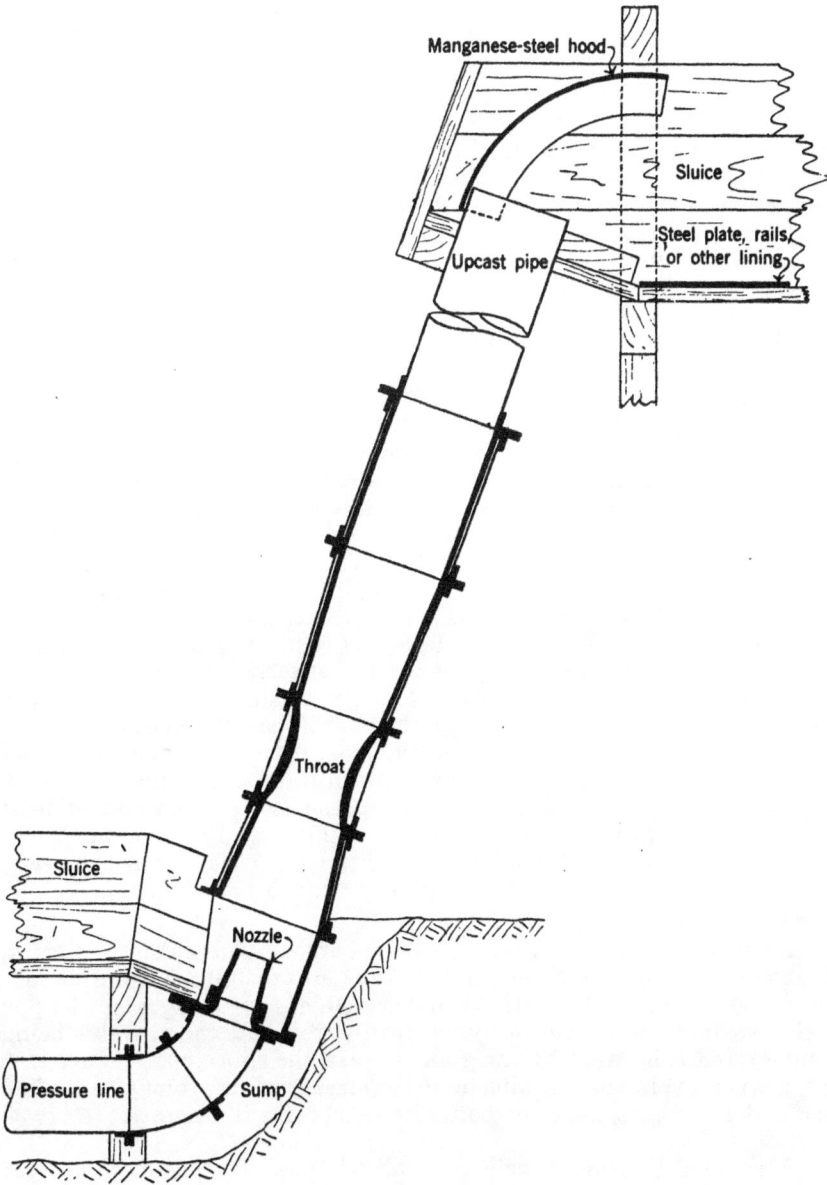

FIG. 38. Hydraulic elevator.

for high banks, as the giants must be set far enough away that caving gravel will not injure the workmen when the banks are undercut. The extreme range of the heads on the giants at the mines was 40 to 450 feet. The usual range was 100 to 300 feet.

In at least 75 percent of the 40 or more operating hydraulic mines visited in 1932 water was conveyed in ditches dug by the early miners. Often, old pipe lines or salvaged pipe were utilized. Some of the present lines are built of pipe first installed 50 years ago. Water was pumped at four mines. Pumping water for hydraulicking, however, has not been generally successful.

The pipe ranged in size from one line with an intake diameter of 46 inches, which was reduced by stages to 24 inches in diameter at the pit, to lines of 10-inch pipe.

Reservoirs where used ranged in size from 0.01 to 15 acre-feet.

Duty of Water. The duty of a miner's inch of water in hydraulicking is defined as the number of cubic yards of gravel which it can break down and send through the sluice in 24 hours. The factors affecting this duty are so varied that it can be compared directly at few mines. An average duty of a miner's inch can not be calculated for the same reason. The duty of water appears to be highest in large-scale operations. Tight or cemented gravel is difficult to break down; a high bank takes less pressure water per cubic yard than a low one; a flat bedrock requires an excessive quantity of water for sweeping; angular rock and gravel with flat or large boulders requires more water to move it than does small-size, rounded material; clay-bound gravels require excessive washing to free the gold; high water pressure is more effective than a low one for cutting or sweeping; and the grade and size of sluices govern the daily yardage that can be washed through them. The calculated duty of water at the mines operating in 1932 ranged from 0.4 to 4.3 cubic yards per miner's inch. In these calculations by-wash water is included.

Conditions at the mines range from the most difficult to at least average. Wimmler[7] reports a duty of as high as 10 cubic yards per miner's inch at some Alaskan placer mines; the usual range, however, was about 0.5 to 1.7 cubic yards per miner's inch per 24 hours.

Piping

After a mine is opened up the gravel bank is undercut by the giant, which allows the overlying material to cave into the pit. The fall breaks the gravel to some extent; it is further reduced by being played upon by the stream from a giant or by by-wash water. As the gravel is being disintegrated it is swept by the giant toward the sluice box. Where the gravel is clay-bound or contains lumps or streaks of clay it may be washed back and forth across the pit bottom one or more times until free from the clay.

A smaller-diameter nozzle generally is used for cutting than for sweeping. As an example, a quantity of gravel may be brought down with a giant with a 4½-inch nozzle. Then the water will be shut off and a 5-inch nozzle put on the giant for driving the gravel to the sluice, or a separate giant with a 5-inch nozzle can be used. Usually two or more giants are set up in a pit even when only enough water is available to run one at a time. One large giant will do more work than two small ones using the same quantity of water. The giants are placed at the

---

[7] Wimmler, Norman L., Placer-mining methods and costs in Alaska: U. S. Bur. Mines Bull. 259, p. 139, 1927.

most strategic points both to cut the bank and wash the gravel to the
sluice box. Where two giants are used at a time one may be used for
cutting and the other for sweeping. The cutting giant is set on an angle
to the face. At the old La Grange mine the streams from two 9-inch nozzles were used together for both cutting and sweeping. Giants may be
set up at the lower end of the sluice to stack the coarse material in the
tailings where the grade is not sufficient for it to be disposed of naturally.

Sometimes a pit is laid out so that all of the gravel washed in one
season is swept to the head of the sluice. After the clean-up the boxes
are extended through the washed-out pit and set up for the next year's
work. At other places the boxes are extended upward as room is made.

When a pit is started a cut is taken across the channel, after which
a diagonal or square face is advanced upstream. In wide channels or
bars two or more parallel cuts may be taken. One pit may be worked
while boulders are handled or bedrock is cleaned in the other. At the
Ruby Creek mine at Atlin, British Columbia, the channel was 250 feet
wide; two 125-foot cuts were made and worked alternately.[8] Wing
dams of timber, logs, or boulders generally are built to guide the water
and gravel into the head of the sluice.

Occasionally the form of the deposit and the contour of the bedrock
are such that the gravel is washed over the side of the boxes rather than
into the end. Then the sluiceway is sunk into bedrock.

At some mines overburden containing little or no gold may be mined
separately. This system has an advantage when dump room at the end
of the main sluice is limited, as the higher material may be disposed of
elsewhere. At one mine, the Salmon River, the light top material was
stripped after the water supply was too low for working the heavier
gravels but was still sufficient to supply one giant. The usual practice,
however, is to mine the full thickness of gravel at one time. The admixture of the top soil and light gravel with the heavier material from near
bedrock may permit moving a larger proportion of boulders to the sluice
than otherwise.

The number of giants used at one time in the mines operating in
1932 ranged from 1 to 4. The size of the giants ranged from nos. 1 to 6
and the diameter of the nozzles from $1\frac{1}{2}$ to 7 inches. A larger nozzle is
used for sweeping than for cutting in about half of the mines and the
same size in the other half. In one mine, the North Fork placer, water
used for sweeping came from a separate source under a lower head; a
smaller nozzle was used than for cutting where the pressure was higher.
The nozzles used in the elevators ranged from $3\frac{1}{4}$ to 4 inches. The distances that the material was elevated were 25, 44, 54, 17, 30, 9, and $19\frac{1}{2}$
feet. The distances the coarse material was elevated by Ruble elevators
were 14, 25, and 11 feet.

Handling Boulders

Where the size and grade of sluices permit, all boulders that can be
moved by the giant are run through the boxes. As stated before, the
upper limit in size at present mines ranged from 4 to 20 inches in
diameter. At some of the early-day large producers boulders weighing
3 or 4 tons were successfully put through the sluices.[9]

[8] Lee, C. F., and Daultin, T. M., The solution of some hydraulic mining problems
on Ruby Creek, British Columbia: Am. Inst. Min. Met. Eng. Trans., vol. 55, p. 90, 1917.
[9] MacDonald, D. F., The Weaverville-Trinity Center gold gravels, Trinity County,
Calif.: U. S. Geol. Survey Bull. 430, pp. 48-58, 1910.

Fig. 39.   Handling boulders with derrick.   *Photo by C. V. Averill.*

In ground sluicing any boulder that can be washed into the sluice by the water usually goes through without trouble. In hydraulicking, however, boulders too large to run through the sluice may be swept into it with a large giant using a high head of water. Boulders too large to be moved by the giant or to run through the sluice are handled in various ways, depending mainly upon the number and size of the boulders encountered and the magnitude of the operations.

In small-scale operations boulders may be rolled by hand to one side or onto cleaned-up bedrock, or dragged away by teams. Occasionally, a boulder too large to handle may be left standing on the floor of the pit and bedrock cleaned up around it. The usual custom when the proportion of boulders is small, however, is to break them up by means of hammers or by blasting and wash the fragments through the sluice. In the larger operations with relatively shallow gravel, as at the Salmon River mine, the boulders may be pulled from the pit by winches or moved by a derrick mounted on a tractor, as at the Salyer mine. At the Diamond City mine a drag line with an orangepeel bucket handled boulders very cheaply under the existing conditions. A relatively narrow cut was being run. The drag line was operated on a bench above the cut and piled the boulders on the bench back of the drag line. The most common method of handling boulders, however, is by means of a derrick. The boulders that can be rolled by hand are loaded onto a sling or a stone boat and hoisted from the pit. Large ones are hoisted by means of chains. At some mines few boulders that can not be moved by the giant are encountered; derricks are used at the head of the sluice for removing those too large to go through. Stumps are handled in much the same manner as boulders.

Cleaning Bedrock

Bedrock usually is cleaned by piping. As much as 2 feet of bedrock may be cut by the giant and the material washed through the sluice. Occasionally a fire hose with a small nozzle may be used for the purpose. When the bedrock is hard and contains crevices, it must be cleaned by hand. The crevices and soft seams are dug out by means of small, flat tools made for the purpose.

Sluice-Boxes and Riffles

Sluice-boxes were laid on bedrock at the most of the mines being operated in 1932. At a few, where high channels were being worked, cuts had been run to bedrock to permit an adequate grade for the sluices. In one mine, a tunnel was used.

Individual boxes were 12 feet long at the majority of places. In a few districts 16-foot boxes were preferred, and occasionally a 10- or 14-foot box was used. The length of the sluice at various mines ranged from 32 to 5,000 feet. The long sluices generally are used only when they are necessary as tailraces. The width of sluice boxes at these mines ranged from 12 to 60 inches. The extreme range in the grade of boxes was from $\frac{1}{8}$ inch to $1\frac{1}{2}$ inches to the foot (1.0 to 12.5 percent). The usual range was from $\frac{1}{4}$ to $\frac{3}{4}$ inch to the foot (2.1 to 6.2 percent).

Riffles serve a twofold purpose, they protect the bottom of the sluice and catch the gold. Both strength and wearing qualities are required in large-scale hydraulic operations where boulders up to a ton in weight may be put through the boxes. Wooden blocks, rails, rock paving, and iron castings, in the order named, were used at the larger mines operated in 1932. When the service was not so severe, poles, angle iron, and Hungarian-type riffles were used. The Hungarian riffles usually were made of wood and were protected from wear on top by strap iron.

At all mines most of the gold was caught in the first few boxes of the sluice. The top boxes were cleaned up twice a season, monthly, weekly, or even oftener. In long sluices the lower boxes were cleaned only at the end of the season or when repairs were needed. At the time of the general cleanup worn riffles were replaced and the sluices repaired if necessary. Quicksilver was used in the sluices at the largest mines, but at the majority it was used only in cleaning up.

Although the sluice is an efficient gold-saving device some gold gets away, especially if the gold is very fine and the gravel carries a relatively large proportion of black sand. To further recover the gold, undercurrents were used at 10 mines. The term "undercurrent" in placer mining is used to designate a device for catching the gold contained in the fine material drawn out through a grizzly in the bottom of the sluice. The undercurrent usually is placed near the lower end of the sluice. At most mines it is not possible to draw all of the material small enough to go through the grizzly to the undercurrent, as not enough water would be left in the sluice to dispose of the coarse material. The quantity drawn off is controlled by the area of the grizzly and the openings between the bars. The grizzly bars are $\frac{1}{8}$, $\frac{1}{2}$, $\frac{3}{8}$, $\frac{3}{4}$, or $1\frac{1}{4}$ inches apart. Undercurrent boxes, or tables as they are sometimes called, are relatively wide to permit a shallow depth of the sands.

The same type of riffle generally is used on undercurrents as in sluices where a screened product is treated. Hungarian riffles, usually similar to those used on dredges, were favored. Steel matting or wire screen over burlap was used at two mines; planks with holes bored in them, angle iron, and stone paving were used at one mine each; and a variety of riffles was used at another mine (Salyer). Quicksilver was used on undercurrents at nearly all of them. An important function of an undercurrent in placers where quicksilver is used in the main sluice is

FIG. 40. Sluice box at hydraulic mine. *Photo by C. V. Averill.*

FIG. 41. Forking boulders along sluice at hydraulic mine. *Photo by courtesy of Roy McGain; reprinted from California Journal of Mines and Geology, October 1941, p. 514.*

Fig. 42.   Stacking coarse tailing with giant; sluice is under grizzly. *Reprinted from California Journal of Mines and Geology, January 1941, p. 50.*

to catch quicksilver or balls of amalgam that may get away in the sluice. As much as 10 percent of the recovered gold may be saved on the undercurrent, but in most places less than 5 percent is so obtained.   At three mines where an estimate was made, 3, 5, and 8 percent, respectively, of the total gold recovered was saved on the undercurrent.   At two places so little gold reached the undercurrents that they were not cleaned up at the end of the 1932 season.

'The sluice-box serves a double purpose in placer mining; it collects the gold or other heavy minerals sought within the riffles of the sluice and conveys the washed material to a dumping ground.   It is an efficient gold saver and is universally used in hydraulicking and ground-sluicing. The principle of the riffled sluice is used for recovering most of the gold on dredges and in other forms of placer mining where the gravels are excavated mechanically.

Sluices are built in accordance with the service to be demanded of them.   Riffles are of varied forms and are made of different materials. Although the form of riffle is chosen largely to fit particular conditions custom in various districts and materials at hand have a bearing upon the practices followed.

The following discussion has a general application and is not confined to any region or method of mining.

Construction of Sluice-Boxes

Sluice-boxes are rectangular in section and are nearly always built of lumber; steel or iron sluices, however, were used at a few washing plants operated in 1932.   The construction of a wooden sluice-box depends somewhat upon the size and service expected of the box; a number of types, however, may be used satisfactorily.

The important features in design are sturdiness and simplicity of construction.   Large flumes may have to withstand severe battering and vibration from the passage of heavy boulders, hence they must be strongly constructed and well braced.   In small flumes this feature is less important, but the use of lighter lumber increases the difficulties of maintenance and prevention of leaks.

The bottom of a narrow sluice should be a single plank if lumber of the desired width is obtainable; for wider boxes two or more bottom planks must be used. The bottom joints may be made tight by the use of soft-pine splines, by batten strips nailed on the outside, or by caulking with oakum or other material. Bowie[10] recommends half-seasoned lumber as most suitable for the construction of boxes. Where local timber is used it is common practice to cut the plank during the dry season or before snow is off the ground. It is not customary to use surfaced lumber for boxes, although a smooth bottom facilitates the clean-up. The lumber should be clear and of uniform size.

For any but small, temporary installations the sides of sluice-boxes should be lined with a wearing surface of rough lumber or sheet iron. Otherwise the entire box must be replaced when the sides are worn out. Board lining is easier to place and replace than sheet iron. In early Californian practice some of the side linings were made of wide, thin blocks nailed on so as to present the endgrain to the wear. Worn iron or steel riffles are used for side lining at some places. Usually only the lower half or third of the side of the box needs this protection, and a single 2-inch board may serve not only for lining but as a cleat to hold down the riffles. False bottoms of planed or rough boards may be used to save wear on the box proper.

Each box should rest on three or four sills, equally spaced. The sills and upright members at the ends of the box serve as battens to prevent leakage at joints. The practice of tapering the box enough to permit a telescope joint is very convenient in small sluices, especially if the boxes must be moved occasionally. Small, three-board boxes may be braced with ties across the top, although this hampers shoveling and clean-up operations. Larger boxes should be braced externally from the ends of the sills. Sills should be weighted with rocks to check any tendency of the sluice to rise. If the sluice is placed in a bedrock or other cut, water under it or at the sides has a strong lifting effect. Moreover, the vibration caused by boulders rolling through the sluice permits fine gravel to be washed under the sills placed on the ground.

As mentioned, the side lining plank may serve as a cleat under which the riffle sections can be wedged to the bottom of the sluice. Otherwise some other provision must be made as the riffles must be held securely. In small boxes it is customary to lay long, narrow boards on edge on top of the riffles and against the sides of the sluice. These boards are wedged down tightly under cleats nailed permanently to the sides of the box. The practice of nailing riffles to the bottom of the box, or using any device that requires driving nails in the bottom or sides, should be avoided as it results in leaks and eventually damages both sluice and riffles. Wooden blocks are the most difficult to secure in place but can be held by the method described in the following section. Rock pavement depends on its weight, on being packed tightly, and sometimes on the slight downstream tilt of the individual stones to resist the shifting action of the current.

Maintenance of Sluice-Boxes

Maintenance work on sluice-boxes consists chiefly in aligning and bringing to grade any boxes that have moved out of position, replacing

[10] Bowie, A. J., Hydraulic mining in California, 3d ed., p. 220, New York, Van Nostrand Company, 1889.

linings, and plugging leaks. Attention to this work at clean-up time will be repaid by greater capacity and freedom from break-downs when the water again is turned into the sluice.

Size

As previously shown, sluice-boxes seldom are built less than 10 inches wide for strictly mining purposes. Eight-inch boxes, however, may be used in sampling or cleaning up. The quantity of water, with its accompanying load of gravel, that will run through a sluice of a given size depends upon a number of factors. The practice at the majority of about 75 hydraulic and ground-sluice mines visited in the preparation of this paper indicates that the carrying capacities of sluices of various widths are within the following limits:

Width of box, inches	Miner's inches of water	
	From	To
12	25	100
18	100	300
24	200	600
36	500	1,300
48 to 60	1,000	3,000

These limits probably represent good practice.

More trouble is experienced from clogging of boxes that are too wide, because the depth and velocity of water are insufficient, than from failure of boxes to carry their load because they are too narrow.

The current velocities required to transport different sizes of material have been studied; works of various authorities are cited by Gilbert.[11] The following table is based chiefly on Dubuat's figures for competent velocity; the figures are adjusted to approximate mean velocity instead of bed velocity. The last three figures are taken from Van Wagenen.[12]

Size of material moved	Mean velocity, approximate feet per second
Sand:	
Fine	0.5
Coarse	1.0
Gravel:	
Fine	1.5
1-inch	2.5
Egg-size	4.0
Boulders:	
3- and 4-inch	5.3
6- to 8-inch	6.7
12- to 18-inch	10.0

    [11] Gilbert, G. K., The transportation of debris by running water: U. S. Geol. Survey Prof. Paper 86, p. 216, 1914.
    [12] Van Wagenen, T. F., Manual of hydraulic mining, p. 88, New York, Van Nostrand Company, 1880.

Well-rounded pebbles are easier to move than angular ones, and rock of low specific gravity is appreciably easier to wash than heavy, dense rock such as greenstone or basalt.

Gold has a better opportunity to settle and be caught in riffles in a wide, shallow stream than in a deeper and narrower stream of the same volume; the wider sluice, however, usually must be set on a steeper grade.

Small- or medium-size boxes generally are roughly square in cross-section; large boxes usually are one-half to two-thirds as deep as they are wide. The water in a sluice should always be more than deep enough to cover the largest boulder that may be sent through. In practice, the depth of the stream in the main sluice at hydraulic mines usually is a fifth to a half the width of the box so as to prevent spills if the box is temporarily plugged by boulders or sand. Where screened gravel is being washed, as in undercurrents or on dredges, wide and shallow streams are necessary for the recovery of fine gold. In "booming" operations the boxes usually are run full in order to handle the relatively large volumes of water that flow for short periods only, and the sluices commonly are about as deep as they are wide. It would be desirable but impracticable to decrease the depth of water by using wider sluices, as flows of 5,000 to 10,000 miner's inches are not unusual when the gate of the reservoir suddenly is opened wide.

Grade of Sluice

Usually the grade of the sluice depends upon the slope and contour of the bedrock. If the gradient of bedrock, however, is too low to permit sufficient fall for the sluice, cuts or tunnels may be run in the bedrock to overcome this difficulty. Very short sluices of only 1 or 2 boxes sometimes are set nearly flat where there is a drop at the end of the box, the gravel being forced through the sluice by the initial velocity and the head of water in the pit.

The opinion of most operators is that about 6 inches in 12 feet is the best grade for average conditions. Grades as flat as 3 inches in 16 feet can be used but only at great loss of capacity. At the Depot Hill mine, where a grade of 3 inches in 14 feet is used, all rocks over 5 to 6 inches in diameter must be left in the pit. Because of the greater friction and the consequent lowering of velocity, steeper grades are needed for small sluices than for large ones; some operators favor grades of 12 inches to 12-foot box. For maximum gold-saving efficiency, as well as for economy in the dump room, grades should be as flat as possible without lowering the velocity to such an extent that the riffles pack with sand. Any increase in slope from that adjustment will increase the capacity of the sluice, increase the wear on the sluice, and decrease the efficiency of the riffles, resulting in gold losses if carried to extremes or if the gold is very fine. If water is scarce, gold recovery may well be sacrificed to capacity. Bowie [13] states that grades of 10 to 24 inches were used in some Forest Hill Divide (California) mines for this reason. Increasing the proportion of water to solids decreases the tendency of riffles to pack with sand.

Sluice capacity increases with grade but more rapidly; that is, doubling the grade of sluice boxes will more than double the quantity of gravel that can be put through the boxes by a given flow of water. The absolute increase cannot be predicted closely as coarseness of gravel,

―――――――
   [13] Bowie, A. J., A practical treatise on hydraulic mining in California, 3d ed., p. 220. New York, Van Nostrand Company, 1889.

velocity, and shape of the box appear to have some bearing on the relation of capacity to slope. For instance, Bowie [14] cites a mine at which changing the grade from 3 to 3½ inches in 16 feet increased the quantity of gravel sluiced through the same boxes with the same flow of water by about one-third.

The established grade should not be decreased anywhere along a sluice, otherwise gravel may accumulate where the current loses velocity. If the water and gravel, however, enter the first box with considerable speed, say, from the discharge of a hydraulic elevator, the first boxes may be placed on less than the regular grade. Bends or curves are undesirable as they complicate construction and induce clogging and running over. When a curve is unavoidable it should be as gradual as possible, the outside of the sluice should be elevated a fraction of an inch, and the grade should be increased perhaps an inch per box at and immediately below the curve. Similar rules apply to turn-outs or branches, and drops of 3 to 4 inches should be provided at junctions to check the deposition of gravel at these points. Such drops occasionally are inserted in straight sluices if the grade is available, particularly if the gravel is a difficult one to wash or if heavy sand tends to settle to the bottom. A drop of even a few inches from one box to the next has a disintegrating effect and mixes the material passing through the sluice, thus assisting gold recovery. At one place where drops were provided at intervals between different types of riffles, 25 percent of the gold recovered in the sluice was found at the drops.[15]

Theory of Gold-Saving by Riffles

The function of riffles is to hold back the gold particles that have settled to the bottom of a flowing stream of water and gravel. Any "dead" space in the bottom of a sluice-box, where there is no current, fills quickly with sand and thereupon loses most of its value as a gold saver, unless the sand remains loose enough to permit gold to settle into it; therefore, the shape of riffles is important, regardless of the fact that under some conditions, as with coarse gold and free-washing gravel, all forms of riffles are almost equally efficient. The riffles should be shaped so as to agitate the passing current and produce a moderately strong eddy or "boil" in the space behind or below it, thus preventing sand from settling there and at the same time holding the gold from sliding farther down the sluice. In other words, riffles, for maximum efficiency, should provide a rough bottom that will disturb the even flow of sand and gravel, will retain the gold, and will not become packed with sand. Where grade is lacking the riffles must be relatively smooth, so as not to retard the current unduly; under these conditions the sluice must be long enough to compensate for the loss in gold-saving efficiency of the individual riffles.

Natural stream beds act as gold-saving sluices, not because they are particularly efficient as such but because most gold is "hard to lose" and the streams are long.

[14] Bowie, A. J., op. cit., p. 266.
[15] Theller, J. H., Hydraulicking on the Klamath River: Min. and Sci. Press, vol. 108, pp. 523-526, March 28, 1914.

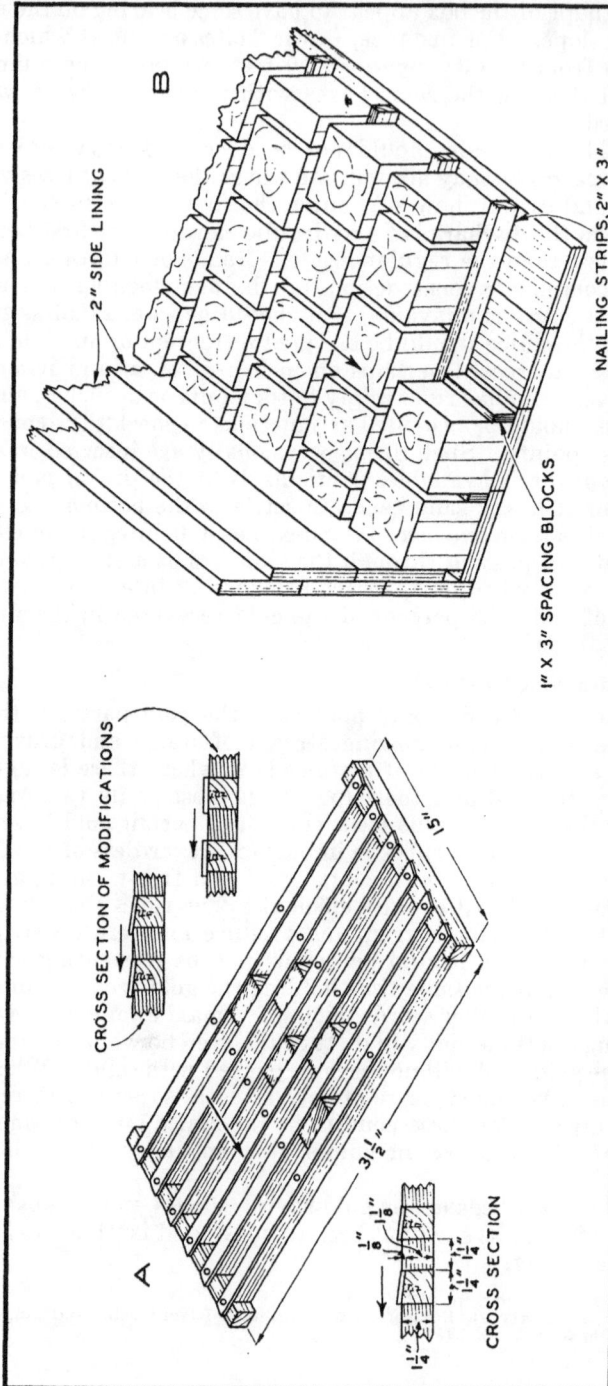

FIG. 43.   Dredge-type riffles and wooden block riffles.

Types of Riffles

Riffles, of course, should be designed so as to save the gold under the existing conditions. They should also be cheap, durable, and easy to place and remove. Not all these qualities are found in any one type.

Sluice-box riffles may be classified roughly as transverse, longitudinal, block, blanket, and miscellaneous roughly surfaced ones, or, according to material, as wood block, pole, stone, cast iron, rail, angle iron, fabric, and miscellaneous. Usually more than one type of riffle is used, although in California very long sluices have been paved entirely with wood-block riffles.

Of about 80 hydraulic, ground-sluice, and mechanically worked placer mines visited in 1932 by the authors, approximately 25 percent used riffles of the transverse variety, loosely termed "Hungarian," consisting generally of wooden crossbars fixed in a frame and sometimes capped with iron straps. About 20 percent used the longitudinal pole type, 15 percent wooden blocks, and 15 percent rails, the last being placed crosswise or lengthwise. Angle-iron riffles, wire-mesh screen or expanded metal on carpet, blankets, or burlap, rock paving, and cast-iron sections together made up the remaining 25 percent. The only general rule observed was that the size of the riffles was roughly proportional to the size of the material to be handled and that for fine material, particularly the screened gravel washed in most of the mechanically operated plants, the dredge-type riffle found most favor.

For a small or medium-size sluice (if lumber is costly and a plentiful supply of small timber, such as the lodge-pole pine so common in many Western States, is available) peeled pole riffles are perhaps the most economical and satisfactory of the various types. Those of transverse variety may have a somewhat higher gold-saving efficiency, but undoubtedly they retard the current more and wear out faster. Poles 2 to 6 inches in diameter may be used, spaced 1 or 2 inches apart. Such riffles are cheap but wear out rapidly. The sections should be a third or half the box length for convenience and 1 or 2 inches narrower than the sluice. At the Golden Rule mine 6-inch pole riffles had to be replaced every 10 days or after each 1,200 cubic yards had been sluiced. The sluice was 30 inches wide and had a grade of 8 inches in 12 feet. At other mines poles last several times as long.

If sawed lumber can be obtained cheaply, riffles similar to the one described may be made of 1- by 2-, 2- by 2-, or 2- by 4-inch material. The top surfaces of the riffles may be plated with strap iron. Transverse riffles of this type may be slanted downstream and the top surfaces may be beveled to increase the "boiling" action, as with the dredge riffles. The effectiveness of this practice is not known, and the authors know of no conclusive tests having been made. Longitudinal riffles of 2- by 4-, 3- by 4-, or 2- by 6-inch material are used at some places.

Wooden-block riffles are held by Bowie[16] to be unexcelled in regions where the material is available cheap. The blocks are 4 to 12 inches thick and of corresponding diameters or widths. They may be round, partly squared, or cut from square timber. One- or two-inch wooden strips separate the rows of blocks, and they are held securely in place by nails driven in both directions. Wooden-block riffles are perhaps the hardest of all types to set because of their tendency to float away. They

---

[16] Bowie, A. J., op. cit., p. 225.

must be nailed to the spacing strips, as stated, and wedged securely at the sides.   The spacing strips are held down at either end by the side lining of the sluice.   Wooden-block riffles are durable, can be worn down to half their original thickness or less, and if made of long-grained wood (such as pitch pine, which "brooms" instead of wearing smooth) may catch some gold in the endgrain.   When discarded, they are commonly burned and the ashes panned to recover any gold so caught.   The life of 10- or 12-inch wooden-block riffles may be a few months to several seasons and, according to Bowie, ranges from 100,000 to 200,000 miner's inches of water; that is, with a flow of 1,000 inches one would last 100 to 200 days.   The grade of the sluice apparently has much to do with the life of block riffles.   At the Superior mine where the sluice was 48 inches wide and had a grade of $2\frac{3}{4}$ inches in 12 feet a set of blocks lasted two seasons, during which time 140,000 cubic yards was sluiced.   At the Salmon River mine the grade was 7 inches in 12 feet and the width of the boxes 30 inches.   Here block riffles lasted 60 to 70 days, during which time about 18,000 cubic yards was washed.   On account of differences in the wearing rates only one variety of wood should be used in a section of a sluice. Douglas fir wears longer than other native western conifers.

Stone riffles are durable and fair gold catchers.   Stones ranging from the size of cobbles to 8 or 10 inches in diameter are packed closely on the bottom of the sluice.   They may be held at intervals of a few feet by transverse wooden strips.   In some instances the stones are roughly hand-shaped and set similarly to street paving.   Stone riffles are difficult to set and generally are not used in portions of a sluice that are cleaned up frequently.   Their main advantage is their long life.   Because of their roughness, stone riffles require a steeper slope than wood blocks, a feature that sometimes would prohibit their use.

Where large quantities of gravel are put through sluices, iron or steel riffles generally are preferred.   Their superior wearing quality as compared with that of wood permits longer runs without stopping to replace the riffles.   Their durability may more than compensate for their higher cost.

Steel rails and angle iron are common riffle materials used in various ways.   Old rails or angle iron can often be obtained cheaply in mining districts or near railroads.   Various other steel products such as pipe and channels have been utilized for riffles.   Cast iron is also used and has the advantage of a lower first cost than steel rail or angle iron.

Iron or steel riffles should not be used in units too long to be handled readily.   Rope blocks on movable tripods have found favor at some places for lifting heavy riffle sections.

When used as transverse riffles lengths of steel rail usually are set upright, the flanges almost touching or not more than 1 or 2 inches apart. Where grade is lacking and gold saving is not particularly difficult, longitudinal rail riffles make excellent paving for a sluice as they provide a smooth-sliding bottom for the gravel and boulders.   The rails ordinarily are bolted together by tierods passing through wood, pipe, or cast-iron spacing blocks, forming riffle sections the width of the sluice and any convenient length.   At La Grange mine in Trinity County, California, 40-pound rails costing about $125 per ton proved more satisfactory than wood riffles.[17]   When 16- by 16- by 13-inch wood blocks were used the

---

[17] MacDonald, D. F., op. cit.

riffles tended to "sand up." Moreover, the blocks had to be replaced every 2 or 3 weeks. Lengthwise rails 8 inches apart lasted 2 months and rails 5 inches apart, 4 months. Strangely enough, transverse rails 5 inches apart lasted 6 months. The rails were spaced by cast-iron lugs and set right side up on timber sills. When the head of the rail was worn off the remainder was used for side lining. This sluice was handling a flow of about 4,000 inches of water and 1,000 cubic yards of material per hour, boulders as large as 7 tons being washed through. The eddies behind the rails were believed to be the cause of the improved recovery as compared with that using block riffles. The lower part of the branching sluice line was cleaned up every other season only. The combination of steel rails and wooden sills used at La Grange mine appears to make an excellent gold saver, and modifications have been used at many large mines.

Angle iron is commonly used for making riffles. Many methods of assembling the lengths of angle iron into riffle sections are in use, and no one method can be said to excel. The irons may be set with flat upper surfaces or inclined slightly to increase the riffling action. Usually the gap between the riffle bars is $\frac{1}{2}$ to 1 inch. The effectiveness of this type of riffle is believed by some operators to depend largely on the vibration of the riffles under the impact of boulders, which keeps the sand trapped under the angles in a loose condition favorable to gold saving.

Cast-iron riffles of all shapes and sizes have been used. If available at low cost they are very economical, as they wear slowly, can be quickly and securely placed, and are efficient gold savers if designed so as not to pack with sand. In an undercurrent at the Indian Hill mine, California, cast-iron riffles were in use that were 4 feet long, shaped like angle irons, and had equal $3\frac{1}{2}$-inch legs $\frac{7}{8}$ inch thick.

For shallow sluice streams carrying only fine material various gold-saving materials are used, including brussels carpet, coco matting, corduroy, and burlap. These may be held down by cleats or by wire screen. Fabrics often are used in combination with riffles to catch fine gold and hinder its being washed out of the riffles by eddies. A corduroy woven specially for a riffle surface is used by some large Canadian lode-gold mines to catch their "coarse" gold before flotation or cyanidation. As such gold would be considered fine by most placer miners it seems probable that such a fabric would be useful for treating finely screened placer sands. The corduroy in question has piles about $\frac{1}{4}$ inch wide and $\frac{1}{8}$ inch high, spaced about $\frac{1}{4}$ inch apart. The piles are beveled slightly on one side. The cost in Canada is about $1.00 per square yard.

Heavy wire screen such as that used for screening gravel makes an excellent riffle for fine or medium-size gravel in fairly shallow sluice streams, and generally it is used with burlap or other fabric underneath.

Expanded metal lathing and woven metal matting are common types of riffles for fine material and are used with carpet or burlap. If the thin strands of metal slant considerably in one direction, the material should be placed with this direction downstream. Eddies in back of the strands will then form gold catchers, whereas if the recesses face upstream they will at once fill with a tight bed of sand and lose their effectiveness.

Solid-rubber riffles were noted at one washing plant. Sponge-rubber riffle material is on the market, but it was not observed in use and nothing is known by the authors of its merits or cost.

FIG. 44. Looking down from hillside at sluice and undercurrents. Lower end of sluice is in center of photo. *Reprinted from California Journal of Mines and Geology, January 1941, p. 61.*

Another form of riffle often used as an auxiliary to other types is a mercury trap, consisting of a board the full width of the sluice with 1- or 1½-inch auger holes in which mercury is placed. Instead of round holes, transverse grooves or half-moon-shaped depressions, 2 to 4 inches wide and with the rounded, deep side downstream, may be cut in a wide board and partly filled with mercury. These riffles have no apparent advantage over the ordinary transverse-bar type and are suitable only for fine gravel, as large pebbles would splash the mercury out of the traps.

Many ingenious and odd kinds of riffles are encountered in the field, some of which have been patented. It is very unlikely, however, that the advantage of any unusual or freakish design of riffle is sufficient to offset the cost of royalties on patented inventions.

Undercurrents

An undercurrent, as defined before, is a device for sluicing separately a finer part of the gravel passing through the main sluice. The fine material and a regulated quantity of water pass through a stationary grizzly in the bottom and usually near the end of the sluice to one or more wide sluice-boxes, commonly called tables, paved with suitable riffles. If the main sluice is in sections, with drops between, the water and sand may be returned from the undercurrent tables to the main stream, and several undercurrents may be installed at convenient points along a sluice.

The screen or grizzly in the main sluice may present the most difficult problem in building a satisfactory undercurrent. The screen should divert all the undersize yet not take so much water that it causes plugging of the main sluice below the undercurrent. The proper size of opening can be determined only by experiment. A screened or barred opening, the full width of the main sluice and a few inches to a foot or more long, will usually draw off as much water as can be spared. New water may be added to either the undercurrent or main sluice if the screen opening does not take out the right quantity for successful operation. Usually minus ¼- to ½-inch material is desired for the undercurrent, and either punched-plate screen or iron-bar grizzlies may be used to make the separation. Grizzlies should be made of tapered bars or screens punched with tapered holes with the largest openings downward, otherwise they will plug and render the undercurrent ineffective.

Because undercurrents need a wide, shallow stream, grades of 12 to 18 inches per 12 feet must be used, depending largely on the type of riffle. Cobblestone, block, transverse or longitudinal wooden strips, rails, screens, or fabrics may be used for riffles. Often several types of riffles are used on successive parts of one undercurrent. Undercurrents may be a few to 25 or 30 feet wide and 10 to 50 feet long

Most of the gold recovered by undercurrents is so fine that it does not settle in the relatively swift, deep current of the main sluice, but part consists of gold that is freed from its matrix of clay by dropping through the grizzly and rolling over the undercurrent riffles. All coarse gold is saved in the first few boxes of the main sluice unless conditions are radically wrong. Unless the undercurrent is installed at the end of the sluice, or at least below where gold is recovered, not all the saving in the undercurrent should be credited to its installation. In the early days when hydraulicking was at its height undercurrents were much favored, sometimes 5,000 to 10,000 square feet of undercurrent being used along a single sluice line. The gold saved in them occasionally exceeded 10 percent of the total clean-up but more often was less than 5 percent. As this recovery usually was effected by 5 or 10 large tables and as considerable would have been saved by the main sluice without the undercurrents, the economy resulting from their use was perhaps doubtful. Bowie[18] presents details of the use of undercurrents in early Californian practice and indicates that their particular field lay in the treatment of cement gravels. Of the several undercurrents observed by the authors in use in 1932 it is doubtful if many were justifying their installation.

Operation of Sluice-Boxes

Under favorable conditions a properly designed and constructed sluice-box requires little attention other than periodic clean-ups and minor repairs which are made at the same time. Unfortunately, such a combination rarely occurs, and an appreciable part of the miner's operating expense is chargeable to work along the sluice lines.

The best results are obtained when a steady flow of water and gravel passes through the sluice. An excessive flow of clear water through the sluice will bare the riffles, causing some gold to be lost. On the other hand, a continued overload of gravel will plug the sluice at some point so that sluicing must be stopped for the time needed to clear the obstruction; this time lost may be appreciable. If plugging can not be prevented by increasing the grade or the flow of water or reducing the feed, one or more sluice tenders must work along the sluice with forks or shovels to keep it open. This added cost may be serious at small mines. All effort should be directed toward getting the gravel into the box and letting the water do the rest.

Large boulders are another cause of expense and lost time. When the maximum size of boulder that the sluice will carry is known, all boulders larger than this should be prevented from entering the boxes. Relatively little work directed to this end will save hours of delay in clearing plugged sluices and unnecessary wear and tear on the boxes and riffles.

An exception is found in the operation of "booming." A necessary condition of this work is a heavy head of water which usually fills the sluice to the brim. Sometimes little or no work can be done in the pit while the water is on, and the entire crew may profitably patrol the sluice with long-handled shovels to guard against stoppages which might be disastrous because of the large flow of water and gravel. Before each "boom" all oversize boulders should be moved out of the course of the water.

Cleaning Up

Clean-up time should be kept to a minimum. This can be done by cleaning up as seldom as practicable and by using efficient methods. Large hydraulic mines, particularly if the water season is short, clean up only once a season except perhaps the upper one or two boxes. Dredges clean up every 10 days or 2 weeks, because large amounts of gold are recovered in relatively short sluices with attendant possible loss when the upper riffles become heavily charged. This necessary delay is used for routine repairs on the dredge. In ground-sluicing the clean-up period ranges from weeks to months, while in shoveling-in operations the sluice

---

[18] Bowie, A. J., op. cit., pp. 252-262.

may be partially cleaned up daily. The danger of theft from the upper, richer boxes can be lessened by filling them with gravel at the end of each day's work.

The general principle is the same in all clean-up operations, but practice differs widely. Clear water is run through the sluice until the riffles are bare, the stream being reduced enough to prevent washing out the gold. Then the water is turned off or reduced to a very small flow, and the riffles of the first box are lifted, washed carefully into the box, and set aside. Any burlap or other fabric used under the riffles likewise is taken up, rinsed into the box, or placed in a tub of water where it can be thoroughly scrubbed. Then the contents of the sluice are shoveled to the head of the box and "streamed down" with a light flow of water. The light sand is washed away, and rocks and pebbles are forked out by hand. This operation is repeated until the concentrates are reduced to the desired degree of richness. Gold or amalgam may be scooped up, as it lags behind the lightest material at this stage, or all the black sand with the gold, mercury, and amalgam may be removed and set aside for further treatment. Successive boxes are treated similarly, until the sluice is bare. The last step is to work over the whole sluice with brushes and scrapers to recover gold and amalgam caught in cracks, nail holes, or corners. At the Wisconsin mine a small box was set up in the main sluice and the concentrate from the riffles shoveled into it to reduce the bulk. At the Round Mountain mine the concentrate from the lower section of the sluice was treated in a quartz mill.

Use of Quicksilver in Sluicing

Quicksilver is used at nearly all placer mines. If it is not used to catch gold in the sluices at least it is probably used in extracting the gold from the concentrates. The average market price for mercury in 1932 was about $58 per 76-pound flask, but quicksilver purchased in 5- or 10-pound lots from a chemical-supply house cost about $1 per pound. Except in districts where placer mining was particularly active, drug stores or other local retailers charged about double this amount. The price in January 1934 was $67.54 per flask, and in late 1945 was $106 per flask.

The characteristics of quicksilver that make it of value to the miner are: (1) Its power of amalgamating with gold and silver; (2) its high specific gravity (13.5), which causes it to lie safely under a stream of water and gravel, floating off on its surface everything but the native metals; and (3) its relatively low boiling point (about 675° F.) far below red heat, which allows it to be driven off by heat from the gold with which it has amalgamated.

Amalgamation is a process in which mercury alloys with another metal. All metals but iron and platinum amalgamate more or less readily. Clean and coarse placer gold alloys readily, but if the gold is partly coated with iron oxide or other substances (for example, "rusty" gold) it amalgamates with difficulty. The mercury itself should be clean enough to present a smooth, shiny surface; the presence of some gold or silver in the quicksilver, however, is said to facilitate amalgamation, that is, to make it more "active."

Quicksilver is placed carefully in the sluice-boxes, where it finds its way to the many recesses in the riffles and lies in scattered pools, ready to seize and hold any particle of gold that touches it. It is used in this

9—56968

manner in almost all important hydraulic operations, but some operators place it in the boxes only shortly before the clean-up, evidently believing that the added gold saved by its use during sluicing does not compensate for the loss of the mercury that passes through the sluice with the tailings or escapes through cracks or other leaks. In exceptional instances the conditions are such that the mercury "flours," that is, breaks into minute, dull-coated drops. Flouring is aggravated by agitation or exposure of the mercury to air. The common practice of "sprinkling" it into sluice boxes may be condemned on this ground, as well as for the reason given by Bowie[19] that the fine particles formed by careless sprinkling are more readily washed away and lost. Flouring is responsible for the most serious losses of quicksilver with the tailings.

Even under the best conditions, 5 to 10 percent of the mercury used is lost. If steep grades, heavy gravel with consequent severe pounding and vibration, old and leaky sluices, or other adverse conditions exist, the loss of mercury may be 20 or 25 percent.

Only clean mercury should be placed in a sluice; even this tends to become fouled or sluggish and to lose its effectiveness. The best cleansing process is retorting, which is discussed later. However, straining the mercury through chamois or tightly woven cloth removes some of the surface scum and foreign material, or the mercury may be treated with potassium cyanide or other chemicals to dissolve the impurities. It should be handled as little as possible and kept from contact with grease or other organic material.

Wilson[20] suggests a cow's horn, sawed off near the small end to leave a hole that can be stopped with the finger, as a useful implement for charging sluices. Most miners charge the sluice from stoneware or heavy glass bottles such as are used for champagne.

Mercury should be kept or carried only in iron, glass, or earthenware containers because of its tendency to amalgamate with zinc (galvanized iron), tin, or other metals.

The quantity of quicksilver used differs according to conditions and custom. According to Bowie,[21] 200 or 300 feet of 6-foot sluice should receive about three flasks (225 pounds) as a first charge and a 24-foot square undercurrent, 80 or 90 pounds. At the Depot Hill mine one flask is placed in the first four or five boxes each month during the washing season. At another mine two flasks were used in a season during which 100,000 cubic yards was washed. Dredge tables, with areas of 1,000 to 10,000 square feet, are charged with 150 to 3,000 pounds of mercury. According to Janin,[22] a 7-foot dredge with a table area of 2,800 square feet uses about 1,000 pounds on the sluices and in the traps. Probably in common practice the range is $\frac{1}{10}$ to $\frac{1}{4}$ pound per square foot of sluice area.

The sluice should be run long enough to plug all leaks before the mercury is added. Usually only the upper 2 or 3 boxes or a quarter or half of the sluice at most is charged with mercury, as otherwise considerable loss occurs. During a run more mercury is added periodically.

[19] Bowie, A. J., op. cit., p. 244.
[20] Wilson, E. B., Hydraulic and placer mining, 3d ed., p. 230, John Wiley & Sons, 1918.
[21] Bowie, A. J., op. cit., p. 244.
[22] Janin, Charles, Gold dredging in the United States: U. S. Bur. Mines Bull. 127, p. 143, 1918.

Whenever the sluice is run down enough to expose the riffles the mercury can be examined. If it does not show here and there with clean surfaces nearly to the top of the riffles, more is added. As the quicksilver takes up gold near the head of the sluice it becomes pasty and finally quite hard and more should be added to keep it in a fluid condition.

The use of mercury in recovering gold from sluice-box concentrates is discussed in the following section.

Amalgamating plates should be used only in treating fine material, generally well under one fourth inch in size and preferably not coarser than 10-mesh, as larger particles abrade the plates too rapidly and prevent building up of the amalgam. Consequently, the application of plates to placer mining is limited to the stamp milling of some drift-mine gravels and the treating of fine undercurrent or other screened sands. The use of plates in stamp milling is a phase of metallurgy beyond the scope of this paper, and reference is made to any standard text or handbook on gold milling.[23]

None of the other applications of amalgam plates to placer mining is of particular importance, probably because the recoveries seldom have justified the labor and expense. Plates may be set in undercurrents treating finely screened sands, such as beach sands or the Snake River gold-bearing sands. They usually are covered with burlap to assist in retaining the gold until it has come in contact with the amalgam. Many other amalgamating devices have been applied to such material, but none is known to the authors to have been of greater value than properly designed sluices and riffles.

Separation of Gold and Platinum-Group Metals From Concentrates

General

No sluice box or other type of gold saver used in large-scale placer mining makes a clean separation of the valuable minerals. The concentrate obtained must be treated further to make a marketable product. Concentrate obtained in cleaning bedrock in some types of mining is treated similarly to sluice-box concentrates.

The concentrate may be cleaned by panning or rocking in auxiliary sluices or by blowing, or it may be amalgamated in a special type of apparatus. The treatment will depend mainly upon the scale of operations, the proportion of black sand in the concentrate, and the characteristics of the gold. The general methods of cleaning concentrate with pans, rockers, or small sluices are the same as those in small-scale mining, described in a previous paper,[24] except that more care is required and smaller quantities are treated at one time. In treating small quantities of concentrate, however, it should be remembered that colors of gold so fine as to present great difficulty in their separation by panning or rocking are probably of small value, and their loss would be inconsequential.

If precise results are desired for sampling or testing, the concentrates should be amalgamated.

[23] See also Chapman, T. G., Treating gold ores: Arizona Bur. Mines Bull. 133, Univ. Arizona, 1932; a brief, nontechnical description of the methods of treating gold ores.

[24] Gardner, E. D., and Johnson, C. H., Placer mining in the western United States, general, hand-mining, and ground-sluicing: U. S. Bur. Mines Inf. Circ. 6786, pt. 1, 73 pp., 1934.

Panning

Panning is the simplest method of separating the valuable constituents from the worthless material and generally is used in small-scale operation.   The method, however, is tedious if the gold is very fine and the concentrate contains much black sand.   Mercury may then be used in the pan to collect the gold.

Rocking

Larger quantities of concentrate may be treated in a rocker and the resulting semifinal product cleaned further in a pan.   A final or almost final product, however, can be made in a rocker, the flat, smooth bottom of which, set on a gentle grade with screen and canvas baffle removed, offers an ideal surface for the purpose.

The concentrates are placed at the upper end, and a small stream of water is poured over the sand while the rocker is swayed gently back and forth.   The lighter material is washed down to the riffle at the lower end, and the coarser particles of gold are left behind.   These are picked up with a scraper, and the operation is repeated, a portion of the concentrates presently being discarded with each washing until at length all gold of appreciable value has been recovered.   This method is satisfactory with ordinary concentrates, but if the gold is very fine, flaky, or particularly light, porous, or angular, the separation is tedious and unsatisfactory, and amalgamation is to be preferred.

The same general method may be used in the mine sluice to recover the bulk of the gold amalgam.

Auxiliary Sluices

Sometimes an auxiliary sluice is used to reduce the volume of concentrate from the mine sluice or to treat concentrate after it is amalgamated.   The small sluice in turn must be cleaned up.   At one mine a 12-inch box was set up in the main sluice into which was shoveled the riffle concentrate from below.

Blowing

The grains of sand remaining in an almost final product may be removed from the gold by blowing.   A flat metal or paper sheet, such as a piece of drawing paper or a large flat tin about 2 feet square with the edges bent up about one-half inch, is best for the purpose.   However, with care and skill the operation can be performed in a common gold pan, as is done by many prospectors, particularly when cleaning dry-washer concentrates.   The material should be perfectly dry.   Much effort is saved by using a magnet to take out any magnetite sand in the concentrates; often this mineral comprises as much as 90 percent of the material.   A piece of paper folded around or held against the end of the magnet will keep the magnetite from sticking to the metal.   When all the magnetite is removed, blowing gently on the remaining sand and gold will drive the former to the farther edge of the sheet, leaving the gold behind.   In most instances the loss of a few fine colors is not serious.

Amalgamation

In Ordinary Gold Pans.   A small quantity of quicksilver, ranging from an ounce to a quarter of a teaspoonful, will catch all the gold from a pan of sluice concentrates.   The mercury is simply placed in the pan with about 5 pounds of concentrates and agitated under water until no

more free gold can be observed.   Then the sands are panned off, care
being taken not to lose any of the amalgam or fine drops of mercury,
which gradually will run together into a single mass.   If the concentrates
are nearly all black sand only a small quantity should be washed at a time,
but if much light sand or rock is present larger quantities can be washed.

Copper-plated pans or pans with steel rims and copper bottoms are
available and are useful for saving fine gold in concentrates.   The copper
is coated with mercury by first cleaning it with emery paper, then rub-
bing clean, bright mercury or amalgam on it until it presents a smooth,
shiny surface.   The gold in the material being treated is picked up
quickly by the amalgam surface.   Only fine sand can be treated to advantage as coarse sand or gravel will scour the amalgam off the copper.   As
fast as amalgam accumulates on the copper it is scraped off with a smooth,
dull-edged iron scraper such as a putty knife.   More mercury may then
be added to keep the surface bright and in a "receptive" condition.

Amalgamators.   In nearly all large-scale operations most of the gold
is amalgamated in the sluice boxes or on the riffle tables, and the amalgam
is separated from the sands during clean-up operations or from the concentrates by rocking or panning.   Tarnished or rusty gold or very fine
gold, however, does not amalgamate readily because it is difficult to make
contact between the gold and quicksilver.   Such gold, generally included
in a black-sand concentrate, requires agitation in the presence of quicksilver or, if rusty, grinding to remove the interfering coat for satisfactory
amalgamation.

Mechanical amalgamators are used to treat such materials.   Occasionally all of the concentrate from the sluice will be treated in an amalgamator, particularly if it contains rusty gold.   The charges for the
amalgamator should be kept clean; grease especially interferes with amalgamation.

A common type of amalgamator is the clean-up pan, which consists
of a cast-iron, cylindrical, flat-bottomed barrel or tub 1 or 2 feet in diameter for small-scale work and 4 to 6 feet in diameter for mill service.   The
concentrate with 1 or 2 percent quicksilver by weight is placed in the pan
with sufficient water to make the mass fluid and agitated by a revolving
spider.   The quantity of water added should be sufficient only to permit
agitation without too great strain on the machine.   The pulp should be
thick enough to hold particles of mercury in suspension.   Shoes on the
lower end of the spider arms slide on a flat, circular race in the bottom of
the barrel, thus adding some grinding to the agitation.   After running
for 1 or 2 hours the batch may be emptied through a drain plug in the
bottom of the barrel and the mercury and amalgam separated from the
sand by panning.   Some pans are provided with side drain plugs at various elevations.   The rotation may then be slowed from its usual speed of
about 60 r.p.m., the shoes raised enough to stop the grinding, and water
added.   This will settle the quicksilver and amalgam; the waste sludge
can then be flushed out through the upper drain plugs and almost complete cleaning of the amalgam and mercury made in the pan itself.

Another device, the so-called amalgam barrel, generally is used at
large stamp mills and occasionally is employed in placer operations, particularly in dredging, to treat accumulated black sands, scrap metal, and
other possible gold-bearing material from clean-up operations.   It is
merely a cast-iron or steel drum revolving on a horizontal axis like a ball

mill and fitted with suitable drain plugs, handholes, manholes, or removable ends, depending on its size and use. The material to be treated is placed in the barrel with quicksilver, water, and a few iron balls, and the barrel is turned slowly for an hour or several hours. The barrel may then be flushed with water from a hose to wash away the lighter products of grinding, turned over, and emptied into a tub, the amalgam and mercury being recovered by panning. Potassium cyanide sometimes is added to brighten the gold; only enough is used to make a very weak solution.

An amalgamator that occasionally is used, especially if a part of the gold is attached to particles of quartz, is the Berdan pan, which is relatively simple in construction and cheap to operate. The pan consists of a revolving cast-iron bowl, usually 3 to 5 feet in diameter, with a raised central hub for the drive shaft, giving it the form of a circular trough. The bowl is supported either by the drive shaft or by rollers and is set with a tilt of about 20 or 30° from the horizontal. It is driven at 10 to 30 r.p.m. either by a crown gear on the inclined shaft of the bowl or by a ring gear on the bottom of the bowl. One or two large cast-iron balls roll in the trough as the bowl revolves. Quicksilver is placed in the bowl with the charge, and as the device revolves a stream of water is directed into it and overflows at the lowest point of the rim. The material to be amalgamated may be added in batches or, if it is to be ground as well as amalgamated, by an automatic feeder, the slimes and fine material overflowing to waste; the bowl then acts as a classifier. For placer concentrates the batch process is used, 100 pounds or more being treated at a time. Too large a quantity of sand lessens the grinding effect of the balls.

A 1- or 2-cu.ft. hand- or power-driven concrete mixer is a convenient amalgamating device for the small- to medium-scale placer miner, particularly if part of the gold is rusty. It costs only $20 to $30, excluding the small gasoline engine, and can be obtained from hardware stores or mail-order houses. The charge for such a machine is two or three pails of concentrates, 1 or 2 pounds of quicksilver, a few round cobblestones 3 or 4 inches in diameter, and water. About a 1-hour treatment will amalgamate practically all of the gold. The charge is emptied into a settling tub and then washed in a pan or small sluice box to recover the amalgam and mercury.

Regardless of the amalgamator used, too violent agitation of the mercury must be avoided, otherwise excessive flouring hinders amalgamation and makes it difficult or impossible to recover the quicksilver.

Cleaning Amalgam. The mixture of quicksilver and amalgam from sluice-box clean-ups usually contains much more mercury than amalgam. It can be freed from sand, scraps of iron, and other solid impurities by careful panning and by washing with a jet of clean water. The amalgam can then be separated from the quicksilver by straining the mixture through buckskin, chamois skin, close-woven canvas, or other strong, tight cloth. This generally is done by hand, preferably under water to prevent scattering of the mercury. The quicksilver thus filtered off contains at the most only about one-tenth percent of gold; this mercury is desirable for recharging the boxes as the small amount of gold makes it more active. The amalgam, after squeezing, still contains some mercury, part of which may drain off if the mass is suspended for several hours in a funnel or other similar container. With or without this last refinement, which one dredge operator used with success, the stiff, pasty amalgam is now ready for fire treatment to separate the gold.   It contains 25 to 55 percent, commonly about a third by weight of gold and silver.

Extracting Gold From Amalgam

Heating

Although retorting is the common method of separating the gold from the quicksilver in amalgam at dredges and other large-scale operations, the mercury in small quantities of amalgam may be volatilized by simple heating.   A common method is to heat the amalgam on a clean iron surface over an open fire or forge, or in a furnace, until all the mercury is driven off.   This is the usual expedient of the single miner or small operator who does not object to the loss of the small quantity of quicksilver involved.   The mercury vapor may appear as heavy white fumes.   Whether visible or not, mercury vapor is exceedingly poisonous, and the work must not be done except where a draft can be depended on to carry all the vapor away from the operator.   As stated elsewhere, mercury boils at 675° F., a temperature about halfway between the boiling point of water and the first visible red heat of iron.   However, it volatilizes at the boiling point of water enough to be dangerous to the health of persons exposed to it.   Consequently, it should be handled carefully, particularly to avoid inhaling its vapors.

In another method of recovering the gold from small amounts of amalgam, a potato is used as a condenser.   This is a device popular with prospectors because it is very simple, yet saves part of the mercury that would be lost by the method previously described.   A large potato is cut smoothly in half, and in the flat surface of one-half a recess is hollowed which should be considerably larger than the amount of amalgam to be treated.   The amalgam is placed on a clean sheet-iron surface, the half potato is placed over it, and the whole is set over a hot fire.   For convenience it may be done in a frying pan or the scrap of sheet iron put on a flat shovel so that it can be withdrawn readily from the fire.   Some mercury vapor will escape under the edges of the potato, and, as before, these fumes must be avoided.   After 15 or 20 minutes of strong heating the potato may be lifted off for inspection.   If all the mercury is gone from the gold the potato may be crushed and panned, and a considerable part of the mercury will be recovered.   It may be desirable to heat the gold further to anneal it; this can be done without removing it from the iron plate.   Any tinned or galvanized metal intended for use in this process should be heated redhot and then scoured to remove all traces of the coating so that a clean iron surface will be presented.

A laboratory method of separating the gold is to put the amalgam in a small beaker and dissolve the mercury in a 1 to 1 solution of nitric acid and water.   When all the mercury is dissolved, the gold may remain as a sponge, which can be washed gently in water and annealed in a small porcelain crucible.   More frequently the gold will be recovered as a fine dust, which also can be washed and annealed but is less easy to handle.

Retorting

A very small amount of amalgam can be retorted quickly and easily in a laboratory in a glass tube 18 to 24 inches long, sealed at one end and bent 2 or 3 inches from that end to a slightly acute angle.   A large tube three-fourths inch in diameter is best.   The amalgam is broken into pieces small enough to be dropped into the closed end where it is then heated, the fumes condensing in the long open end of the tube. The gold can be annealed by heating the tube to redness after all mercury is driven off.

A retort for treating a few ounces at a time can be made cheaply of ⅜-inch pipe, pipe connections, and a large grease cup. The lower and open end of the ⅜-inch pipe is inclosed in a larger pipe. Cooling water is poured through the space between the two pipes from an open connection in the top of the outer one. The charge of amalgam is placed in the grease cup cover which is then screwed into place; graphite lubricant is placed on the threads to make a tight joint. Heat is applied to the grease cup, and the quicksilver is condensed in the lower end of the pipe. The method of using and the general arrangement of the device are similar to those of the next retort described.

The typical quicksilver retort for placer mines is a cast-iron pot with a tight-fitting cover in which a hole is tapped to accommodate the condenser pipe. The capacities of such retorts range from a few to 200 pounds of amalgam, or about a quarter pint to 2 gallons. They are listed in chemical-supply catalogs at prices ranging from $4 to $30, not including the condensers. The condenser commonly used with this type of retort is an iron pipe 3 or 4 feet long leading from the hole in the retort cover at a downward angle of 20 to 30°; it is encased for most of its length in a considerably larger pipe through which cooling water is circulated. When heat is applied to the charged retort the mercury vapor enters the condenser pipe where it cools and condenses; it trickles down the pipe into a vessel placed under the open end of the pipe. In the treatment of a large amount of amalgam the temperature of the pipe might be raised to a point where some of the vapor would escape; therefore, a cooling device is necessary.

The retort may be heated over a large bunsen burner, by a gasoline blow torch, in a forge, or in one of several types of furnaces built for the purpose. Very high temperatures are unnecessary, and a wood fire is considered better than a coal fire. The flame should cover as much of the retort as possible.

A rigid, strong stand for the retort and condenser should be constructed if the apparatus is to be used regularly.

The retort should be coated on the inside with chalk, or painted with a thin paste of chalk, clay, mill slimes, or a mixture of fire clay and graphite and thoroughly dried before putting in the charge. This prevents the gold from sticking to the iron, which sometimes causes trouble. A lining of paper serves the same purpose but tends to form an objectionable deposit in the condenser pipe.

The retort should not be filled over two-thirds full of amalgam (a third or half full when retorting liquid mercury), otherwise there is danger of some of the contents boiling over into the condenser tube. The amalgam is broken into pieces and piled loosely. Then the cover is put on and clamped tightly with the wedge or thumbscrew provided, first making sure that the attached condenser pipe is clean and free of obstructions. The ground joint between the cover and body of the retort is seldom tight enough to prevent leakage and should be luted with clay or some sealing compound. One satisfactory cement is made readily by moistening a mixture of ground asbestos and litharge with glycerin.

A low heat is applied at first, then after 10 or 15 minutes the temperature is increased just enough to start the mercury vaporizing and condensing. Too rapid heating harms the retort, and only enough heat should be used to maintain a steady trickle of quicksilver from the condenser. When no more mercury appears the temperature should be increased for a few minutes to red heat to drive the last of the quicksilver out of the retort; then the fire should be withdrawn from the retort and the latter allowed to cool. Some mercury vapor always remains in the retort, and the operator should take care not to breathe these fumes upon taking off the cover.

The likelihood of dangerous amounts of mercury vapor passing through a long cold pipe without condensing is very small. However, if much amalgam is to be retorted, or if the operation is of daily or frequent occurrence, it usually is desirable to provide some form of water seal at the end of the condenser tube to prevent the escape of such fumes. Many miners have followed the dangerous practice of submerging the end of the condenser pipe in the bucket of water used to receive the condensed mercury. This should not be done, as a slight cooling of the retort would cause water to be sucked into the pipe, and if the water reached the retort an explosion would follow. Such an experience has taught more than one "oldtimer" the danger of this practice.

If the volume of the receptacle is very small compared with that of the condenser pipe and if the discharge pipe is barely submerged the danger is avoided, as any large rise of water in the pipe would lower the water surface enough to break the suction. At some properties the end of the condenser pipe is in a large sheet-iron cylinder, a few inches in diameter, open at the lower end, which may be placed 2 or 3 inches into the water in a receptacle of only slightly larger diameter, thus making a good water seal yet avoiding the danger of explosions.

The simplest method is that recommended by Louis;[25] it consists merely of tying a piece of cloth such as canvas or burlap around the end of the condenser pipe and letting it dip in the water 2 or 3 inches below, forming a damp filter which will condense any escaping vapor yet not be tight enough to permit water to be sucked into the retort.

Large gold mines use cylindrical retorts, usually set horizontally in specially built furnaces. Such installations probably would be needed in placer mining only by large dredging companies. The operation is similar to that of a pot retort, except that the amalgam usually is placed in several small iron trays, rather than on the floor of the retort proper, and charged through a door or removable cover at one end of the retort, while the condenser is attached at the opposite end.

Separation of Platinum-Group Metals From Gold

In several localities in the Western States sluice concentrates from placer mining are likely to contain platinum or its associated metals in sufficient quantities to be of economic interest. The separation of these minerals from gold is difficult. Their specific gravity is too near that of gold to permit a separation by panning. Coarse platinum particles can be picked out of the gold by hand, but most placer platinum is exceedingly fine. Although platinum does not amalgamate, quicksilver can be made to coat and hold platinum particles by treatment with chemicals; thus it is possible to separate successively the gold and platinum from the concentrates.

---

[25] Louis, Henry, A handbook of gold milling, p. 386, London, 1894.

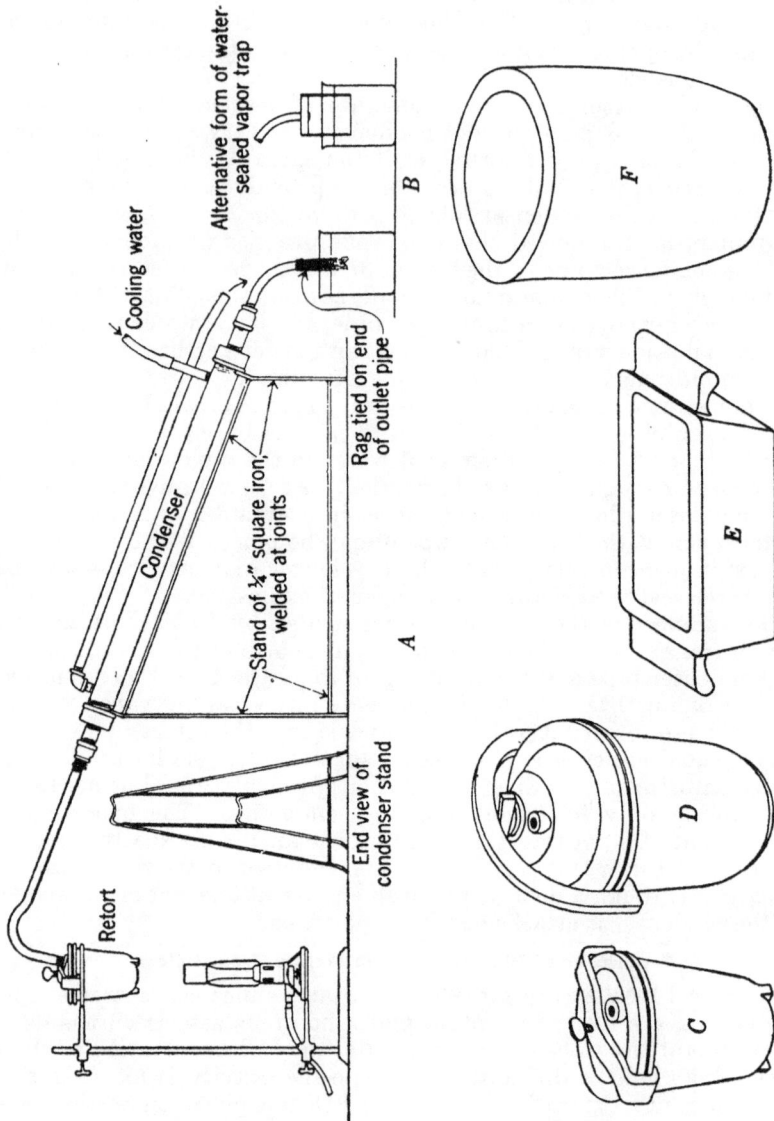

FIG. 45.   Amalgam retorts, bullion mold, and crucible. A. Apparatus for retorting; B. methods of keeping end of pipe out of water; C, D, retorts; E. bullion mold; F. crucible.

One dredging company in California which recovers platinum metals uses the following clean-up procedure :[26]

In cleaning up, the riffles are removed from the sluices, starting at the head end, carefully washing them off and washing the sluice down with water from a hose. This washes away the light sands and concentrates the amalgam and heavy sands, which are carefully scooped up into buckets and carried to a "long tom" for further treatment. In the long tom most of the mercury and amalgam and some of the platinum-group metals are caught in the upper box. Most of the platinum, some rusty gold, scattered particles of mercury and amalgam, and the sand and refuse are washed out over riffles where the heavier components are caught. The sand finally passes through a screen at the end of the tom, into a sand box, and the gravel goes to waste. The mercury and amalgam from the upper box are transferred to a bucket, in which the gold amalgam settles to the bottom; the lead or other base-metal amalgams float on top. The latter is partially cleaned by panning, which separates some metallic platinum, then retorted. The gold amalgam is squeezed free of mercury and likewise retorted.

The gold amalgam, usually containing about 55 percent gold and silver, is retorted in a standard make of gasoline-fired retort. The mercury condenses in a water-jacketed pipe and drains into a bucket of water. The gold remaining in the retort is transferred to a crucible and fused in the same furnace. It is then poured into molds, producing bars which are shipped to the Selby smelter. The bullion averages 890 parts gold, 90 parts silver, and 20 parts impurities per 1,000.

The riffle concentrates and sand from the end of the long tom are placed in small batches in a steel barrel mill 4 feet long and 2½ feet in diameter. Mercury is added and the batch ground for 1 or 2 hours. Then the amalgam is removed by panning and added to the other base amalgam for retorting. Further panning and rocking reduce the remaining sand and concentrates to a product containing about half black sand and half platinum, by volume. This is treated by the addition of water, mercury, zinc shavings, and sulphuric acid; this causes the platinum metals to be coated and held by the mercury, so that a final separation from the sand is possible. The final concentrate is then washed with water and drained to remove acid and excess mercury, after which treatment with nitric acid dissolves the mercury, leaving a final residue of platinum, iridium, and osmium.

The base amalgam, which includes shot, bullets, and small particles of copper and brass scrap, as well as some precious metals, is retorted to recover the mercury, melted, and poured into molds to form bars for shipment to the smelter. These bars range in value from $1 to $8 per troy ounce.

The United States Mint does not now buy platinum or pay for the platinum content of gold shipments. The following buyers of crude platinum reported purchases in 1930 :[27]

American Platinum Works, 225 New Jersey Railroad Avenue, Newark, N. J.
Baker & Co., Inc., 54 Austin Street, Newark, N. J.
J. Bishop & Co. Platinum Works, Malvern, Pa.
Sigmund Cohn, 44 Gold Street, New York, N. Y.

---

[26] Patman, C. G., Method and costs of dredging auriferous gravels at Lancha Plana, Amador County, Calif.: U. S. Bur. Mines Inf. Circ. 6659, pp. 12-13, 1932.
[27] Davis, H. W., Platinum and allied metals in 1930: Mineral Resources U. S., 1930, pt. 1, p. 105, 1931.

Thomas J. Dec & Co., 1010 Mallers Building, Chicago, Ill.
Kastenhuber & Lehrfeld, 24 John Street, New York, N. Y.
Pacific Platinum Works, Inc., 814 South Spring Street, Los Angeles, Calif.
Schwitter, Clover & Startweather, Inc., 312 Passaic Avenue, Newark, N. J.
Wildberg Bros. Smelting & Refining Co., 742 Market Street, San Francisco, Calif.

Lots ranging from less than an ounce to hundreds of ounces ordinarily are marketable but preferably not less than 2 ounces. Settlement is based on assay, either by the buyer or, for large lots, by both parties. The price paid in 1930 for domestic crude platinum ranged from $30 to $40 per troy ounce;[28] the average quotation for the refined metal was $45.

Melting Gold [29]

The spongy mass of gold left after retorting can be sold to the mint or other agencies just as it comes from the retort, but generally it is melted and poured into molds to form bars or ingots for marketing.

The melting generally is done in graphite crucibles placed in a special furnace. In small operations the crucible is usually heated in a blacksmith forge in which coke is used for fuel. The graphite crucible must be dried thoroughly before it is used by being warmed gradually for several hours.

Small quantities of gold frequently are melted without fluxes in make-shift devices such as dented frying pans; in most instances, however, some flux is desirable. If the gold is fairly pure, that is, has a bright yellow color, it may be melted with only a small quantity of borax glass for flux. If, however, it contains impurities and is grey or black in color, the melt requires larger quantities of flux to take up these impurities. Sometimes niter, sodium carbonate, or silica is used to remove specific impurities. The flux is melted first, then the gold is placed in the crucible and likewise melted. Enough flux is used to form a covering about one half inch deep over the molten metal.

In large-scale operations the melted gold is poured from the crucible into cast-iron molds holding 50 to 1,000 ounces. A mold should be larger at the top than at the bottom so that the bullion will drop out readily when it is inverted. A mold 3 inches by 12 inches at the top, 1 inch narrower and shorter at the bottom, and 3 inches deep holds about 1,000 ounces. The common practice is to smoke the mold over an oil flame, then to heat it before pouring the gold. Another practice is to coat the mold with graphite or oil or to pour a quarter inch of vegetable oil in the mold and heat it to boiling, then to pour the gold into the oil.

When the gold has just set in the mold and before the slag has hardened, the contents of the mold are tipped into water. This granulates most of the slag, and any particles still adhering to the gold usually can be brushed off. Tightly adhering slag can be loosened by washing the gold with nitric acid.

The bar of bullion may be stamped with identifying marks or names, or these may be cast in reverse in the bottom of the mold. The bar is then ready for market.

Sampling and Weighing Gold

Sampling

There are several methods of sampling gold bullion. The most accurate one is to dip a sample from the melted bullion before casting it.

---

[28] Davis, H. W., op. cit., p. 105.
[29] Under *Gold Reserve Act of 1934*, a license from a U. S. Mint is required for melting gold. A copy of this act is available from Superintendent of Documents, Government Printing Office, Washington, D. C., for 10 cents.

A graphite rod suitably shaped at one end to dip up the desired amount of gold, usually 1 to 5 grams, is heated redhot, stirred about in the melt, and lifted out with the sample. The sample is then poured into an oiled mold or into a shallow bath of heated oil. This method has been used by a few mining companies and is said to eliminate slight inaccuracies to which other methods are subject. It is impracticable for small amounts of gold and is inconvenient in that the sample is not obtained in a form convenient for assay; except for bullion containing large quantities of base metals, simpler methods generally are sufficiently accurate.

Other methods depend on taking samples from the solid cast bar of bullion. Chips can be cut with a cold chisel from the surface of the bar, at one or more places, hammered thin, and trimmed to the desired weight for assay, or holes can be drilled and the drill cuttings used for samples. The latter method is used most. Holes an eighth inch or less in diameter are drilled a quarter to a half inch into the bar, usually one on top and one on the bottom of the bar, on the center line a short distance from the opposite ends. Two diagonally opposite corners, one on top and one on the bottom, are sometimes preferred, although the difference probably is negligible. It has been found that in bullion containing base metals there is a strong tendency for the base elements to segregate at the bottom of the bar and for the top surface to be above the average fineness. Special methods of sampling then must be used. However, for most placer-mine bullion a sample of the desired weight, obtained in almost any convenient fashion, will be sufficiently accurate. Drill samples taken as described above usually check the mint or smelter return within 5 parts per thousand.

When gold is to be shipped to the mint, assaying the bullion is a needless expense as there is no recourse from the mint assay returns.

Weighing

Analytical balances suitable for weighing small amounts of gold with great accuracy cost from $150 to $300, and balances that will weigh large amounts of gold, such as the gold bars shipped to the mint by mining companies, with sufficient accuracy so that their value can be calculated to the nearest cent, are very costly. Balances capable of weighing a few ounces of gold to the nearest cent can be purchased for $20 to $30, and convenient pocket scales, either of the hand-balance type or arranged to be set up on the cover of their cases as mounted balances and capable of weighing 3 or 4 ounces to the nearest cent, are sold by most chemical-supply houses at prices ranging from $2.50 to $15, including weights. With little or no expenditure a set of hand balances can be made, which will weigh 1 or 2 ounces of gold to within a grain or the nearest 5 cents. The balance beam may be of wood, 6 or 8 inches long, suspended by a pin, needle, or bent wire hook through a hole in the exact middle of the beam. The pans can be made of tin, cut $1\frac{1}{2}$ to 2 inches in diameter and hammered dish-shaped, or of the lids of small tin cans, each suspended by three threads from the ends of the beam by means of bent wire hooks. No pointer is necessary, as the beam can be leveled closely enough by eye. It is not necessary that all parts be of exact weight, as the balance beam or the pans can be trimmed to make the assembly balance. For even approximate accuracy, however, it is necessary that the pans be suspended from points on the beam the same distance from the center bearing, and for stability the end bearings must be slightly lower than the center one. The nicer the construction and the more nearly frictionless the method of suspension, the greater the accuracy; but even with very little attention to these points a sensitivity of less than a grain is obtainable when weighing as much as 2 ounces. Weights can be purchased in convenient sets for 50 cents or more or can be made or improvised from bits of wire or sheet metal cut to match any available standards.

Marketing Placer Gold

Five classes of buyers usually are available to the miner who wishes to sell gold dust, retort sponge, or bullion bar: (1) Individual gold buyers; (2) local stores; (3) local banks; (4) smelting companies; and (5) United States mints and assay offices. If the miner has base bullion or concentrates the smelter or custom mills are usually his only market.

Local stores are the principal buyers of small amounts of gold, ranging in value from a few cents to $50 or more. The merchant, who is often the chief retailer of supplies to the miners, finds it brings him trade to act as a commission buyer of gold, making it possible for the prospector and miner to convert their winnings promptly into cash. If his commission is fair, this is satisfactory, as it saves the miner much distasteful annoyance in preparing his gold for shipment, filling out various registration and report forms, and then waiting several days for his check. It likewise makes possible the sale of less than 1 ounce of fine gold at a time, which is the least amount of retort sponge, gold dust, or nuggets the mint will accept. The discount of the merchant ranges from $1 to $2 per fine ounce. The miner must remember that no placer gold is pure and that the merchant has only his judgment to tell him how much the mint will pay for his gold. Not all gold from a district assays the same degree of fineness, and the merchant is not to be blamed for staying on the safe side.

In most mining districts there are assayers, company officials, jewelers, metal brokers, or other individuals who for profit or for the convenience of employees, lessees, or customers make a practice of buying gold in small lots from prospectors and miners and paying cash for the value of the estimated weight of fine gold, less certain charges. Likewise banks in many districts receive gold, either purchasing it outright on the basis of their own or commercial assayers' analyses, or merely acting as shipping agents, receiving the gold, shipping it to the mint, and paying the miner upon receipt of mint returns. In the latter case a commission of about 1 percent usually is charged, for which the bank assumes all risk and trouble otherwise taken by the miner himself.

A few smelters or refineries buy gold or silver metals; the melting and refining charges probably will closely approximate those of the mint. Most smelter or refineries handling precious metal ores buy gold-bearing concentrate. Smelting charges on such material are variable, and an inquiry, accompanied by a close description or a sample of the material offered for sale, should be made in advance.

Gold can be shipped by express or by mail. If by express, the parcel can be insured with the express company for its full estimated value. United States mail shipments usually are sent as registered first-class mail and should be insured. The mail registry system provides insurance in graduated amounts from $5 to $1,000 at a cost of 15 cents to $1, including the registration fee but excluding postage. If regular mail shipments of considerable value are being made, it is possible to secure commercial insurance for them. However, for amounts greater than a few ounces the first-class postage rate of 3 cents per ounce becomes so costly that express shipments are advisable. All shipments must be prepaid.

The best container for shipping gold, either by mail or express, is a lead-sealed canvas sack, securely tagged with the addresses of the sender and addressee. Gold bars may be wrapped securely in canvas and packed in wooden boxes.

When a shipment of gold is sent to the United States Mint a letter should be sent separately containing the prescribed affidavits. Form TG-19, for a person shipping gold that he has mined himself, and form TG-21, for gold buyers, can be obtained by writing to a United States mint or assay office. Form TG-19 need not be sworn to if the amount of gold is 5 ounces or less. Since January 1934 the mints have paid $35 per troy ounce of fine gold, less one-fourth of 1 percent, as compared with the former price of $20.67+ per ounce.

The mint charges $1 for melting any deposit of 1,000 ounces or less and 10 cents additional for each 100 ounces over this amount. An extra charge of $1 or more is made for melting gold dust or gold containing non-metallics if the loss of weight in melting is more than 25 percent.

If the gold is 992 fine, or finer, no charge is made for parting and refining. If less fine, or if more than 50 parts base metals are present per 1,000, charges of 1 cent to 5 cents or more per ounce are imposed for parting and refining. Bullion less than 200 fine is not accepted.

Current market prices are paid for the full silver content; however, if necessary forms are submitted for silver qualified under Executive Proclamation of December 21, 1933 the depositor will receive the number of silver dollars that can be coined from one half of the fine silver content. No other constituent in the bullion is paid for.

Laws Regulating Ore Buyers

California laws require ore buyers to take out licenses. The California Ore Buyer's License Act, passed in 1925, includes as buyers all persons sampling, treating, or buying gold dust, gold or silver bullion, gold or silver specimens or ores, or concentrates of these metals and gives the State Mineralogist the duty of licensing such persons.[30] The license fee is set at $15 per year if the gold and silver treated or purchased in a year exceed $1,000 in value or at $2 if less. The licensee is required to keep on record the names of the sellers, the amount and description of each lot purchased, the stated source of each lot, and other data and to report all purchases monthly to the State Mineralogist. Provision is made for the issuance of licenses, recovery of stolen metals, and penalties for violation of the act. The latter are severe. No regulation, of course, is placed upon the gold buyer as to terms of purchase, nor is this a matter of record under the act; however, one effect of the act, which primarily is intended to prevent the ready sale of stolen gold or silver, doubtless is to improve the chances of the small producer getting fair treatment from ore buyers by driving dishonest dealers out of business.

[30] Ricketts, A. H., American mining law: California Div. Mines Bull. 123, pp. 659-664, 1943.

DEBRIS DAMS

When hydraulic mining was at its height, annual production of gold by that method reached $15,000,000, and averaged $10,000,000 per year for the 30 years ending in 1884.[1] The best estimates[2] of the amount of gravel that was mined to recover this gold indicate that 1,295,000,000 cubic yards was washed into tributaries of the Sacramento alone. This is equivalent to a body of gravel 10 miles long by 1 mile wide by 125 feet deep.

The immense scale of these operations caused serious silting of the river channels and partial blockage of streams tributary to them, so that in flood times much damage was done to farm lands adjacent to the rivers. In consequence there were many actions at law in the State courts, which resulted in injunctions against the mines. Immediately prior to the general closing of hydraulic mining the annual yield is estimated to have been reduced to $5,300,000.[3]

Finally the decree of Judge Sawyer of the United States Circuit Court restrained the North Bloomfield Mining Company from discharging debris into the streams. This was in 1884. The court reserved the power to modify or suspend the injunction upon the defendant's showing that conditions had been so changed that the discharge of debris might be conducted so as not to continue the nuisance complained of. That is to say, hydraulic mining was not declared to be illegal, but the discharge of debris into the streams was forbidden. The provision of a dam to restrain the debris might make the necessary change in conditions to allow mining to be resumed.

Mine after mine was closed by injunction based upon this decision and the federal legislation passed in 1893 (the California Debris Commission Act, or Caminetti Act) embodied the principles declared by these injunctions, but enacted more severe penalties for their non-observance. The Caminetti Act is printed in full in the appendix of this bulletin. It applies only to the territory drained by the Sacramento and San Joaquin Rivers.

Under the Caminetti Act, charges for debris storage behind dams contemplated by the law were set at 3 percent of the gross proceeds of a mine. As this was inadequate to amortize the cost of such dams, the Caminetti Act was in effect more than 40 years before any debris dams were built. An amendment approved June 19, 1934, set the charges on the basis of cubic yards of gravel mined behind the dam. The matter of building such dams by the Federal Government was reviewed by the California Debris Commission, and a report[4] containing recommendations against the building of debris dams was submitted to the Board of Engineers for Rivers and Harbors under date of February 13, 1935. A hearing in protest of this report was held before the Board of Rivers and Harbors in Washington on April 15, 1935, and Walter W. Bradley, California State Mineralogist, presented before that board a brief to show the value of placer gravel that would be made available for mining behind

---

[1] Jarman, Arthur, An investigation of "the feasibility of any plan or plans whereby hydraulic mining operations can be resumed in this state"; California State Min. Bur., Mining in California, State Mineralogist's Rept. 23, pp. 54-116, 1927.

[2] Lindgren, Waldemar, The Tertiary gravels of the Sierra Nevada of California: U.S. Geol. Survey Prof. Paper 73, p. 21, 1911.

[3] Haley, C. S., Gold placers of California: California State Min. Bur., Bull. 92, p. 16, 1923.

[4] Jackson, T. H., Irvine, E. S. J., and Drinkwater, J. G., Report of the California Debris Commission: 74th Cong., 1st sess., H. Doc. 50, pp. 9-34, 5 maps, 1935.

such dams. This brief was published in the July 1935 issue of the California Journal of Mines and Geology.[5] A map shows locations of proposed dams.

The Board of Engineers for Rivers and Harbors[6] under date of May 23, 1935, reported in favor of the construction of four dams at a cost of $6,945,000. The dams recommended were the Upper Narrows dam on the Yuba River about 1½ miles above its confluence with Deer Creek near Smartsville; Dog Bar dam about 7 miles up the Bear River from the existing Combie dam or 3 miles southwest of Colfax; the North Fork dam on the North Fork American River about 2 miles above its confluence with Middle Fork American River near Auburn; and Lower Ruck-A-Chucky dam on Middle Fork American River a mile below the mouth of Canyon Creek.

An appropriation was provided by Congress, and two of these dams were built in the late thirties, and are now ready for use, the Upper Narrows dam and the North Fork dam. In an announcement dated January 5, 1942 the California Debris Commission established a tax rate of 3.11 cents per cubic yard for storage of hydraulic mining debris in the North Fork Reservoir on the North Fork of the American River, and a tentative tax rate of 2.30 cents per cubic yard for storage of debris in the Upper Narrows Reservoir on the Yuba River. Power is generated at the Upper Narrows, and this helps to pay for the dam, as provided in an amendment to the Caminetti Act approved June 25, 1938. This amendment is printed in the appendix.

Work was started on the Lower Ruck-A-Chucky dam, but the rock-formations at the proposed site were found unsuitable to support the dam, and the site was abandoned. If a dam is built to restrain debris on the Middle Fork of the American River, a new site farther up the stream will probably be selected. On the Bear River, where the Dog Bar dam was to be constructed, a conflict exists with regard to the generation of power, and trouble has been caused by mud and silt in the water at the existing Combie dam. This conflict of interests makes plans uncertain for a debris dam on the Bear River. On the Yuba River, a mile below the mouth of Willow Creek, near Camptonville, the Bullard Bar dam owned by the Pacific Gas and Electric Company has been available for storage of hydraulic mining debris for many years. The charge per cubic yard in 1941 was 2 cents.

Although the two debris dams mentioned above have been built and are ready for use, some serious problems still remain to be solved before hydraulic mining can be resumed on a large scale. One of these was brought out by case no. 60,474 in Department 4 of the Superior Court of Sacramento County, Carmichael Irrigation District and others vs. Lost Camp Mining Company and others. Carmichael Irrigation District alleged that damage was being caused by mud and silt from hydraulic mining above the North Fork dam on North Fork American River near Auburn. A temporary restraining order was issued in 1939 against hydraulic mining during the irrigation-season. This order allows hydraulic mines on North Fork American River to operate only between November 15 and April 30. McGeachin Placer Gold Mining Corporation

---

[5] Bradley, Walter W., Dams for hydraulic debris: California Jour. Mines and Geology, vol 31, pp. 345-367, 1935.
[6] Pillsbury, G. B., Control of mining debris on the Sacramento River and tributaries: 74th Cong., 1st sess., H. Doc. 50, pp. 3-8, 5 maps, 1935.

and The Mayflower Gravel Mining Company filed answers in this case in March 1943 alleging that the United States Government is a party in interest and that the case should be transferred to a Federal court.

A second serious problem is that of water-supply. Many of the old water-rights, and even some of the ditches formerly used by hydraulic mines have been acquired by power companies and irrigation districts. Where the old water rights are no longer available, the only practicable method of replacing them is by the storage of surplus water by means of large reservoirs high in the mountains. This is, of course, expensive. Even where water-rights have been retained by hydraulic mining companies, many of the ditches and flumes require rebuilding.

Because of these difficulties, much consideration is being given by owners of large gold-bearing gravel deposits to mechanical methods of handling gravel with tailing storage at the mine as a substitute for hydraulic mining. Such methods are rapidly being reduced in cost, and may soon become available. Dredging may be used for some of the deposits, but many of the large deposits are too deep for any dredge yet constructed.

SECTION II

GEOLOGY OF PLACER DEPOSITS

NEW TECHNIQUE APPLICABLE TO THE STUDY OF PLACERS *

BY OLAF P. JENKINS **

OUTLINE OF REPORT

* Reprinted from April 1935 *California Journal of Mines and Geology*, pp. 143-210.
** Chief Geologist, California State Division of Mines.

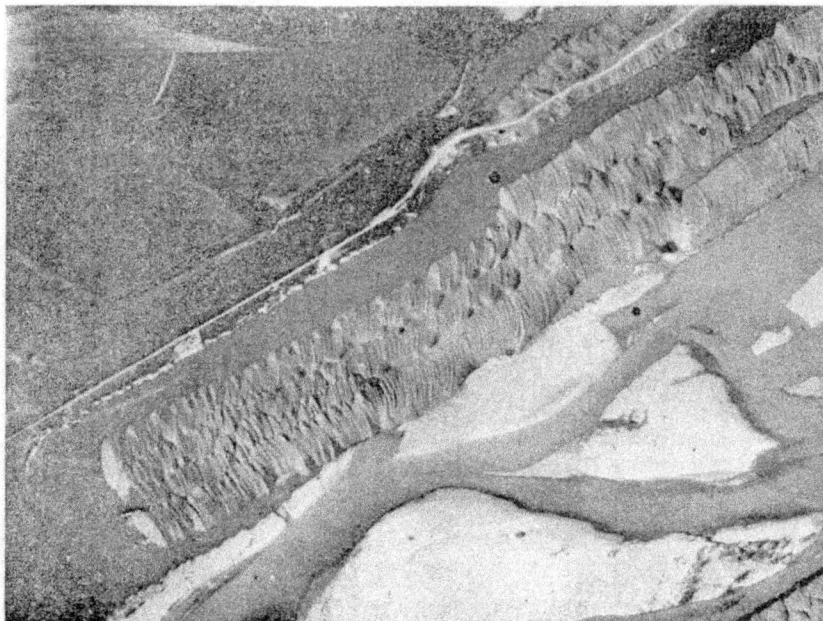

FIG. 46.   Aerial photograph of a dredged strip along the Yuba River; note the dredge at work.   *Photo by Russell; reprinted from Engineering and Mining Journal.*

Abstract

The exploration of placers is a problem involving nearly all phases of the science of geology, especially physiography and stream sedimentation, neither of which has been given sufficient consideration in connection with the economic problems concerned. The use of aerial photography is a great aid in the study of placers, both ancient and modern.   The use of geophysics, when applied as a part of a geologic program of exploration, can materially assist in guiding drilling operations and underground prospecting, with the result that the cost of expensive development work can be greatly reduced and hastened to an earlier completion.

Placers are classified according to the way they are formed: residual, eluvial, stream, glacial-stream, bajada, eolian, and beach.   Since the ordinary stream placer is by far the most important, various phases of stream study are discussed in this paper. A need for further scientific study and the development of systematic working criteria is apparent.

The depleted placers of California consist largely of Recent and Pleistocene stream gravels and uncovered or buried, but easily accessible Tertiary channels, while the large reserves lie in more remote positions.   These are exemplified by hidden buried bench gravels not connected with the surface nor with channels already worked, and by the lower untouched channels that lie on true bedrock beneath the 'false bedrock' of the dredged areas along the western foot of the Sierra Nevada.   Some Pleistocene gravels still lie in pockets beneath the waters of the larger rivers where faults have caused down-dropping of the stream bed.   Benches still lie in isolated regions such as the Klamath Mountains.   The desert affords several types of deposits: stream placers buried by alluvial fans, reworked older placers, gravels interbedded with lavas, and the more recent bajada placers.   Marine placers of Cretaceous, Eocene, Pleistocene, and Recent periods exist in the state, which may in places be worth investigating.

The largest of all these possible reserves in California probably lies in the remaining buried Tertiary stream channels of the Sierra Nevada.

FIG. 47. A mined-out segment of the Tertiary Central Hill channel near Murphy, Calaveras County. White rhyolitic ash, which covered this channel, is exposed on the left.

Introduction

Object of the Report

The gold-bearing gravel deposits or placers of California that still remain untouched lie, for the most part, in more obscure positions than the depleted gravels which formerly produced vast wealth for the state. The depleted gravels, which were once readily accessible but are now nearly mined out, fall into three principal classes:

Depleted placers
1. Recent and Quaternary stream gravels.
2. Uncovered channels of Tertiary age.
3. Buried Tertiary channels, easily located.

About one billion dollars worth of gold came from these three sources, the first having produced twice as much as the other two put together.

The placers which still remain to be sought out and worked, offer a challenge to the ingenuity of the exploration geologist. The problems involve the following types of deposits:

Placer reserves

1. Deep gravel deposits lying immediately beneath several large rivers, such as the Feather and Klamath.
2. Isolated high benches such as those found in the Klamath Mountains.
3. Ancient gravels that lie beneath 'false bedrock' (interbedded volcanic layers) of the dredging areas along the western foot of the Sierra Nevada.
4. Gold-bearing gravels occurring in the 'shore' deposits of the Ione (Eocene) formation, and the Chico (Cretaceous) formation.
5. Buried Tertiary channels and associated benches located in the known gold-bearing districts of the state.
6. Buried Tertiary channels and benches in the lava-covered district which lies between the Sierra Nevada and the Klamath Mountains.
7. Bajada placers, or desert alluvial-fan deposits, where gold is derived directly from the original mineralized bedrock source.
8. Desert placers, where the gold is reconcentrated from more ancient gold-bearing streams.
9. Buried desert stream placers.

The scope of these problems indicates the great need of an understanding of the geological principles involved. In no kind of mining is geology more applicable than in the exploitation of these more obscure placer deposits. For this reason, the following discussion is written.

Significance of Improved Exploration Methods

A widespread geologic study of the ancient Tertiary gold-bearing stream channels of the Sierra Nevada, the gravel deposits of which are found to a large extent buried beneath a mantle of volcanic materials, was concluded by the United States Geological Survey a quarter of a century ago. Lindgren's "Tertiary Gravels"[1] summed up, in a splendid manner, in 1911, these various geologic studies. His data were drawn from his own careful observations, from those of his associates, H. W. Turner, F. L. Ransome, and J. S. Diller (issued largely in folio form), and from such early sources as J. D. Whitney, W. H. Storms, and Ross E. Browne.

Lindgren's Colfax folio,[2] published in 1900, was the last great detailed field study of this kind in the Sierra Nevada. By no means, however, is this folio confined to the subject of stream channels, for it deals with every phase of the geology of the quadrangle. Everything of importance which it was possible to accommodate on a map of the small scale used—two miles to the inch—was recorded. At the time this field work was done, the best equipment and finest technique of the day were employed, and very little escaped Lindgren's keen observation, each feature being scrutinized and shrewdly interpreted by his masterful mind.

Since then, however, considerable advance has been made in exploration technique; other and different points of vantage are now available; and a greater degree of refinement of study is therefore in order. Furthermore, mining itself has many advantages today over the earlier methods. With the recent enormous advances in development of motive power, pumps, and other machinery, the drift-mining industry could easily undergo a greatly accelerated development, if guided by intelligent exploration carried on in advance of the attempt to extract 'pay' gravel.

[1] Lindgren, W., The Tertiary gravels of the Sierra Nevada of California: U.S. Geol. Survey Prof. Paper 73, 226 pp., 1911.
[2] Lindgren, W., U.S. Geol. Survey Geol. Atlas, Colfax folio (no. 66), 10 pp., maps, 1900.

Aerial Photography. From the air, regional photographs are systematically taken by qualified aerial photographers.[3] The pictures are then examined under the stereoscope, or used in constructing topographic maps [4] which show the most amazing completeness of detail. Many surface features never before realized are thus simply unfolded before the eye. Geologic truths in great numbers are revealed, and many important problems solve themselves. Used as a base for location of field observations and surface mapping, these aerial photographs are unexcelled They are undoubtedly the greatest practical aid which has yet reached the hands of the geologist; besides they give secrecy, speed, and low-cost surveying to the program of modern exploration.

Geophysical Surveying. Added to this regional view from the air is the greater insight into the very interior of the earth itself afforded by several types of geophysical instruments, now well-tried and standardized. Peculiar characters of rock structure and composition are not only revealed but measured with precision by skillful engineers. Since the proper interpretation of all results thus obtained requires sound geologic reasoning, it is important that a better and more detailed background of geology should be drawn, and this is made more effective by aerial photography.

Physiography. The subject of physiographic geology or geomorphology has in recent years made notable progress in developing sound, scientific principles concerning the history and origin of the present surface configuration of the earth. Since these principles are directly applicable to the more ancient earth surfaces of the Sierra Nevada, over which flowed the Tertiary streams now extinct, Tertiary physiography is the key to the ancient channel problem.

Study of Desert Processes. Recent study of the geologic processes at work in the desert has led to a better understanding of the desert placers, which offer a practically virgin field for exploration, holding a potential wealth not yet known.

Stream Sedimentation. Furthermore, the study of sedimentation has now reached a refined stage of development. There are today available various methods of technique which may be, but have not yet been, extensively applied to stream deposition. This sort of research includes the critical study of texture and structure of strata, as well as the microscopic examination of their mineral grains. It should yield a wealth of practical information concerning the processes involved in the accumulation of gravels, in the nature and direction of stream flow, in knowledge of what to expect as regards the concentration of gold and other heavy minerals, and in the correlation of channels of the same period or of the same system.

The more obvious criteria of stream sedimentation have long been used by the experienced miner, who, by examining the gold particles

[3] Lee, Willis T., The face of the earth as seen from the air; a study on the application of airplane photography to geography: Am. Geog. Soc., Special Pub. 4, 1922.
English, Walter A., Use of airplane photographs in geologic mapping: Am. Assoc. Petroleum Geologists Bull. 14, pp. 1049-1058, 1930.
Eliel, Leon T., Aerial photography and its importance to California geologists: California Div. Mines, Mining in California, State Mineralogist's Rept. 26, pp. 64-71, 1930.
Eliel, Leon T., Aerial photography proves its importance to California geologists: Oil Bull., vol. 15, pp. 1177-1181, 1253, 1929.
[4] Haquinius, E., and Shuster, E. A., Fr., Construction and operation of the Hugershoff aerocartograph: U.S. Geol. Survey, Section of Photo-Mapping, 1929.

Fig. 48.　View northeast toward Table Mountain, Toulumne County, showing how this resistant lava flow, which filled a canyon cut in earlier volcanic materials, now stands out in relief, while the surrounding country is eroded from its sides. *Photo by Burt Beverly.*

under the simple hand lens, infers whether they were robbed from earlier channels or whether they came directly from a vein.　Also he uses the 'shingling' of gravels to tell him the direction of stream flow.　These and a few other working criteria now used by the miner, however, have not all undergone a thorough scientific test, and at present there is a wide difference of opinion as to their interpretation.

The advances in technique and sound geologic interpretation should, therefore, be made to serve as wide practical aids in channel exploration, and much more definite and conclusive results should be gained now than were possible 25 years ago.　Since it is well known that many channels still lie buried, undoubtedly containing a potential gold wealth not yet half exhausted, why should not new discoveries now be made if the new technique in placer exploration is employed?

Certainly the more advanced methods of technique in exploration were called upon by the petroleum industry, resulting in the flooding of the country with oil.　Now the application of some of the same methods is taking a leading part in various types of mining, bringing further success to other mineral industries.　If the gravel mining industry takes full cognizance of the importance of the new exploratory methods of the science, successes for it should indeed be assured.

Usefulness of Contouring an Ancient Surface

The most elucidating method of tracing in detail and showing graphically the position and course of an ancient stream valley is by preparing a contour map [5] of the old drainage surface.　The contours may be superimposed over a base map which should also show the present surface topography and areal distribution of the geologic formations, as well as any mine workings, drill holes, etc.

With the old surface contours superimposed on the present surface contours, an accurate estimate may be made of the thickness, extent, and

[5] Lindgren, W., The gold-quartz veins of Nevada City and Grass Valley districts, California: U.S. Geol. Survey, 17th Ann. Rept., pt. 2, pl. II, 1896.　(Contour map of Neocene bedrock surface.)

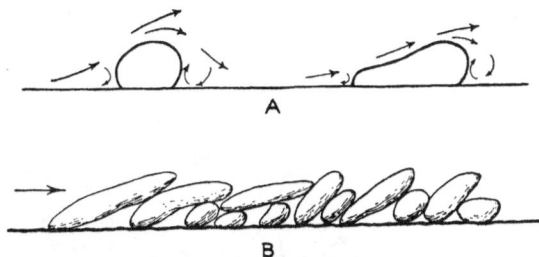

FIG. 49.   A, Effect of stream-bed irregularities on water current. *After Chamberlin and Salisbury.* B, 'Shingling'. The arrow points in the direction of stream flow. This arrangement of flat stones may be found in the bed of a mountain stream and is used by the drift miner to tell which way the ancient stream was flowing at that particular point. *After Geike; originally from Jamieson, T. G., Quarterly Journal, Geological Society of London, vol. 16, 1860.*

yardage of the intervening channel-filled area.   The points which are to be used in preparing an old-surface contour map should be secured through careful study of geologic and physiographic conditions, and obtained during the surface and underground survey.   Drill-hole data and the essential results of geophysical observations should also be represented on this map.   Careful enough study should be made so that the points used represent only one period of erosion.

If the geologic work is done prior to a contemplated geophysical survey and drilling program, much time and money may be saved in the location and number of points of observation, as well as in the number of drill holes needed.   An approximate old-surface contour map may generally be constructed as a preliminary step by a skillful geologist, though he is limited only to surface data.   This map should locate the general trend of the ancient valley to which later detailed work should be confined, thus eliminating much unnecessary and more costly work in adjoining areas unlikely to be productive.   The most accurate elevations are those taken on bedrock where it is in direct contact with the older gravels, or with the pipe-clay and other volcanic materials of the oldest period of the area.   Thus geological investigation should limit the extent of the geophysical work, which in turn defines the area to be drilled and later to be explored by underground methods.

The basis for such mapping may effectively be prepared as part of a plane-table survey of the surface topography and geology.   Aerial photography, however, has now become an almost indispensable aid to such work.   Enlarged aerial photographs may be used as plane-table sheets, thus adding to the final map an enormous amount of useful information and physiographic detail.

Provinces of the Ancient Channels in California

The so-called 'buried channels' occur for the most part in the northern Sierra Nevada, where, during the Tertiary period, millions of years ago, volcanic outbursts with their mud-flows, covered the then existing network of stream courses.   Thus entrapped and preserved, this old surface of the earth, probably Eocene in age, with all its forest-covered hills, beautiful valleys, and winding streams laden with gold, became completely hidden.   Not until the area was lifted by mountain-making forces did the later rivers cut their canyons through the heavy mantle and expose whatever was beneath it.   But even then, further volcanic mudflows, cobble-washes, and newer streams of lavas, filled and refilled the

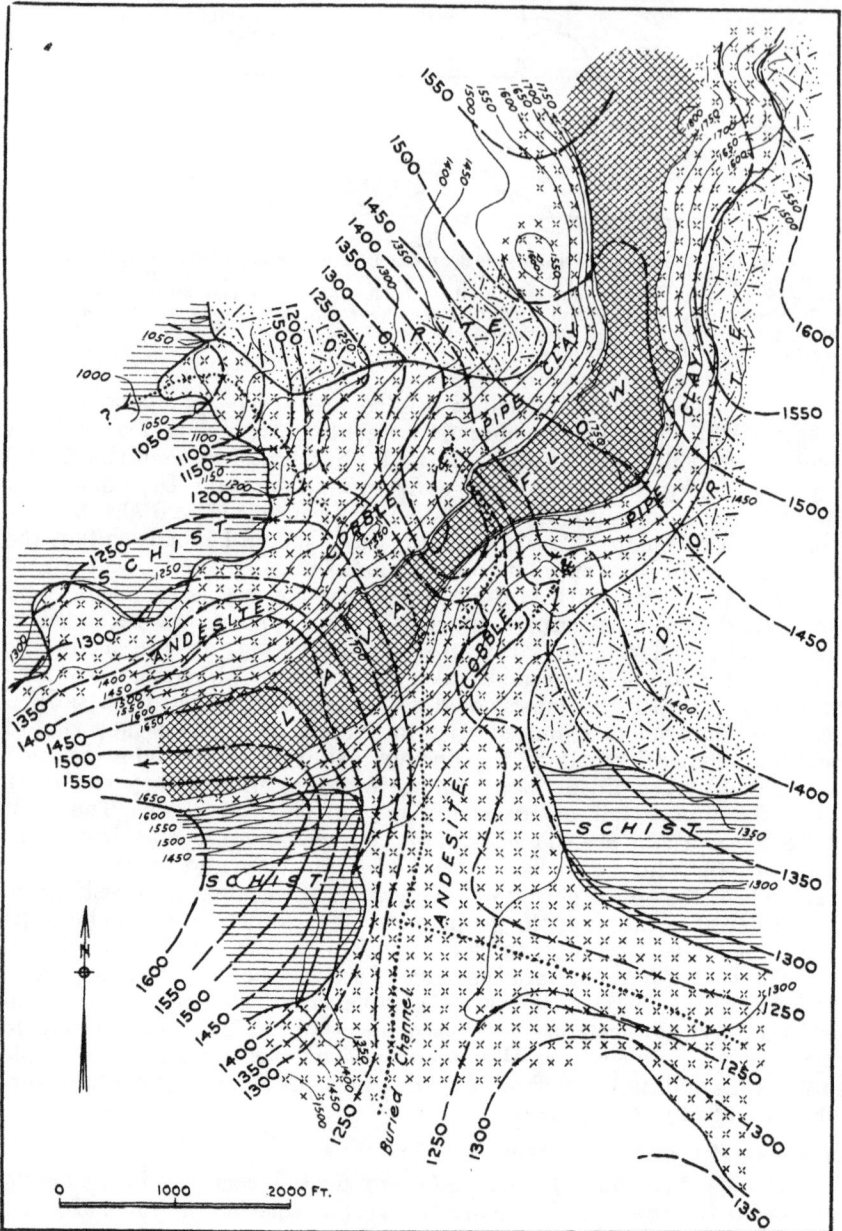

FIG. 50. Example of a geologic map showing the earlier prevolcanic topography in contrast with the present topography. Old-surface contours (dashed lines) are drawn over present-surface contours (fine, full lines). A major early Tertiary valley is thus reconstructed, and the general position of its buried stream channel is indicated, lying beneath volcanic materials of two later geologic periods, i.e., (1) Miocene cobble and pipe-clay (mud flows and lake beds), and (2) Pliocene lava flow which followed a canyon cut in the andesitic materials. The older channel cut in bedrock is gold-bearing. Its course was controlled by bedrock, for it followed the contact between hard diorite and softer schist. The gradient of the lava flow is not as steep where it is directed south as where it turns toward the west. This is explained by the westward tilting of the Sierra Nevada.

valleys formed. Finally, modern canyon-cutting trenched the whole area deeply, leaving remnants of flat-topped ridges between.

Though no volcanic mantle ever covered the ancient streams of the Klamath Mountains, the region was uplifted in late Tertiary and Quaternary times, and the rivers were therefore entrenched. As a result, benches of gravel were left at various elevations along the sides of the canyons. Faulting also played an important role in the history of the uplift, trapping gold-bearing gravels which date back to the Eocene.[6] The whole subject of the surface features of the region, past and present, represents a series of very interesting problems not yet entirely solved, though they are now being carefully studied.

In the Mojave Desert,[7] Tertiary gold-bearing streams may have once flowed south from the region which is now the Sierra Nevada; but during the Pleistocene period their deposits became so broken and disrupted by faulting, and so extensively covered by later desert wash, that now there remains little to be easily recognized or traced.

The problem of the 'dry placer' is one of considerable importance, and when more is learned about it, an immense potential gold wealth may be discovered which has not yet been glimpsed. At present, the lack of water[8] is largely responsible for limited operations. It is quite possible, however, to find and develop a sufficient underground water supply in many places for dredging operations. The finding of such water supplies may also be greatly assisted through the use of geological knowledge and geophysical surveying.

In the Peninsular Ranges of San Diego County, some placers have been mined. In the Poway (Eocene)[9] marine conglomerate is found the Ballena placer,[10] described as early as 1893.

Economic Significance of the Tertiary Gravels

Lindgren[11] in 1911, made the statement that "about $300,000,000 is a conservative guess for the amount obtained from the Tertiary gravels," including hydraulic, drift, and other forms of placer mining. Just how much has been taken from the buried channels by underground methods, alone, is difficult to determine. J. M. Hill[12] states:

"Drift mining on the ancient buried channels was started at Forest Hill, Placer County in 1852 and was an important source of placer gold, by 1866. During the period beginning about 1876 and continuing to about 1890, drift mining was most productive."

[6] Hinds, Norman E. A., Mesozoic and Cenozoic eruptive rocks of the southern Klamath Mountains, California: Univ. California, Dept. Geol. Sci., Bull., vol. 23, p. 368, 1935.

[7] Hulin, C. D., Geologic features of the dry placers of the northern Mojave Desert: California Jour. Mines and Geology, vol. 30, pp. 417-426, 1934.

[8] Simpson, E. C., Geology and mineral deposits of the Elizabeth Lake quadrangle, California: California Jour. Mines and Geology, vol. 30, p. 409, 1934.

[9] Donnelly, Maurice, Geology and mineral deposits of the Julian district, San Diego County, California: California Jour. Mines and Geology, vol. 30, p. 369, 1935.

[10] Merrill, Frederick J. H., The counties of San Diego, Imperial: California Min. Bur., State Mineralogist's Rept. 14, p. 652, 1914.
Fairbanks, H. W., Geology of San Diego County, also portions of Orange and San Bernardino Counties: California Min. Bur., State Mineralogist's Rept. 11, p. 91, 1893.

[11] Lindgren, W., The Tertiary gravels of the Sierra Nevada of California: U.S. Geol. Survey Prof. Paper 73, p. 81, 1911.

[12] Hill, J. M., Historical summary of gold, silver, copper, lead, and zinc produced in California, 1848 to 1926: U.S. Bur. Mines, Econ. Paper 3, p. 6, 1929.

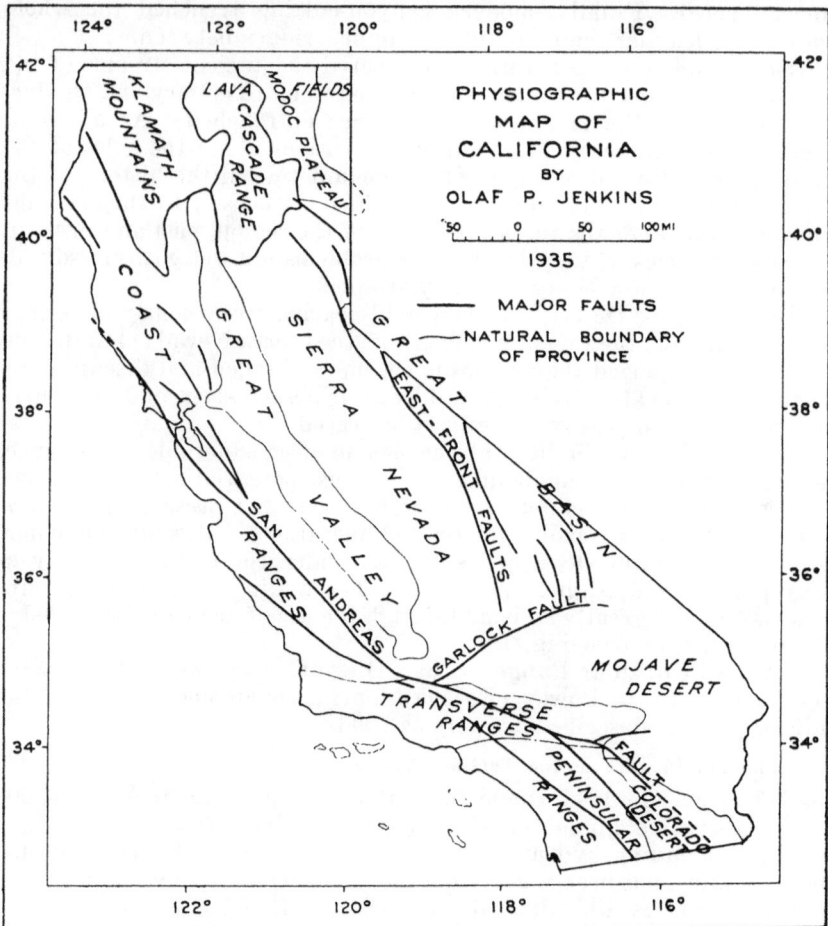

FIG. 51. Map of California showing the principal physiographic provinces and the major faults or fault zones of the State. The principal gold provinces occur in the northern Sierra Nevada, the Klamath Mountains, the Great Basin, and the Mojave Desert. Gold has been mined in the Peninsular Ranges, and a small amount has been found in the southern Coast Ranges. It was first discovered in 1831 in the Transverse Ranges east of Los Angeles; but not until James W. Marshall's discovery on the American River in the foothills of the Sierra Nevada, January 24, 1848, did the occurrence of gold in California become significant.

From 1848 to 1933 California's gold production amounted to nearly two billion dollars, of which one and a quarter billion came from placer gravel. The ratio of gold produced from placers to gold produced from lodes is shown in the following table:

Ratio of placer production to lode production

Period	Per-cent	Period	Per-cent
1848 to 1850	100	1891 to 1900	20
1851 to 1860	99	1901 to 1910	35
1861 to 1870	90	1911 to 1920	44
1871 to 1880	70	1921 to 1930	44
1881 to 1890	25		

The decline in this ratio was caused by the closing down of hydraulic mines by the Sawyer Decision of 1884 together with the Caminetti Act of 1893, as well as by the increased development of lode mining. The later increase of the ratio was due to the use of the dredge, which was introduced in California in 1898.

Haley[13] has estimated that "In the neighborhood of $600,000,000" worth of gold (on the basis of the old price) is still recoverable from the old Tertiary channels, referring to both hydraulic and drift-mining properties. There are undoubtedly many unaccounted-for channels, such as those that lie buried beneath the extensive lava flows which occur in the region lying between the Sierra Nevada and Klamath mountains. Though most of these are not today available for mining, the channels along the margin of this area may some day be reached. It is conceivable, however, that the reserves in buried channels throughout the Sierra Nevada, workable only through underground methods, should be reckoned at least in terms of hundreds of millions of dollars.

The economic significance of the buried channel and the importance of its exploration should not, therefore, be slighted, for it represents a vast future source of gold wealth for California. The revival of the channel-mining industry is made still more interesting by an understanding of the detailed geologic features, the possibilities they suggest for certain of these buried stream courses, and the realization of the vast extent of the region yet to be explored.

Since an estimate of the potential wealth of the desert placers depends upon further exploration and mining of them, it is not yet possible to evaluate their place in this study.

History of Development

The discovery of the Tertiary channels followed shortly after the discovery of gold in the present stream courses. In places, remnants of Tertiary channels were found lying exposed high up on 'flats,' stripped of their volcanic covering by Pleistocene erosion. Starting from these flats, the miners followed the uncovered channel to the point where it was covered by the lava capping, the top of which formed another kind of 'flat.' Water was nearly always encountered, which led to the construction of long and expensive drainage tunnels. To place these tunnels at the proper elevation to serve their best purpose was the most serious problem, for the old-timers did not have the powerful pumps which we use today.

The ideas developed concerning the courses and positions of the ancient channels were many and varied. Misconceptions of Tertiary physiography led many persons astray, and millions of dollars were spent in vain. In spite of the millions gained, the losses were so great as to stamp this form of mining as hazardous in the extreme.

It is most instructive to follow carefully the recorded history of mining in a given district and to consider its relation to the geologic condition of that area. The two are so closely related as to provide a guide to the probable history which might be expected to be found in another area if the geology were known, or vice versa; the recorded history reflects what geologic conditions may be expected.

---

[13] Haley, C. S., Gold placers of California: California Min. Bur. Bull. 92, p. 5, 1923.

FIG. 52. Old work of the Valdor Dredging Company in the bed of the Trinity River, Junction City. Bench gravel exposed on the opposite side of the canyon shows the effect of early hydraulic mining.

FIG. 53. Ancient river terraces of Canyon Creek, tributary of the Trinity River, partly hydraulicked.

The geologic history and structure of the buried channels are so complex that the best engineers have been baffled by them. Fragmentary benches and segments of rich gravel deposits which still rest in positions completely hidden from the surface, or even from the underground passages which enter into the lower, main channels, afford alluring possibilities to the geologist and geophysicist as well as to the prospector. A three-dimensional surface, complex and irregular in the extreme, is the problem to be faced.

The key to the solution is geology, aided by aerial photography, followed by new geophysical surveying, and finally by directed prospecting through means of the drill, shaft, incline, or tunnel. To be effective, all these methods should be coordinated into one unified exploration program.

Classification of Placers

Outline of Classification

A systematic geological study of placers calls for an orderly classification dividing them into genetic types, which indicate how they were first formed. The following classification is based upon the fundamental conditions of deposition:

Fundamental classification of placers

A. Residual placers or 'seam diggings.'

B. Eluvial or 'hillside' placers, representing transitional 'creep' from residual deposits to stream gravels.

C. Bajada placers, a name applied to a certain peculiar type of 'desert' or 'dry' placer.

D. Stream placers (alluvial deposits), sorted and re-sorted, simple and coalescing.

E. Glacial-stream placers, gravel deposits transitional from moraines, for the most part valueless.

F. Eolian placers, or local concentrations caused by the removal of lighter materials by the wind.

G. Marine or 'beach' placers.

Of these seven types, the ordinary stream placer is by far the most important. All types are, however, more or less interrelated and intergradational; they are all subject to deformation or burial, and they may be formed during any geological period.

Most of these various types of placer deposits have been well described and classified by Brooks,[14] Mertie,[15] and others[16] in their studies of Alaskan geology. Webber[17] recently proposed the name, "bajada placer" for peculiar desert, or so-called 'dry' placers, which he carefully analyzed. The concentration of gold by the agency of wind in Western Australia has been described by Rickard[18] and Hoover.[19]

[14] Brooks, Alfred H., The gold placers of parts of Seward Peninsula, Alaska: U.S. Geol. Survey, Bull. 328, pp. 111-139, 1908. . . . Mineral resources of Alaska, report. on progress of investigations in 1913: U.S. Geol. Survey, Bull. 592, pp. 27-32, 1914.

[15] Mertie, J. B., Jr., The occurrence of metalliferous deposits in the Yukon and Kuskokwim regions: U.S. Geol. Survey Bull. 739, pp. 160-165, 1923.

[16] Wimmler, Norman L., Placer mining methods and costs in Alaska: U.S. Bur. Mines Bull. 259, p. 11, 1927. (Classification of placers.)

[17] Webber, Benjamin N., Bajada placers of the arid southwest: Am. Inst. Min. Met. Eng., Tech. Pub. 588, pp. 3-16, 1935.

[18] Rickard, T. A., The alluvial deposits of Western Australia: Am. Inst. Min. Met. Eng., Trans., vol. 28, pp. 490-537, 1898.

[19] Hoover, H. C., The superficial alteration of Western Australian ore-deposits: Am. Inst. Min. Met. Eng., Trans., vol. 28, pp. 758-765, 1898.

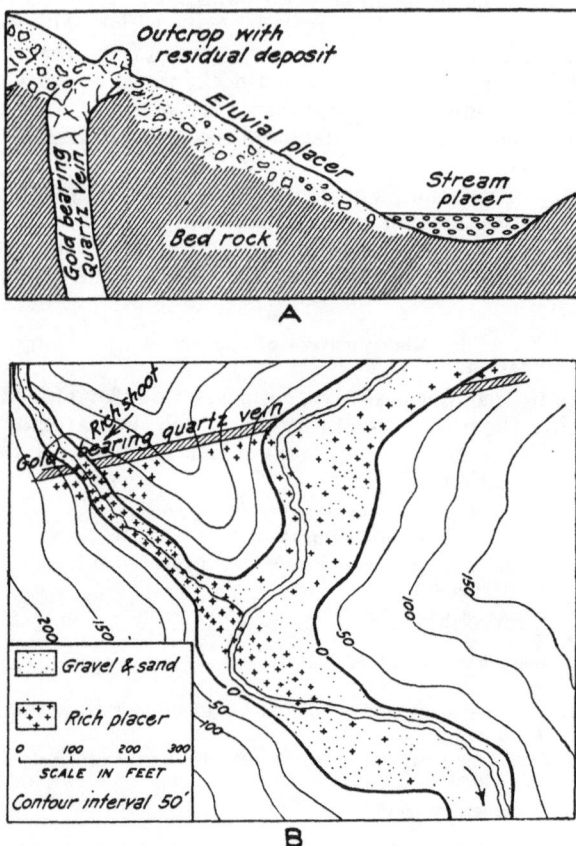

FIG. 54. *A*, Diagrammatic cross-section showing the transitional stages in the development of placer deposits: First, the quartz vein; second, disintegration at the outcrop to form a residual placer; third, formation of eluvial placer by 'creep' of residual material down the hill slope; fourth, deposition of water-worked material as alluvium, forming an auriferous gravel deposit, or stream placer. *B*, Sketch map showing development of rich placers broken down directly from the disintegration of a gold-bearing vein; *after Lindgren, Mineral Deposits.*

General Statement

In valuating a placer, one of the first considerations should be the determination of how it was formed. It is recommended, therefore, that the exploration engineer should classify genetically each gravel deposit to be prospected. This calls for an understanding of the historical geology of the region and of the processes which have been responsible for the formation of the deposit, as well as how it came to be preserved or modified from its original form. The actual sampling of a deposit is carried on in a much more intelligent and satisfactory manner when a clear understanding of the geologic set-up has been acquired.

Characteristics of the Principal Types of Placers

Residual Placers

In order that gold may become released from its original source in bedrock, the encasing material must be broken down.   This is most effectively done by long-continued surface weathering.   Disintegration is accomplished by persistent and powerful geologic agents, which effect the mechanical breaking-down of the rock and the chemical decay of the minerals.

The surface portion of a gold-bearing orebody will become enriched during this process of rock disintegration, because some of the softer and more soluble parts of the rock are carried away by erosion, leaving the remaining portion of higher tenor.   The name 'residual placer' is applied to this type of deposit.   After the residual portion is mined away by comparatively inexpensive methods, the harder mineralized rock is encountered, and the mining methods must be changed to accommodate another type of deposit, i.e., the lode.

The so-called 'seam diggings' are in weathered, gold-bearing quartz stringers, occurring along fracture zones of disintegrated schists.

Eluvial Placers

After gold is released from its original bedrock encasement through agents of rock decay and weathering, the whole weathered mass may 'creep' down the hillside (in some regions partly because of frost heaving) and may finally be washed down rivulets into gulleys.   Lindgren [20] states:

"When the outcrops of gold-bearing veins are decomposed a gradual concentration of the gold follows, either directly over the primary deposits or on the gentle slopes immediately below.   The vein when located on a hillside bends over and disintegration breaks up the rocks and the quartz, the latter as a rule yielding much more slowly than the rocks; the less resistant minerals weather into limonite, kaolin, and soluble salts. The volume is greatly reduced, with accompanying gold concentration.   The auriferous sulphides yield native gold, hydroxide of iron, and soluble salts.   Some solution and redeposition of gold doubtless take place whenever the solutions contain free chlorine. The final result is a loose, ferruginous detritus, easily washed and containing easily recovered gold.   This gold consists of grains of rough and irregular form and has a fineness but slightly greater than that of the gold in the primary vein."

On its way down the hillside, gold is sometimes concentrated in sufficient amount to warrant mining.   Such deposits are classified as eluvial placers.   They are transitional between residual and stream, or alluvial, deposits.

There have been a number of residual and eluvial gold deposits mined in both the Sierra Nevada and Klamath Mountains; for example, the 'seam diggings' of Georgia Slides,[21] El Dorado County, and Scott Bar,[22] Siskiyou County.

Stream Placers

By far the most important type of placer is the ordinary alluvial gravel or stream placer.   So far, it has been the source of most of the placer gold mined; but now its supply is nearing depletion, save for values remaining in those ancient channels which lie deeply buried beneath a cover of lava or rock debris.

[20] Lindgren, W., Mineral deposits, p. 213, McGraw-Hill Book Co., 1919.
[21] Logan, C. A., Mother Lode gold belt of California: California Div. Mines Bull. 108, pp. 43-44, 1935.
[22] Logan, C. A., Siskiyou County: California Min. Bur., Mining in California, State Mineralogist's Rept. 21, p. 474, 1925.

FIG. 55.   Quartz Hill mine, Scott River, Siskiyou County. Disentegrated material formed in the upper part of gold-bearing veins, by residual weathering, has been removed by hydraulic mining. Note the terrace-like profile of the hill, indicating that an older, less rugged surface existed at the time this deep weathering took place. Uplifting of the region entrenched the drainage, and much of the gold released by weathering was washed by Pleistocene erosion into Scott River where rich gravel bars have been mined.

FIG. 56.   Residual-weathered part of a gold-bearing orebody, King Solomon mine, Siskiyou County.

Deposits by streams include those of both present and ancient times, whether they form well-defined channels or are left merely as benches. Stream placers consist of sands and gravels sorted by the action of running water. If they have undergone two or more periods of erosion, and have been re-sorted, the result will in all probability be a comparatively high degree of concentration of the heavier mineral grains.

Quoting from Mertie:[23]

"All bench placers, when first laid down, were stream placers similar to those of the present stream valleys. In the course of time the stream gravels, if not reworked by later erosion, may be left as terraces or benches on the sides of the valley, if the local base-level is lowered and the stream continues to cut down its channel. Such deposits constitute the so-called bench gravels. On the other hand, if the regional or local base-level is raised, the original placer may be deeply buried and a second or later placer deposit may be laid down above it. * * * If the local base-level remains practically stationary for a very long period, a condition seldom realized, ancient and recent placers may form a perfectly continuous deposit in a long valley, for the deposition of a gold placer is known to occur at that point in a valley where the stream action changes from erosion to alluviation, and such deposits are therefore formed progressively upstream.

"Where several parallel and contiguous streams that are forming placers emerge from their valleys upon an open plain, perhaps into some wide valley floor, a continuous or coalescing placer may be formed along the front of the hills. If the streams empty into some lake or estuary, a delta placer, genetically the same but perhaps different in some minor respects, may be formed. Manifestly such compound placers may be formed by either present or ancient streams and may be elevated or buried in the same way as simple stream placers."

In order to understand thoroughly the subject of stream placers, streams themselves must be studied in regard to their habit, history, and character. The effects of existing and changing climates, the relation to surrounding geologic conditions, and the effect of movements of the earth must also be considered. A special chapter in this report has therefore been devoted to a brief outline of the fundamental geologic features of streams.

Glacial-Stream Placers

It is a frequent fallacy of the placer miner to attribute the deposition of gold-bearing gravels to the action of glaciers. Contrary to such a belief, glaciers do not concentrate minerals; the streams issuing from melting ice, however, may be effective enough in sorting debris to cause placers to be formed under certain especially favorable conditions. In California, glaciers occurred throughout the high Sierra during the Pleistocene, but in Tertiary times they were wholly lacking. The Pleistocene streams cut through the earlier channels, robbing them of much of their gold.

Quoting from Blackwelder:[24]

"Since it is the habit of a glacier to scrape off loose debris and soil but not to sort it at all, ice is wholly ineffective as an agency of concentration for metals. Gold derived from the outcrops of small veins is thus mixed with large masses of barren earth. Attempts to mine gold in glacial moraines, where bits of rich but widely scattered float have been found, are for that reason foredoomed to failure.

"If a glacier advances down a valley which already contains gold-bearing river gravel, it is apt to gouge out the entire mass, mix it with much other debris and deposit it later as useless till. Under some circumstances, however, it merely slides over the gravel and buries it with till without disturbing it.

[23] Op. cit., pp. 161-162.
[24] Blackwelder, Eliot, Glacial and associated stream deposits of the Sierra Nevada: California Div. Mines, Mining in California, State Mineralogist's Rept. 28, pp. 309-310, 1932.

FIG. 57.   Hydraulicking a high bench-gravel deposit of the Trinity River, near its confluence with the South Fork.   Location: Salyer mines, pit No. 5.

"On the other hand, the streams born of glaciers or slowly consuming their moraines have the power to winnow the particles of rock and mineral matter according to size and heaviness.   Such streams may form gold placer deposits in the well-known way by churning the load they carry and allowing the heavy minerals to sink to the bedrock.   Placers may therefore be found in the deposits of glacial rivers if there are gold veins exposed in the glaciated area upstream.   Nearly all the gravel which has been dredged for gold along the foothills of the Sierra Nevada was deposited by rivers derived in part from glaciers along the crest of the range, but most of the gold was probably picked up in the lower courses of such rivers.   Since glacial rivers choke themselves and build up their channels progressively, their deposits are likely to be thicker and not so well concentrated as those of the more normal graded rivers which are not associated with glaciers."

A few gold-bearing deposits in re-worked glacial till may be found along the eastern front of the Sierra Nevada, as, for example, in the region just north of Mono Lake on the road to Bridgeport.

Bajada Placers

The name 'bajada' was first used by Tolman [25] for confluent alluvial fans along the base of a mountain range.   Recently Webber [26] named and described the bajada type of placer deposit as follows:

"Bajada is the Spanish term for slope and is used locally in the Southwest to indicate the lower slope of a mountain range, the portion consisting of rock debris and standing at a much lower angle than the rock slope of the range proper.   *   *   *

"The total production of gold from bajada placers in the southwestern United States is necessarily small, probably not over ten million dollars.   *   *   *

"Most all bajada placer gravels are Quaternary and the larger part are recent. *   *   *

"The genesis of a bajada placer is basically similar to that of a stream placer except as it is conditioned by the climate and topography of the arid region in which the placer occurs.   *   *   *

"Erosion, transportation and deposition in a region of extreme aridity present some phenomena not encountered in more humid areas.   Practically all the work of running water is strongly conditioned by aridity.   *   *   *

[25] Tolman, C. F., Erosion and deposition in southern Arizona bolson region: Jour. Geology, vol. 17, pp. 136-163, 1909.
[26] Webber, B. N., op. cit., excerpts from pp. 3, 4, 6, 8, 10, 11.

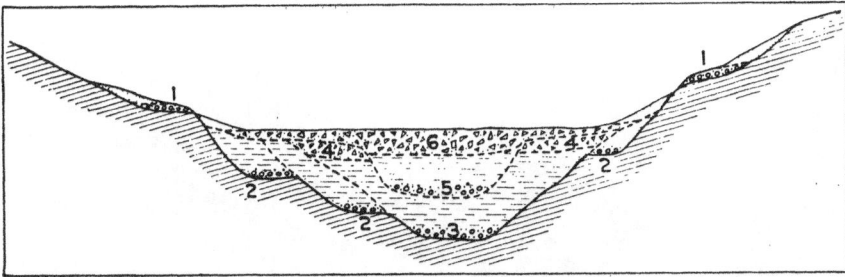

FIG. 58.   Cross-section of a gold-bearing desert stream valley (Manhattan, Nevada), showing the results of several periods of stream deposition from the oldest (1) to the youngest (6).   *After Ferguson, U. S. Geol. Survey Bull. 723.*

"Rock-floored canyons through which rock fragments are moved by infrequent torrential floods should constitute excellent pebble mills for the further reduction of the material, but the amount of attrition accomplished seems to be slight, as fragments, large or small, on the bajada slope are decidedly angular and show little effect of attrition.   *   *   *   Probably a small percentage of the gold is freed during this phase of the movement of gravel.   The gradient of these intermont drainage channels is too high to permit lodgment of the finer gravel.   When a small amount of gravel is temporarily lodged in one of these channels, the deposit displays most of the characteristics of stream gravel.

"As debris reaches the bajada slope a rapid diminution in volume of water due to seepage and an extreme decrease in the grade of channel causes deposition of debris, and either (1) an alluvial fan or (2) a gravel-mantled pediment may be formed.   If detritus is supplied to a bajada slope much faster than it can be removed, an alluvial fan is the result.   *   *   *   If rock debris is supplied to the bajada slope in considerable volume but not in excess of the quantity capable of transference to the center of the basin by the existing agencies, a gravel-mantle pediment results.   *   *   *

"The bulk of the gold that has been released from its matrix on the journey from lode outcrop to bajada slope is deposited on the bajada slope close to the mountain range.   The gold is dropped along the contact of the basin fill and bedrock; this is referred to hereafter as the lag line and is coincident with the line of contact of bajada gravels lying at a low angle and the rock slopes of the range standing at a high angle.   *   *   *

"The heaviest deposition of gold is on bedrock at the lag line, and since the lag line is moving in the direction of the crest of the range, values on bedrock may be distributed over a large area of which the longest dimension is parallel to the foot of the range.   Because bulk concentration does not operate as in a river channel, and a certain percentage of the gold is still locked in fragments of matrix, to be partly released by further disintegration on the bajada slope, there is a strong tendency for less gold to reach bedrock and for more to remain erratically distributed throughout the detritus than in the case of stream gravels."

There are probably many examples of typical bajada placers in the Mojave Desert and Great Basin regions of California, but recognition of them as such has not yet reached publication.   Reconcentrations from former gold-bearing streams in the Mojave Desert have been described by Hulin [27] and dry placers in southern California have been listed by Sampson. [28]

Eolian Placers

Webber [29] states that: "Bajada placers usually show an appreciable and even considerable enrichment on the surface due to removal of lighter material by wind and sheet floods."   This applies to some of the dry

[27] Op. cit.
[28] Sampson, R. J., Placers of southern California: California Div. Mines, Mining in California, State Mineralogist's Rept. 28, supplement, pp. 245-255, 1932.
[29] Op. cit., p. 15.

FIG. 59. 'Placer mining in the beach sands at Santa Cruz, California,·1933.

placers of California, though no commercial eolian gold deposits such as those mined in Australia, previously referred to, are known in this state.

Wind action, however, is responsible for the removal of large amounts of fine detritus in the desert. The process involved has been called 'deflation' and its results described by Blackwelder.[30] It is quite likely that it will be found to play an important part in the surface concentration of desert placers.

Spurr[31] described "auriferous sand dunes" in the Nevada desert seven miles south of Silver Peak, 18 miles from the California boundary line.

Beach Placers

Concentrations of heavy minerals occur in various places along the Pacific Coast as a result of the action of shore currents and waves, which tend to sort and distribute the materials broken down from the sea cliffs or washed into the sea by streams. The heavy minerals consist for the most part of magnetite, chromite, ilmenite, monazite, and zircon, with occasional fine particles of gold and platinum. Beach placers are of two kinds, (a) present beaches and (b) ancient beaches. An elevated coast line is often found overlaid with terrace gravels which were deposited at a time when the coastline was depressed. The beach placers of economic importance are those that have been reconcentrated over and over again.

Excellent descriptions of the geologic processes involved may be found in reports on beach placers of Nome,[32] Alaska, and of the coast

- [30] Blackwelder, Eliot, The lowering of playas by deflation: Am. Jour. Sci., vol. 21, 5th ser., pp. 140-144, 1931.
[31] Spurr, J. E., Ore deposits of the Silver Peak quadrangle: U.S. Geol. Survey Prof. Paper 55, pp. 96, 97, 1906.
[32] Moffit, Fred H., Geology of the Nome and Grand Central quadrangles, Alaska: U.S. Geol. Survey Bull. 533, pp. 109-123, 1913.

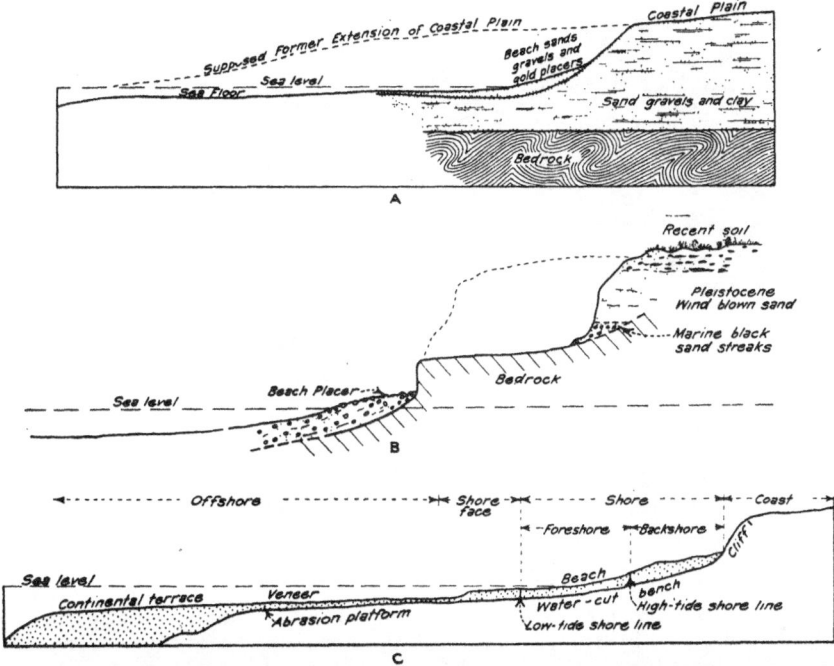

Fig. 60. *A*, Diagrammatic cross-section illustrating the formation of beach placers in Alaska; *after Collier and Hess, U. S. Geol. Survey Bull. 328.* *B*. Cross-section of a typical beach placer (Oregon); *after Pardee, U. S. Geol. Survey Circ. 8.* *C*, Diagrammatic cross-section of a coast, showing shore zones in an advanced stage of development; *after Johnson; see Pardee, U. S. Geol. Survey Circ. 8, 1934.*

of Oregon[33] and California.[34] In discussing the origin of the gold in the Oregon beach placers, Pardee[35] says:

"Some of the miners believe that the gold of the beaches comes up out of the sea, an idea suggested by the fact that after a storm a formerly barren stretch may be found to be gold-bearing. This notion is true so far as the immediate source of some of the gold is concerned. Materials composing the foreshore are carried out in the offshore zone at one time and returned to the beach at another. In the process a shift up or down the coast may occur. * * * Soundings of the Coast and Geodetic Survey show black sand to occur in the offshore zone at the present time. Gold and other minerals are doubtless present also. * * * For the beaches that border retreating shores, however, the most of the gold and other minerals come directly from the rocks that are being eroded by the waves."

The economic possibilities of mining the black sands of the California coast for their gold content have long been discussed.[36] Although gold and platinum have been the only minerals in the black sands which have been mined at a profit, much study has been given to the possible economic value of the other constituents.[37]

Gold-bearing gravels of marine origin occur in the Chico (Upper Cretaceous) sediments of northern California. That they are marine in

[33] Pardee, J. T., Beach placers of the Oregon coast: U.S. Geol. Survey, Circ. 8, 1934.
[34] Hornor, R. R., Notes on the black sand deposits of southern Oregon and northern California: U.S. Bur. Mines Bull. 196, 1918.
[35] Pardee, J. T., op. cit., pp. 29, 30.
[36] Edman, J. A., The auriferous black sands of California: California Min. Bur. Bull. 45, 1909.
[37] Day, D. T., and Richards, R. H., Black sands of the Pacific slope: Mineral Resources U.S., 1905, pp. 1175-1258, 1907.

FIG. 61. Andesite 'cobble-wash' and breccia overlying volcanic-ash beds. Material of this sort once covered most of the foothill belt of the northern Sierra Nevada. Location: 2 miles south of Knights Ferry, Tuolumne County.

origin and not fluvial is shown by their content of abundant fossil sea-shells, as well as by the character of their strata. They were formerly wrongly classed as "the gravel-filled channel of a Mesozoic river."[38] Gold-bearing gravels have also been reported from marine sediments of the Lower Cretaceous of northern California.

Since the gravels of the Eocene rivers of the Sierra Nevada were richly gold-bearing, it is to be expected that some of the gold reached the sea. The sedimentary deposits of this Eocene sea are known as the Ione formation. They occur along the western foot of the Sierra Nevada.

Lindgren says: [39]

"At the mouth of the rivers which descended from the Tertiary Sierra Nevada extensive delta deposits were accumulated, and it is thus difficult in many places to draw any exact line between the Ione formation and the river gravels proper. The gravels in the formation are locally auriferous, though generally poor, because spread over large areas."

The deposits contain quartz gravels and finely divided quartz grains; they are closely connected with the oldest river-channel deposits; they occur along the extreme western foot of the Sierra Nevada. Therefore, as stated by Allen,[40] the Ione sediments

"* * * indicate delta deposits formed at the mouths of many westward-flowing streams. The presence of marine fossils in the upper part of the Ione formation shows that they accumulated on the shores of an Eocene sea."

The processes involved in the distribution and concentration of gold in the marine strata of both the Cretaceous and Eocene formations have never been carefully studied. If these marine and delta placer deposits have any particular economic significance, it certainly has not been adequately demonstrated.

---

[38] Dunn, R. L., Auriferous conglomerate in California: California Min. Bur., State Mineralogist's Rept. 12, pp. 459-471, map p. 461, 1894.
[39] Lindgren, W., op. cit., Tertiary gravels, p. 24.
[40] Allen, Victor T., The Ione formation of California: Univ. California, Dept. Geol. Sci. Bull., vol. 18, p. 348, 1929.

Preservation of Placers

Placers are preserved if something keeps them from being eroded away. Since streams are continually changing their positions, fragments of their deposits are often left isolated. In cutting a deeper channel, a stream leaves 'benches' or 'terraces' at different intervals along its valley sides; but erosion tends to destroy them, unless they are protected in some way.

Burial is the most effective way in which a placer may be preserved. The name 'buried channel' has often been restricted to streams covered deeply by lavas; mud-flows, ash-falls, etc., all of which were very common during the Tertiary period in the Sierra Nevada. There are, however, other means by which burial may be effected.

1. By covering with landslide material. (An example occurs in Canyon Creek, Trinity County.)

2. By covering with gravel, caused by the faulting-down of a part of the river system. (Examples are believed to occur along the Klamath, Trinity, and some of the larger rivers of the Sierra Nevada.)

3. By covering with like deposits. (Many of the buried Tertiary channels were covered first by lake sediments, called 'pipe-clay,' before lava or mudflows poured over them.)

4. By covering with gravels when the stream is choked. (Examples are common along stream systems.)

5. By covering with gravel when the stream course is lowered below the general base-level of erosion. (Examples of this case are found along the western foot of the Sierra Nevada.)

6. By covering of the bedrock surface of down-faulted blocks (graben) by sediments of various sorts. (Many examples, especially in the Great Basin and Mojave Desert regions.)

7. By covering of older stream courses with alluvial fan material, as conditions favorable to stream existence fail. (Many examples in the Great Basin and Mojave Desert regions.)

8. By covering with glacial till. (Examples may be looked for in the glaciated areas of the Sierra Nevada.)

9. By covering of beach placers with marine sediments as the fluctuating coast is submerged, but later elevated. (Such as the present elevated beach placers which are in places covered with other marine sediments.)

10. By covering of one geologic formation with another, through the processes of earth deformation and thrust-faulting. (In a geologically active region such as California, examples of this case might very well be found.)

11. By submergence of river canyons [41] to great depth beneath the ocean. (Off the coast of California many submarine channels have been discovered and mapped by the U. S. Coast and Geodetic Survey. One extends over 75 miles from the shore and attains a depth greater than 10,000 feet. These channels are too deep, therefore, to explore for placer gold. A reported recent project to operate a sea-going gold dredge would probably have to do only with off-shore marine deposits and not submerged stream placers.)

To find gold-bearing stream channels, buried and preserved in such a manner that they may be profitably mined, is the challenge to the exploration geologist.

Modification of Placer Deposits

Placer deposits may be greatly modified in form and structure by earth deformation. The gravel content may also become firmly cemented

[41] Davidson, George, The submerged valleys of the coast of California: California Acad. Sci., Proc. (3) vol. 1, no. 2, 1897.

Reed, Ralph D., Geology of California (with references), p. 3, Am. Assoc. Petroleum Geologists, 1933.

Shepard, Francis P., Investigation of California submarine canyons (abstract): Geol. Soc. America Proc. 1933, pp. 107-108, 1934.

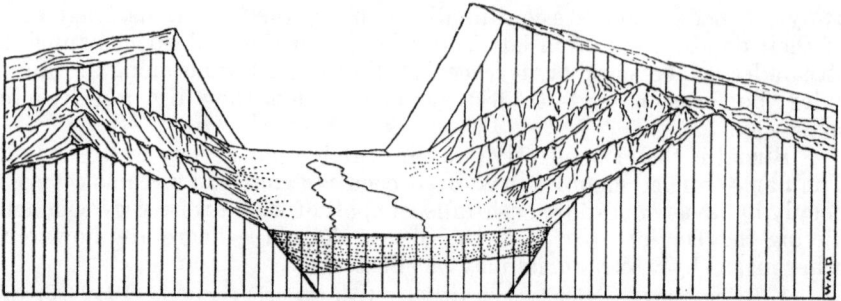

FIG. 62. Diagram showing a down-dropped fault block (graben) between two uplifted fault blocks. Erosion covers the one with materials derived from the others. Streams cut in bedrock prior to faulting may thus be buried under the alluvium of the graben. *After Davis, State Mineralogist's Rept. 29.*

FIG. 63. Ideal sketch showing how a landslide may dam up a mountain valley to form a lake. Silt, sand, and gravels deposited in and on the edge of this lake will cover the stream gravels in its bottom. *After Davis, State Mineralogist's Rept. 29.*

by interstitial deposition of mineral matter, such as by lime and iron carbonate, or silica, through the action of infiltrating solutions.   The older the placer, the more apt it is to have been modified in these ways from its original form and attitude.

The regional tilt of the Sierra Nevada has increased the gradient of the Tertiary channels [42] (where they lie in the direction of the tilt) from 20 or 30 feet to the mile to twice, three times, or even several times that amount.   Locally, tilting has been even greater.   In places where the ancient channels lie in opposite direction to the tilt the original gradient may have been reversed.   Often steep tilting is accompanied by local faulting of a few feet to several hundred feet.   Generally the downthrow is on the east side of the fault plane.   In form, this is a replica of the action which took place and still is taking place along the eastern escarpment of the Sierra Nevada.   Such displacements and changes in channel-gradient as well as in actual position of the channel, are important factors which greatly influence mining procedure.   They should be understood so far as surface data will permit, before actual mining is started in a given area.

In the Mojave Desert and Great Basin region, faulting and tilting have been extremely active, greatly affecting streams antedating late Tertiary and early Quaternary periods.

The flow of ground water through stream gravels, the former channels of which have been blocked, cut off, tilted, or folded by earth movements, is a factor of considerable consequence when it comes to mining such placers.

Gold in Placers

Original Source of Gold

The particles of gold found in placers originally came from veins and other mineralized zones in bedrock, from which they were released through surface weathering and disintegration of the rock matrix. Though the original source may not in every case have been a deposit which could today be mined at a profit, the richer placers usually indicate a comparatively rich source.   A long period of deep weathering, resulting in separation and release of large quantities of gold from the bedrock, followed by a more active period of erosion, generally due to uplift, is an ideal condition for gold to be swept into stream channels and there to be concentrated into rich placers.   Still richer deposits may be formed through reconcentration from older gold-bearing gravels.

These are the most important geologic conditions which have been found to exist in the various gold belts of the world, and particularly in the Sierra Nevada of California.

For the most part, the original source of gold is not far from the place where it was first deposited after being carried by running water. This is certainly true in both the Sierra Nevada and Klamath Mountains. The streams, flowing through regions of metamorphic and intrusive igneous rocks threaded throughout by gold-bearing veins, were found by the early miners to contain auriferous gravels.   But the more recent streams which have had only barren lavas to pass over, as in the volcanic covered area between the Sierra and Klamath regions, have proved to be barren.

Lindgren [43] states:

        [42] Lindgren, W., op. cit., Prof. Paper 73, pl. X, *Profiles along Tertiary channels of the Sierra Nevada*.
        [43] Lindgren, W., Mineral deposits, p. 213, McGraw-Hill Book Company, 1919.

"The great majority of gold placers have been derived from the weathering and disintegration of auriferous veins, lodes, shear zones, or more irregular replacement deposits. * * * In many regions the rocks contain abundant joints, seams, or small veins in which the gold has been deposited with quartz. * * * It is often stated that gold is distributed as fine particles in schists and massive rocks and that placer gold in certain districts is derived from this source. Most of these statements are not supported by evidence, though it is not denied that gold may in rare instances be distributed in this manner."

Release of Gold from Bedrock

Without some widespread process of release from the quartz veins and rocks—vaults in which the metal was originally firmly held—gold particles could not have escaped to be transported as such by running water. Therefore, extensive rock weathering and decay over a long period of time is a primary factor of extreme importance. It has permitted the original source to contribute gold particles, large and small, to placer accumulations. The same geologic processes which form residual and eluvial or 'hillside' concentrations of commercial merit operate in the general release of gold from bedrock.

The factors of prime importance in weathering are solution, changes of temperature, depth of water-table and therefore depth of oxidation, action of rain, effect of gravity, growth of vegetation, nature and composition of material acted upon, and degree of topographic relief. Rock weathering, especially complete disintegration down to the water-table, rather than deep disintegration, is often more favored by tropical climates. This, however, is only one factor, and large areas of deep secular weathering are found in the north, in such places as Alaska, where placers are abundant.

The processes which take place in the separation of gold from bedrock are described in detail by Brooks,[44] who says:

"The breaking down of the rock and the accompanying chemical changes of the constituent materials set free the gold, one of the relatively indestructible minerals, and this becomes intermingled with the other insoluble material. Clay dominates in the residual mass, but if the parent rock contained quartz, this, too, usually remains, being probably the most refractory of all the common minerals toward purely chemical agencies. Mineralized vein quartz very commonly carries easily decomposed minerals, such as pyrites, and is therefore readily broken up, allowing the insoluble ingredients of the ore body, such as gold, to be set free. This process is hastened by purely physical agencies, such as frost and changes of temperature, which break up the insoluble rock constituents. * * *

"As a rule, the changes in a rock mass brought about by weathering result in a very material reduction in its bulk.[b] The loss of material by weathering among siliceous crystalline rocks, according to Merrill,[c] amounts to more than 50 per cent, and in the purer forms of limestone it may reach as high as 99 per cent. Pumpelly[d] has estimated that in the limestone areas of the Ozark Mountains the residual material represents only from 2 to 9 per cent of the original rock mass. Such reductions in volume necessarily result in more or less concentration of any insoluble material that may have been disseminated in the parent rock. This concentration will be materially greater in the case of substances of high specific gravity, such as gold, than in that of the lighter minerals, for the former will have a constant tendency to settle to the bottom of the loose material. On declivities gravity will accelerate the process and help to sort the material, producing in some places a rough

[44] Brooks, Alfred H., The gold placers of parts of Seward Peninsula, Alaska: U.S. Geol. Survey Bull. 328, pp. 125-127, 1908.
[b] Merrill has shown that in certain changes by hydration there is an increase in bulk. He estimated that in the conversion of granite into soil (District of Columbia) there had been an increase in volume amounting to 88 percent. Compare Merrill, G. P., Principles of rock weathering: Jour. Geology, vol. 4, p. 718, 1896.
[c] Merrill, G. P., Rocks, rock weathering, and soils, p. 234, New York, 1897.
[d] Pumpelly, Raphael, The relation of secular rock disintegration to loess, glacial drift, and rock basins: Am. Jour. Sci., 3d ser., vol. 17, p. 136, 1879.

stratification.[e] This is a secular process and will proceed as long as the rocks continue to disintegrate. * * *

"It is evident that the effectiveness of all these agencies is proportional to the length of time in which they are operative. A land mass must remain stable relative to sea level, for a long period of time to permit the accumulation of any considerable amount of residual material. Uplifts bring about renewed activities of the watercourses, and the residual mantle is quickly removed by erosion. It is evident that the conditions that are most favorable to the accumulation of residual material are those in which the land mass is at or near base-level when erosion is reduced to a minimum."

It seems that topographic and climatic conditions which existed during the Eocene period in the Sierra Nevada were very favorable for the release of gold. Rejuvenated drainage at the close of the period swept the immense quantities of gold, freed from the enclosing hard matrix into the early Tertiary stream channels; and these, soon after, became buried and preserved by lake sediments, masses of gravel, cobble-wash, volcanic ash, breccia, and lava flows.

Associated Minerals

Mineral grains that are very heavy and resistant to mechanical and chemical destruction accompany the gold in placers. The so-called 'black sands,' generally made up principally of magnetite, are well-known to the miner. A long list of the minerals found in sluice-box concentrates is recorded by the United States Bureau of Mines.[45] Besides magnetite, there are found titanium minerals (ilmenite and rutile), garnet, zircon, hematite, chromite, olivine, epidote, pyrite, monazite, limonite, platinum, osmiridium, cinnabar, tungsten minerals (wolframite and scheelite), cassiterite, corundum, diamonds, galena, as well as quicksilver, amalgam, metallic copper, bird-shot, bullets, hob-nails, penknives, watches, and nails.

Buried deeply in the gravels of the modern Feather River was once found (and someone thought it was an ancient fossil) the remains of a mule's hind leg with hoof and iron shoe nailed to it. What may be found in some of the placers of today may not, therefore, be representative of what was deposited by the more ancient streams.

The presence of quantities of magnetite associated with extremely fine gold particles, makes a difficult metallurgical problem. To the geophysicist, however, the presence of any minerals having a strong effect on the magnetometer is a godsend to effective exploration.

The determination of heavy minerals and their approximate relative percentage has been extensively used in subsurface correlation[46] of sedimentary beds in the oil fields, and tables have been developed for use in determining these minerals.[47] The same method of research could well be applied to the tracing of channels. Though it has not yet been given consideration in California, this interesting field is open for study, with a well-developed technique[48] available.

---

[e] Kerr, W. C., The gold gravels of North Carolina, their structure and origin: Am. Inst. Min. Eng., Trans., vol. 8, pp. 461-462, 1879-80.

[45] Gardner, E. D., and Johnson, C. H., Placer mining in the Western United States. Part I, General information: U. S. Bur. Mines, Inf. Circ. 6786, pp. 15-20, Sept. 1934.

[46] Tickell, F. G., The correlative value of the heavy minerals: Am. Assoc. Petroleum Geologists, Bull. 8, pp. 158-168, 1924.

[47] Tickell, F. G., The examination of fragmental rocks, Stanford University Press, 1931.

[48] Raeburn, C., and Milner, H. B., Alluvial prospecting, the technical investigation of economic alluvial minerals, D. Van Nostrand Co., 1927.

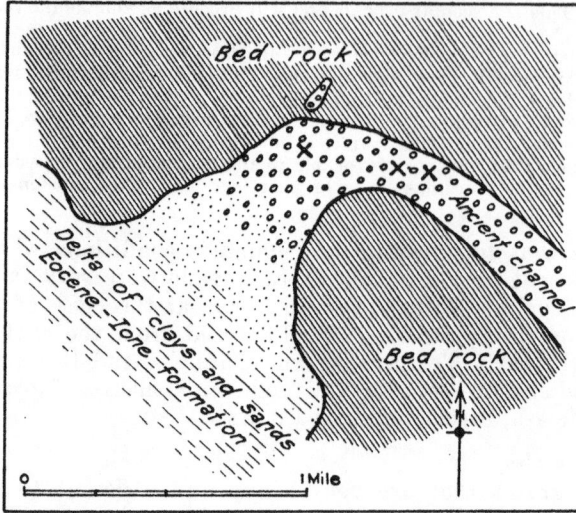

FIG. 64. Sketch map of an early Tertiary channel and its delta in the Eocene (Ione) sea. The crosses indicate where the gold has been mined in the channel—on a bend in the stream, and at the point where a tributary entered. Finely divided gold particles occur interbedded in lenses of quartz pebbles and sand lying above clay layers (false bedrock).

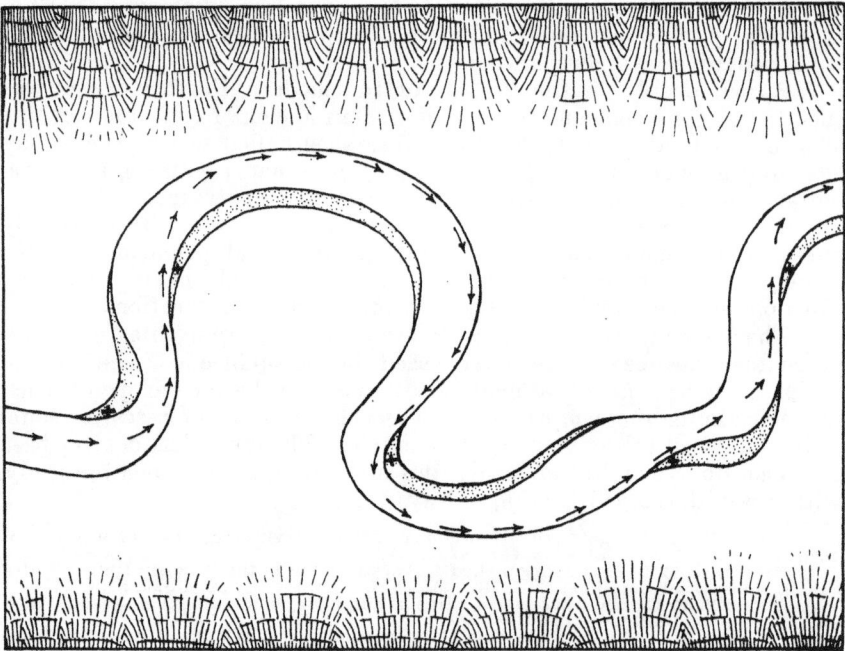

FIG. 65. Diagram to illustrate the course of a river, indicating where gold particles are most likely to become concentrated. *After Spurr, U. S. Geol. Survey 18th Ann. Rept., 1898.*

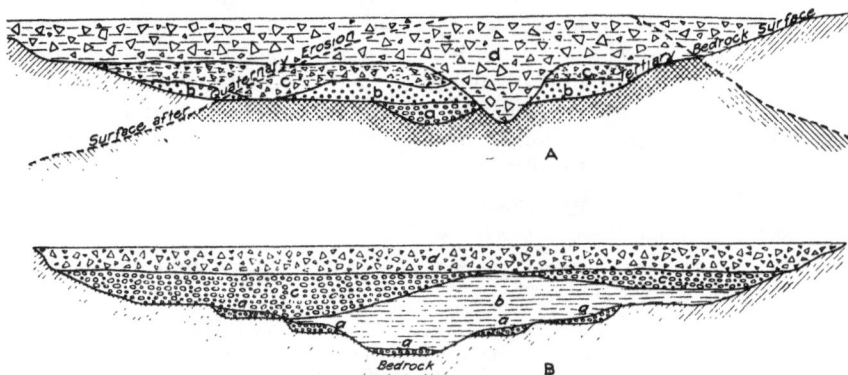

FIG. 66. A, Diagrammatic cross-section showing the four principal epochs of Tertiary gravel deposition in the Sierra Nevada. The deep gravels, a, represent Eocene; b to d are successively younger and probably represent Miocene stages for the most part. The rhyolite period is represented by c and the andesite by d. B, Diagram showing deposits in the Deep Blue lead, Placerville; the older channel and benches of the inter-rhyolitic epoch are represented by a; rhyolite tuff, b; andesite cobble, c, andesite tuff-breccia, d; after Lindgren, U. S. Geol. Survey Prof. Paper 73.

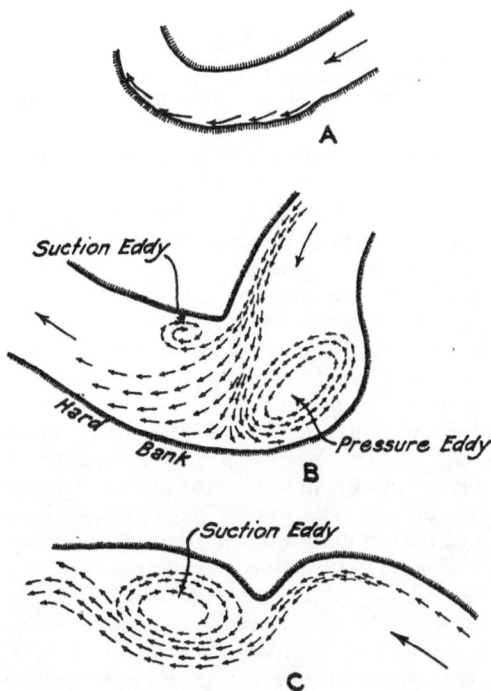

FIG. 67. A, Diagram showing the place of greatest erosion on the bend of a river. B, Diagram to show positions of pressure and suction eddies in a river; gold is more likely to be deposited in the suction eddy than in the pressure eddy. C, Diagram to show how a suction eddy is formed in a river; after Thomas and Watt; see Ries and Watson, Engineering Geology.

12

The source of the minerals deposited and associated with the gold particles lies in the rocks over which the stream has flowed. The source of chromite, platinum, and diamonds is generally attributed to belts of serpentine and related ultra-basic igneous rocks, while garnet, ilmenite and magnetite might come from metamorphic rocks, and monazite, zircon, cassiterite, wolframite, and scheelite would probably have their source in granite pegmatites.

Transportation, Deposition, and Retention

The processes of transportation and deposition of gold in a stream are aptly stated by Brooks :[49]

"The transporting power of a stream is dependent on its velocity, which is a variant determined by the gradient, volume, and load. When a stream is overloaded with sediment, the excess is dropped. When it is underloaded, it erodes. When equilibrium has been established, neither erosion nor deposition takes place. Gradient, volume, and load usually vary in the same stream so that deposition may be going on in one part of its valley and erosion in another. When a stream is eroding, the material within reach of its activity is constantly moved in a downstream direction. All movements of this kind are accomplished by more or less sorting and make for the concentration of the heavier particles.

"Deposition takes place in a stream when the velocity is decreased, either by the periodic changes in volume or by a change of gradient. Where there is a change of grade, resulting in diminished velocity, the gold is laid down with the other sediments. It must be remembered, however, that placer gold may find lodgement in inequalities of the bedrock surface where no considerable deposition of detrital matter has taken place, though extensive placers are, as a rule, not formed because of irregularities in the bed-rock surface alone. The concentration of gold in river bars is analogous to its deposition in stream beds, for it is dropped where the velocity of the current is checked by the formation of eddies, due to the inequalities of the river floor."

A further study of this subject is made herein in connection with the more detailed analysis of stream action.

When the bed of a stream is the actual rock floor of the valley, it is called 'bedrock' in the true sense of the word. Later in its history the stream may flow on an aggraded bed of gravel or other sediment. If the stream gravels become covered with volcanic or other materials, the stream is obliged to flow over this new cover, called 'false-bedrock.' Gold particles are normally deposited on or near the bed of the stream, which is called 'bedrock' or 'false-bedrock' according to whether the bed is the true hard rock floor of the valley or whether it is a superficial layer of clay, volcanic tuff, or some such impervious material overlying previously deposited gravel. It is readily surmised, therefore, that there may be two or more tiers of gold-bearing stream channels, but the upper ones do not necessarily lie directly above the older and lower channels, and may not follow the direction of their courses at all. If the stream cuts clear down to the true bedrock, remnants of the older channels will lie at relatively higher positions instead of at lower horizons. In some cases, where gravel is deposited deeply on bedrock, forming a new aggraded bed, rejuvenation of the stream will stir up the entire mass of gravels, including the gold-bearing layer deposited on top, and the final result will be that most of the gold particles will reach a position very near true bedrock. The complexity of the history of these processes is apparent; so also are the difficulties of the engineer who attempts to do a fair job of sampling.

---

[49] Brooks, Alfred H., op. cit., p. 128.

In excavating for the Boulder Dam a sawed plank of lumber was found 60 feet deep, "lying under gravel on the edge of the inner gorge, a place that it could not have reached in any imaginable way except by burial during some comparatively recent flood." [50]   This case and many others show that the depth of burial by recent rivers does not necessarily mean that a great period of time has elapsed for the accumulation of the deposit.   During high water, the whole mass may be stirred up and even boulders floated in the soupy mixture of heavy rock debris and water. This action gives the gold particles a chance to work their way toward the bottom of the mass.

For thousands of years particles of gold of various sizes, from nuggets to flour gold, were dropped and lodged in the riffles of bedrock along the natural river-sluices of the Sierra Nevada.   Flattened particles are most-easily carried; sometimes, suspended by air-films, tiny scales even float on the surface of the water. [51]   Extremely fine grains of gold were swept by torrents through the canyons and out into the Great Valley. In the present dredging grounds where they are found, they have been easily caught in false-bedrock, which consists of clayey layers of volcanic tuff.

The very high specific gravity of gold, six or seven times that of quartz, with the ratio increasing to nine times under water, is the primary factor which causes this heavy resistant metal eventually to work its way to a point where it may sink no farther.   Once it is caught on bedrock, the stream has great difficulty in picking it up again.

When a stream leaves its mountain canyon and enters a more level country or a still body of water, the materials carried by that stream are deposited in the form of a fan or delta.   At the apex of this fan or delta fine gold may be deposited, and may never reach bedrock.   It may be deposited on top of clayey, 'false bedrock' layers.   The stirring action found to occur in rugged mountainous canyons during time of floods, which permits gold to reach bedrock, does not take place in the delta.

Lindgren state: [52]

"By an odd paradox, gold is at the same time the easiest and most difficult mineral to recover.   It is divisible to a high degree and owing to its insolubility the finest particles are preserved.   A piece of gold worth one cent is without trouble divisible into 2000 parts, and one of these minute particles can readily be recognized in a pan."

Although gold is very malleable, and may be hammered into different shapes by stones hitting it as they tumble along in the stream, different particles are not welded together to form larger nuggets, as some people are prone to believe.   Lindgren [53] has shown that the largest masses of gold have come from lodes, not placers.   Particles of gold may be broken down, however, from another piece.   The more rounded or flattened nuggets have probably gone through more knocking about than the rougher pieces.   Those showing the original crystalline forms have probably not traveled far in the 'free' state.

It is also found that the more ancient placers, and those which have undergone many reconcentrations, contain gold of a higher degree of fineness than those whose source is near by, or in which the gold has not

[50] Berkey, Charles P., Gorge excavation confirms geological assumptions: Eng. News-Rec., vol. 111, p. 762, fig. 2, 1933.
[51] Lindgren, W., Mineral deposits, p. 220, 1919.
[52] Lingren, W., Mineral deposits, p. 220, 1919.
[53] Lindgren, W., op. cit., Tertiary gravels, p. 66.

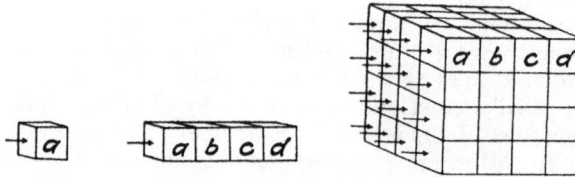

Fig. 68.   Diagram to represent the theoretical
effect of increased velocity (V) on transporting power
(T∝) of a stream; *after Chamberlain and Salisbury.*
$T \propto V^6$, i.e., if the velocity is doubled, the transporting
power increases as much as 64 times.

"A statement more frequently encountered is to
the effect that the quantity varies with the sixth power
of the velocity; and the origin of this assertion is now
in doubt.   It is an erroneous version of a deductive law.
. . . The law, as formulated by Hopkins, is that 'the
moving force of a current, estimated by the volume or
weight of the masses of any proposed form which it is
capable of moving, varies as the sixth power of the
velocity'; and this law pertains not at all to the quantity
of material moved, but to the maximum size of the
grain or pebble or boulder a given current is competent
to move."   *Gilbert, G. K., The transportation of debris
by running water: U. S. Geol. Survey Prof. Paper 86,
pp. 15-16, 1914.*

been deposited for such a long period of time.   This may be due to the removal of alloyed silver by the dissolving action of surface waters.[54]

The solution and reprecipitation of gold in the gravels is shown to be exceedingly rare or nonexistent, commercially.[55]   On the other hand, in some of the Tertiary channels, thin crusts of pyrite or marcasite are found deposited on the surface of the gold particles themselves.

Factors of Concentration

A placer worthy of mining is like any other commercial mineral deposit in that it is a special case of concentration due to several combining natural processes which were all in favor of the accumulation of the one desired mineral.   Source, release, transportation, deposition, reconcentration, and retention of the gold have already been discussed. Three extremely important factors are: (1) Structural control of the stream pattern, so that the streams run along the course of the zones of mineralization; (2) Decay and disintegration at the surface of the mineralized rocks prior to erosion; (3) A change in the cycle of erosion, causing rejuvenated flow of streams and the rapid washing of the released gold into stream channels.

Quoting from a recent paper appearing in the Engineering and Mining Journal:[56]

"Accumulation of gold in an important placer deposit is rarely a mere coincidence; it is rather the fortuitious concurrence of several favorable factors.   In regions where nature has bestowed the advantages of extensive mineralization, rapid rock decay; and well-developed stream patterns, a relatively large amount of gold placer may be formed.   But even in this ideal case, the favorableness of these several important factors must be assumed.

"The general considerations which favor the accumulation of gold in special locations have been frequently discussed.   Physically, the phenomenon is simple; in such locations where the gold has been deposited, the transporting power of the stream has become insufficient to carry away the particles of gold that have settled. The richness of the deposit will therefore depend not only upon the completeness of this loss of transporting power, and on the ability of the bedrock to hold the deposited gold at this point, but also, most importantly, on the general relationship of the gold sources to the stream.   The early miners untiringly sought the 'ledge' or 'mother lode' which furnished certain rich placers.   However, with the presence of mineralized zones as a source of the gold, the richness of a placer is perhaps due more to the efficiency of the stream as a concentrating device than to its uncovering rich lode deposits.

"The ability of a stream to transport materials is essentially dependent upon the velocity of the water and the area and specific gravity of the particles of material being carried.   The transporting power of water varies approximately as the sixth power of the velocity.   This means that even small velocity changes have an enormous effect upon transporting power, ranging rather abruptly from velocities which can not transport an appreciable amount of gold to those which easily transport much of the gold that may enter the stream or be released from the gravels therein.   The velocity of water is a complex relationship of grade, shape, and size of the channel, quantity of water, and other factors.   A grade ranging from 30 to approximately 100 ft. per mile will favor the deposition of gold.   With appropriate conditions of flow, these limits may be somewhat increased or reduced without serious handicap.   When considering the grades of the ancient channels, however, one must remember that faulting and regional tilt often have considerably modified the original grade.

"For the richest accumulations, the erosional conditions must be well balanced, so as to provide a long period of concentration.   Slight uplifts tend to rework and further enrich placer deposits, as do increased volumes of water, inasmuch as both of these factors tend to increase the velocity.   Local variations in the shape of the

---

[54] Lindgren, W., op. cit., Tertiary gravels, p. 68.
[55] Lindgren, W., op. cit., Tertiary gravels, p. 69.
[56] Jenkins, O. P., and Wright, W. Q., California's gold-bearing Tertiary channels: Eng. and Min. Jour., vol. 135, pp. 501, 502, November 1934.

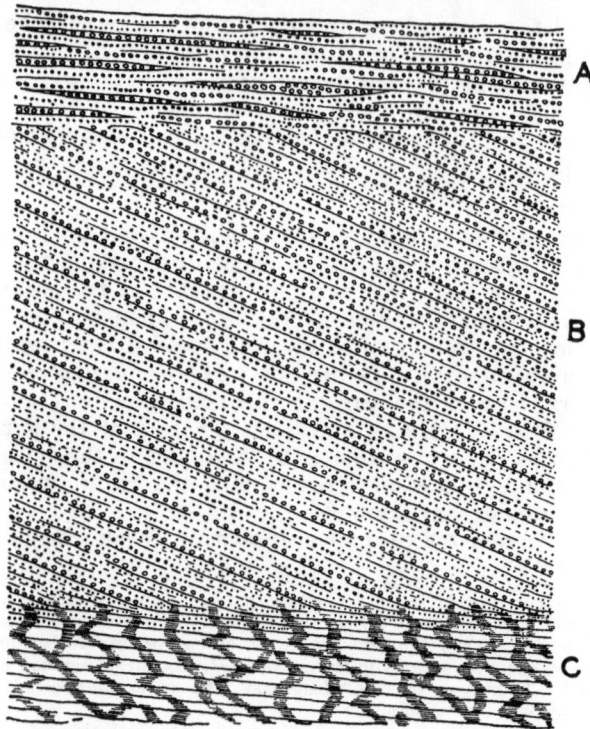

FIG. 69.   Ideal vertical section of a delta, showing in greater detail than figure 70 the typical succes-
sion of strata: A, topset; B, foreset; and C, bottomset
beds.   *After Gilbert, U. S. Geol. Survey 5th Ann. Rept.
1883.*

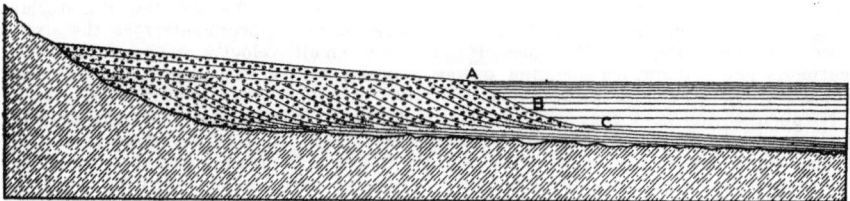

FIG. 70.   Ideal cross-section of a delta showing *A*, topset; *B*, foreset; and *C*, bottomset beds.   *After Gilbert, U. S. Geol. Survey 5th Ann. Rept., 1883.*

channel are of most interest, however, because they are immediately responsible for specific deposits of placer gold. When a stream canyon widens out, deepens, turns, or joins another watercourse, certain zones of concentration will be formed where the water velocities have been somewhat reduced and where eddy currents occur. These reductions in velocity immediately allow gold and heavy mineral particles to separate from the mass of gravel that is being carried and rolled down the canyon. Gold has a specific gravity of approximately six times that of the gravel, but under water this ratio becomes about nine times. This large gravity difference permits the gold quickly to work its way to bedrock and into any crevices therein. Here it remains, requiring excessive erosion to remove it to new locations.

"In order that a major deposition of gold may occur, there must be an abundance of source material which contains more or less gold and which may be more or less easily eroded. A decayed formation of low-grade material could easily furnish more gold than a hard, higher-grade deposit. The decomposed material also supplies more gravel for balanced conditions of stream transportation, providing that overloading or choking is minimized by uplifts or increasing water volumes. Plainly, a stream running along a vein system will have a greater opportunity to accumulate gold than one merely crossing it. Bedrock-controlled streams, therefore, provide a maximum contact with source material.

"A further and very important factor is the ability of bedrock to hold the deposited gold in spite of the scouring action of the stream at higher water stages. A smooth, hard bedrock is a very poor one for placer accumulations. Bedrock formations which are decomposed or possess cracks and crevices are good, and those of a clayey or of a schistose nature are excellent in their ability to retain particles of gold.

"Gold tends to resist most stream transportation. Coarser gold will migrate downstream an amazingly short distance from its apparent source throughout a long erosion period. The fine gold, which the stream can transport, is dropped rather completely within a restricted area at the mouth of the stream canyon."

Age of Placers

Significance

The geologic age of a placer deposit is often a factor of primary economic interest. In the Sierra Nevada, the oldest system of Tertiary channels has proved to be the richest, for it was formed prior to the extensive volcanic activity which resulted in the covering of the mineralized bedrock surfaces as well as the valleys in which gold-bearing streams flowed. These streams, which cut directly through the mineralized zones, had an ideal opportunity to tap the primary gold resources of the region, while those which flowed only over a barren volcanic cover remained themselves barren of gold. In the region of the buried channels of the Sierra Nevada, many stream deposits of different periods are now found intermingled. To decipher their history and relative age is an essential part of the exploration geologist's work, in his search for the channels of greatest possible economic consequence.

Structural Criteria

Various criteria are used in determining the relative age of stream deposits, but the most positive evidence is structural relationship. For example, the cutting channel is younger than the channel which it cuts.

The deepest channel is not necessarily the oldest nor yet the youngest. In the case of the modern canyons of the Sierra—the youngest are the deepest; yet along the western foot of the range the present streams flow over older channels buried beneath. Where a canyon is filled with detritus or with lava, the newer stream flows on top of the deposit and is therefore higher than the old stream-bed beneath, while still more ancient stream terraces or benches may lie at higher and varying elevations on either side of the canyon. Some benches, representing former streams, may have even been left prior to a lava flow covering the deepest channel, while other benches may have been left later.

It is apparent, therefore, that the subjects of historical sequence and relative age are matters of detailed geologic and physiographic study which deserve more than superficial examination. They cannot be classified by dogmatic rules.

Paleontologic Criteria

In order to assign definite geologic periods to the deposits of ancient streams, their age should be related in some way to the well-established epochs of regional geologic history. Fossils, diagnostic in determination of geologic periods, if found in the stream deposits, are of inestimable value in this regard.

In the Sierra Nevada, parts of fossil plants consisting of leaves, logs, etc., have been extensively collected and determined by paleobotanists. Also, some vertebrate bones have been sent from the old drift mines to the Smithsonian Institution and elsewhere for scientific study.

Geologic periods the world over have been established largely on their marine fauna rather than their continental flora or fauna. For instance, marine beds of the Ione formation contain fossil sea-shells definitely assigned by paleontologists to the Eocene, or earliest Tertiary period. The fossils found in the sediments filling the Tertiary valleys of the Sierra are not, however, marine, but are of ancient lands and lakes. Correlation of geologic age by means of these land plants and land animals brings in complications that have not yet permitted the two bases of criteria, marine and nonmarine, to be perfectly coordinated. Besides, most of the fossil leaves occur in tuffaceous lake beds that overlie the gold-bearing quartz-gravel deposits, and therefore do not give much direct evidence as to the age of the latter. Fossil wood, so common in the most ancient of the gold-bearing gravels is not as yet determinable nor diagnostic. In most cases it represents unstudied tropical forms which have been placed in the Eocene by paleobotanists because of the known climate of that period. The older gravel containing this fossil wood has previously been referred to the Cretaceous. In the Smartsville quadrangle, for example, a deposit containing petrified logs of probable Eocene age is described by Lindgren as follows:[57]

"The high, isolated area of well-washed gravel 3 miles north-northwest of Montezuma Hill is noteworthy; it is so much higher than the adjacent gravel channel of North San Juan that it must be assumed to belong to an earlier period; very likely it is of Cretaceous age."

It is a fact, however, that no fossils indicating a Cretaceous age have yet been found in these older gravels. Wherever definite marine Cretaceous beds do occur on the western foot of the Sierra Nevada the oldest stream channels of the vicinity are found to cut the Mesozoic sediments, showing a profound difference in age between the two.

Chaney[58] summarizes the results of paleobotanical study of the fossil plants found in the Sierra Nevada as follows:

[57] Lindgren, W., op. cit., Tertiary gravels, p. 124.
[58] Chaney, Ralph W., Notes on occurrence and age of fossil plants found in the auriferous gravels of Sierra Nevada: California Div. Mines, Mining in California, State Mineralogist's Rept. 28, p. 301, 1932.

"The tuffs and shales in which fossil plants occur interbedded in the Auriferous Gravels range in age from lowermost Eocene to upper Miocene. During this time, there was a climatic trend from subtropical to temperate conditions, which resulted in the elimination of palms and other large-leafed species and the incoming of types of plants similar, in general, to those now living in North America. The Miocene flora indicating a temperate climate, includes genera no longer living in western America, although they occur in eastern America and eastern Asia. The evidences of difference in living conditions in the Eocene and the Miocene make it possible readily to differentiate between the older and the younger floras of the Auriferous Gravels."

Most of the fossil vertebrate bones described from the drift mines of the Sierra Nevada have been collected and sent in to paleontologists by persons who did not record their definite location, so that the exact geologic formations in which the fossils were embedded are unknown.

The paleontologist who works with vertebrate remains is not always apt to apply the same age to beds as that which has been assigned them by the paleobotanist; generally the former assigns a younger age. Many of the well-established Sierran Tertiary as well as later beds containing vertebrate bones were once given the blanket designation of Pleistocene.[59]

Physiographic Criteria

Age correlation has sometimes been done purely on physiographic evidence. Matthes [60] assigns Eocene, Miocene, Pliocene, and Quaternary periods of uplifts to the various old surfaces found in the Yosemite region, tying in the Miocene surface correlation with geologic features of a fossil leaf locality occurring in the Tuolumne Table Mountain region. The fact that old surfaces have been resurrected [61] during the Pleistocene has only recently been given consideration.

"Quaternary erosion resulting in the uncovering of Tertiary volcanic ash and the resurrection of early Tertiary surfaces, formerly cut into pre-Cretaceous bedrock along the western flank of the Sierra Nevada, is a widespread geologic process which has heretofore not received the recognition it deserves. The process involves features of vast economic concern. One key locality for this study is in the region of Table Mountain, Tuolumne and Calaveras counties, California, where a very resistant late Tertiary latite flow has served the purpose of preserving not only fragments of earlier and less resistant geologic bodies consisting of volcanic materials, mud-flows, lake beds, and stream gravels of different ages, but also the underlying bedrock surfaces of earlier Tertiary age. These ancient surfaces, the topography of which appears to have been controlled by bedrock structure, may be found in various stages of resurrection. In this area, the gold-bearing gravels were mined in ancient channels that ran in directions opposite or at an angle to the Table Mountain Channel; the latter apparently never contained any appreciable amount of gold-bearing gravel, contrary to the common belief. Though unmantled and dissected through Pleistocene and Recent epochs, fragments of upland peneplained bedrock and in places gravel-covered surfaces are actually early Tertiary land forms, which have been brought to light after having been buried throughout later Tertiary volcanic epochs."

Lithologic Criteria

The nature and composition of the material filling the ancient channels and valleys also indicate to what geologic period a deposit may belong. Gold-bearing gravels, composed purely of sand and quartz-pebbles, or of the bedrock complex, indicate that the channel

[59] Hay, Oliver P., The Pleistocene of the western region of North America and its vertebrate animals: Carnegie Inst. Wash., Pub. 322-B, 1927.
[60] Matthes, François E., Geologic history of the Yosemite Valley: U. S. Geol. Survey, Prof. Paper 160, 1930.
[61] Jenkins, Olaf P., Resurrection of early surfaces in the Sierra Nevada: California Jour. Mines and Geology, vol. 30, p. 5, 1934.

is of the pre-volcanic period, possibly Eocene in age. Most of the rhyolites were apparently formed during the latter part of this period and in the Oligocene. The most abundant of the volcanic rocks are composed of andesites, which seem to be largely of late Miocene or early Pliocene age. During the late Pliocene and early Pleistocene there were many basalt flows. Tuolumne Table Mountain is composed of latite, probably of late Pliocene age.

In the Recent period much pumice has been expelled from craters and blown over various parts of the Sierra Nevada.

Streams may always be considered younger than the rocks from which their gravel has been derived, though mud-flows may receive their materials from active volcanoes. There is here an opportunity for a petrographic study of both volcanic rocks and the materials of the sediments. Special detailed analysis of stream correlation might be performed by means of the method known as "heavy mineral separation," previously mentioned as widely employed in petroleum geology.

Coordination of Criteria

In order to assign a geologic age to a placer, all criteria should be sought out and used so far as it is possible. A regional study of age relationship should include a coordinated study of all phases: structure, fossils, surface features, and materials deposited.

The principal placers in California occur in the Quaternary, Tertiary, and Cretaceous geologic time divisions, which are grouped in the following manner:

ERA	PERIOD	EPOCH		DURATION in millions of years
Cenozoic	Quaternary	Recent		
		Pleistocene		1
	Tertiary	(Neocene or Neogene)	Pliocene	7
			Miocene	12
		(Eogene)	Oligocene	16
			Eocene	23
Mesozoic	Cretaceous			75
	Jurassic			40
	Triassic			40
Paleozoic				415
Proterozoic				Unknown

Life History and Habit of Streams

Need for Scientific Background

Exploration of placers and various ancient stream channels requires an understanding of the habits and life history of streams in general. This includes, on the one hand, processes of erosion and deposition, and on the other, physiographic history. Each is important; the first, more directly, while the second has to do with regional features, knowledge of which is essential to the exploration geologist. The fundamental science of streams has been outlined in a simple yet splendid manner by G. K. Gilbert [62] in his masterpiece on the Henry Mountains, Utah. Later, I. C. Russell published an excellent text on streams.[63]

Such natural processes as those related to streams are so universal that a study of them in one part of the world may be applied to conditions found in another. Similarly, an understanding of the ancient Tertiary streams of the Sierra may be gained by applying the knowledge of processes found in operation today where conditions and environments would appear similar.

In a province, such as the Sierra Nevada, where the development of the drainage system has been repeatedly interrupted by earth movements or by burial as a result of volcanic mud washes and lava flows, the history of the stream of any one chronological horizon is a separate entity, and may be entirely different in form and pattern from either earlier or subsequent systems. This fact, together with the complexity of any one system presents a problem much involved.

The need for a scientific background in the study of streams is therefore apparent. The more pertinent features of this study, together with its terminology, are outlined in the following pages.

Stream Erosion

Stream or fluvial erosion is complex. It may be divided into the several processes: hydraulic action, abrasion, solution, and transportation. A brief statement of the essential factors which control erosion is quoted from an authoritative textbook[64] as follows:

"The capacity of a stream to erode depends on its volume and velocity. The velocity in turn depends on (1) the slope down which the stream is flowing, (2) volume of water, (3) the shape of its channels, and (4) weight and volume of its load.

"The rate of descent of the bed of a stream is the *stream gradient*. It is ordinarily expressed as so many feet per mile. The gradient changes from place to place along the course of the stream. Velocity increases rapidly with increase of gradient. Thus mountain streams with high gradients erode their valleys much more quickly than lowland streams of comparable size with low gradients. It follows that streams wear high gradients down to low ones by continued erosion, and that as the gradients are worn down the rate of erosion must decrease.

"The volume of water is a variable factor in all streams, largely because of fluctuations in rainfall. Velocity and rate of erosion in any stream are therefore always changing. As a rule, these changes are too slight to be readily noticeable, but in some regions they are great enough to cause streams to dry up at certain seasons, and to rise in floods at others. In other regions fluctuations are less extreme. Every spring the lower Mississippi has a normal rise in water level of 15 to 20 feet. The Nile normally rises 24 feet and the Ganges 32 feet. The erosive effect of such floods is considered below.

"The shape of the stream channel as seen in cross-section also influences velocity. Since friction between water and channel slows a stream down, velocity is greatest in channels with the smallest area in proportion to volume of water. Deep, narrow channels therefore give greater stream velocity than broad, shallow ones.

"A stream continues to acquire a load until it is carrying the greatest possible amount permitted by the gradient, volume of water, and kind of material available."

The "laws of erosive power" concern both transportation and abrasive power of the stream. If all the fragments of rocks had the same specific gravity, then the following definite action would take place.

[62] Gilbert, G. K., Report on the geology of the Henry Mountains: U. S. Geog. and Geol. Survey of the Rocky Mountain Region, Chap. V, Land Sculpture, pp. 99-150, 1877.
[63] Russell, Israel C., Rivers of North America, a reading lesson for students of geography and geology, G. P. Putnam's Sons, 1898.
[64] Longwell, C. R., Knopf, A., and Flint, R. F., A textbook of geology, pt. 1, physical geology, pp. 42-44, John Wiley & Sons, 1932.

"If the velocity of a stream be doubled, the diameters of rock fragments it can move are increased 4 times. In other words, *the maximum diameter of the individual rock fragments a stream can move varies as the square of the velocity.* * * * Calculations have shown that doubling the velocity of a stream increases its abrasive power at least 4 times, and under certain conditions as much as 64 times. In other words, *abrasive power varies between the square and the sixth power of the velocity.*

"These laws not only explain the vastly greater erosion accomplished by swift streams than by slow ones under normal conditions, but they show clearly why exceptional floods, greatly increasing velocity by increasing volume, have such tremendous destructive power. The volume of the Colorado River measured at Yuma, Arizona, during a flood in 1921, was 155 times its normal volume. Again, when the St. Francis Dam near Los Angeles gave way in 1928 and flooded the valley below, huge blocks of concrete weighing up to 10,000 tons each, were moved by the escaping water. In India, during the Gohna flood of 1895, which lasted just four hours, the water picked up and transported such quantities of gravel that through the first 13 miles of its course the stream made a continuous gravel deposit from 50 to 234 feet."

Preparation of Material Removed by Erosion

As previously stated and described, the materials which are removed and washed into the streams are first prepared through weathering processes. Particles are loosened from the outcrop by surface disintegration, consisting largely of oxidation, hydration, and solution.

Since climatic environments were different in the past than they are now, the subject of ancient climates[65] is an important problem in its relation to the development of ancient stream channels. The study of fossils imbedded in the deposits gives the most important clue to the nature of ancient climates. The condition and composition of the sediments themselves give another, as repeatedly pointed out by Reed.[66]

Transportation

The subject of river engineering brought forth at an early date much definite information as regards the carrying power of streams. The following statement is quoted from Stevenson:[67]

"The following are results deducted from experiments made by Bossut, Dubuat, and others, on the size of detrital particles which streams flowing with different velocities are said to be capable of carrying:

3 in. per sec.—0.170 mile per hour, will just begin to work on fine clay.
6 in. per sec.—0.340 mile per hour, will lift fine sand.
8 in. per sec.—0.4545 mile per hour, will lift sand as coarse as linseed.
10 in. per sec.—0.5 mile per hour, will lift gravel the size of peas.
12 in. per sec.—0.6819 mile per hour, will sweep along gravel the size of beans.
24 in. per sec.—1.3638 miles per hour, will roll along rounded pebbles 1 inch in diameter.
3 ft. per sec.—2.045 miles per hour, will sweep along slippery angular stones the size of a hen's egg."

The following table is quoted from F. C. Gilbert[68] to show the maximum diameters of boulders which can be moved in sluices at certain velocities:

[65] Smith, J. P., Ancient climates of the West Coast: Pop. Sci. Monthly, vol. 76, pp. 478-486, 1910. . . . Climatic relations of the Tertiary and Quaternary faunas of the California region: California Acad. Sci., Proc., ser. 4, vol. 9, pp. 123-173, 1919.
[66] Reed, Ralph D., Geology of California, Am. Assoc. Petroleum Geologists, 1933.
[67] Stevenson, David, The principles and practice of canal and river engineering, p. 361, Edinburgh, 1886.
[68] Gilbert, F. C., Design of sluices for gold placer mining: Eng. Jour., Arizona, vol. 16, no. 8, p. 4, 1932.

Diameter (inches)	Velocity (feet per second)
2	3.3
4	5.3
6	6.2
8	7.4
10	8.4
12	9.1
16	10.8
20	11.9
24	13.0
30	13.7

Transportation, as shown by McGee,[69] is done by the carrying of materials in solution, through suspension, and by the process of saltation. The materials thus carried are deposited by precipitation from solution, sedimentation from suspension, and grounding after 'leaping' along by that process called 'saltation.'

It has been shown by G. K. Gilbert,[70] who carried on extensive laboratory experiments with running water, that the materials borne in suspension are easily enough sampled and their quantity measured; but the 'bed load' is much less accessible. This load is carried forward by sliding or rolling along a smooth channel bed, as well as by saltation, which takes place when the bed is uneven and causes the particle to move irregularly in a series of jumps.

Gilbert calls the transportation of the bed load "hydraulic traction" in contrast to "hydraulic suspension." His summary of "Modes of transportation, collective movement" is expressed as follows:

"When the conditions are such that the bed load is small, the bed is molded into hills, called dunes, which travel downstream. Their mode of advance is like that of eolian dunes, the current eroding their upstream faces and depositing the eroded material on the downstream faces. With any progressive change of conditions tending to increase the load, the dunes eventually disappear and the débris surface becomes smooth. The smooth phase is in turn succeeded by a second rhythmic phase, in which a system of hills travel upstream. These are called antidunes, and their movement is accomplished by erosion on the downstream face. Both rhythms of débris movement are initiated by rhythms of water movement."

In showing how complicated a stream's action may be, Gilbert states:

"The flow of a stream is a complex process, involving interactions which have thus far baffled mechanical analysis. Stream traction is not only a function of stream flow but itself adds a complication. Some realization of the complexity may be achieved by considering briefly certain of the conditions which modify the capacity of a stream to transport débris along its bed. Width is a factor; a broad channel carries more than a narrow one. Velocity is a factor; the quantity of débris carried varies greatly for small changes in the velocity along the bed. Bed velocity is affected by slope and also by depth, increasing with each factor; and depth is affected by discharge and also by slope. If there is diversity of velocity from place to place over the bed, more débris is carried than if the average velocity everywhere prevails, and the greater the diversity the greater the carrying power of the stream. Size of transported particles is a factor, a greater weight of fine débris being carried than of coarse. The density of débris is a factor, a low specific gravity being favorable. The shapes of particles affect traction, but the nature of this influence is not well understood. An important factor is found in form of channel, efficiency being affected by turns and curvature and also by the relation of depth to width. The friction between current and banks is a factor and therefore likewise the nature of the banks. So, too, is the viscosity of the water, a property varying with temperature and also with impurities, whether dissolved or suspended."

---

[69] McGee, W. J., Outline of hydrology: Geol. Soc. America Bull. 19, p. 199, 1908.
[70] Gilbert, G. K., The transportation of débris by running water: U.S. Geol. Survey Prof. Paper 86, pp. 10, 11, 15, 16, 1914.

Gilbert classifies streams according to their transportational characters:

"The classification of streams here given has no other purpose than to afford a terminology convenient to the subject of débris transportation.

"When the débris supplied to a stream is less than its capacity the stream erodes its bed, and if the condition is other than temporary the current reaches bedrock. The dragging of the load over the rock wears, or abrades, or corrades it. When the supply of débris equals or exceeds the capacity of the stream bedrock is not reached by the current, but the stream bed is constituted wholly of débris. Some streams with beds of débris have channel walls of rock, which rigidly limit their width and otherwise restrain their development. Most streams with beds of débris have one or both banks of previously deposited débris or alluvium, and these streams are able to shift courses by eroding their banks. The several conditions thus outlined will be indicated by speaking of streams as *corrading*, or *rock-walled*, or *alluvial*. In strictness, these terms apply to local phases of stream habit rather than to entire streams. Most rivers and many creeks are corrading streams in parts of their courses and alluvial in other parts.

"Whenever and wherever a stream's capacity is overtaxed by the supply of débris brought from points above a deposit is made, building up the bed. If the supply is less than the capacity, and if the bed is of débris, erosion results. Through these processes streams adjust their profiles to their supplies of débris. The process of adjustment is called gradation; a stream which builds up its bed is said to aggrade and one which reduces it is said to degrade.

"An alluvial stream is usually an aggrading stream also; and when that is the case it is bordered by an alluvial plain called a flood plain, over which the water spreads in time of flood.

"If the general slope descended by an alluvial stream is relatively steep, its course is relatively direct and the bends to right and left are of small angular amount. If the general slope is relatively gentle, the stream winds in an intricate manner; part of its course may be in directions opposite to the general course, and some of its curves may swing through 180° or more. This distinction is embodied in the terms *direct alluvial* stream and *meandering* stream. The particular magnitude of general slope by which the two classes are separated is greater for small streams than for large. Because fineness is one of the conditions determining the general slope of an alluvial plain, and because the gentler slopes go with the finer alluvium, it is true in the main that meandering streams are associated with fine alluvium."

Commenting on the curvature of a channel, which greatly complicates the transportation and deposition of debris by a stream, Gilbert says:

"In a straight channel the current is swifter near the middle than near the sides and is swifter above mid-depth than below. On arriving at a bend the whole stream resists change of course, but the resistance is more effective for the swifter parts of the stream than for the slower. The upper central part is deflected least and projects itself against the outer bank. In so doing it displaces the slow-flowing water previously near the bank, and that water descends obliquely. The descending water displaces in turn the slow-flowing lower water, which is crowded toward the inner bank, while the water previously near that bank moves toward the middle as an upper layer. One general result is a twisting movement, the upper parts of the current tending toward the outer bank and the lower toward the inner.[71] Another result is that the swiftest current is no longer medial, but is near the outer or concave bank. Connected with these two is a gradation of velocities across the bottom, the greater velocities being near the outer bank. The bed velocities near the outer bank are not only much greater than those near the inner bank, but they are greater than any bed velocities in a relatively straight part of the stream. They have therefore greater capacity for traction, and by increasing the tractional load they erode until an equilibrium is attained. On the other hand, the currents which, crossing the bed obliquely, approach the inner bend are slackening currents, and they deposit what they can no longer carry.

---

[71] The system of movements here described has been observed by many students of rivers. They were demonstrated by the aid of a model channel by J. Thomson, in connection with an explanation which differs somewhat from that of the present text. See Roy. Soc. London Proc., pp. 5-8, 1876, and 356-357, 1877; also Inst. Mech. Eng. Proc., pp. 456-460, 1879.

"It results that the cross section on a curve is asymmetric, the greatest depth being near the outer bank. As the winding stream changes the direction of its curvature from one side to the other, the twisting system of current filaments is reversed, and with it the system of depth, but the process of change includes a phase with more equable distribution of velocities, and this phase produces a shoal separating the two deeps. The shoal does not cross the channel in a direction at right angles to its sides but is somewhat oblique in position, tending to run from the inner bank of one curve to the inner bank of the other. In meandering streams it is usually narrow and is appropriately called a bar. In direct alluvial streams, where bends are apt to be separated by long, nearly straight reaches, it is usually broad and may for a distance occupy the entire width of the channel."

Deposition

The nature of stream or fluvial deposits and their detailed structure and texture is described in various textbooks, but not with sufficient detail to explain all the types of complicated features found in gravel deposits, especially complex deposits such as those of a mountainous region like the Sierra Nevada.

Quoting from Longwell, Knopf, and Flint [72] on "fluvial deposition":

"The constructive process of fluvial deposition goes forward side by side with fluvial erosion. This is a result of the complexity and variability of the stream currents, which constantly drop some rock fragments to the bottom while they pick up others. When a stream is actively eroding its bed at a certain point, it is merely picking up and carrying away more rock material than it is depositing there, and when it is actively depositing the reverse is going on. Therefore, whereas fluvial erosion and deposition are processes physically opposed to each other, they can be separated in practice only by recognizing the preponderance of one over the other."

The arrangement of materials deposited in a delta, however, is well known, and gives a picture which is more or less duplicated whenever the current of a stream is checked by a body of standing water and the materials transported are permitted to drop. The term 'foreset beds' is applied to the deposition on the frontal slope of the growing embankment. 'Bottomset beds' are of finer grain and are formed by the particles carried out beyond the slope and deposited in deeper water. 'Topset beds' are composed of the materials laid down and spread out on top of the other materials by the fluctuating stream.

The material deposited by a stream is called alluvium and makes up fan, floodplain, and delta deposits. The term 'fanglomerate' is used for the gravel materials of alluvial fans.

An alluvial fan is built up at the point of abrupt change in gradient of a loaded stream. A floodplain is a series of coalescing alluvial flats along a valley. A delta is the final deposit by a stream, unloaded as it enters a still body of water.

Overflow of a stream onto its floodplain will cause natural levees to be built up as low ridges bordering the channel. Lateral swinging of a stream causes cutting on the outer sides of the curves and deposition on the inside, which results in the widening of the valley. A meandering stream may develop to the point of straightening itself in places by the cutting off and silting up of the meanders, forming oxbow lakes as a result. A stream which forms a complex interlocking network on its floodplain typifies anastomotic drainage. An overloaded stream on a low gradient, becoming choked, and constantly obliged to cut new channels, develops an intricate network on a floodplain; the process is termed 'braiding.'

---

[72] Longwell, Knopf, and Flint, op. cit., p. 44.

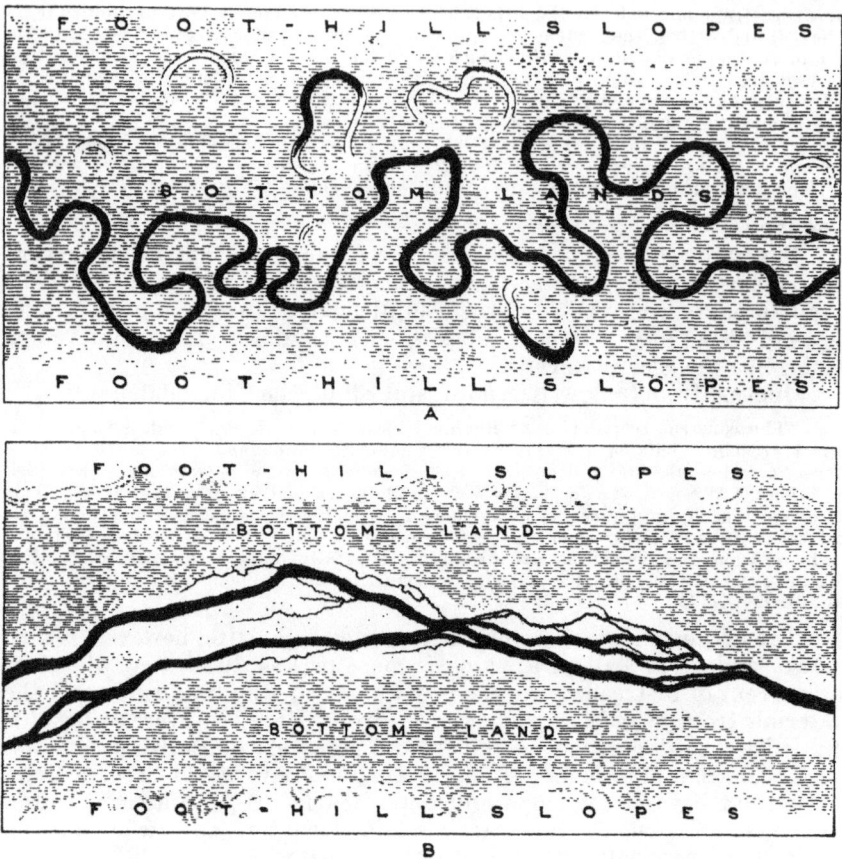

FIG. 71. *A,* Diagram to show a meandering stream with oxbow loops. Such a stream develops in a valley worn down to base-level and not subject to extensive floods. *B,* Diagram to show a subdividing or anastomosing stream in a valley subject to floods; *after Johnson, U. S. Geol. Survey 22d Ann. Rept., 1900.*

If the stream is rejuvenated and therefore cuts deeper, floodplain terraces or benches are formed. The benches may, however, be cut and left in the bedrock and covered with only a film of gravel on the surface.

Streams composed entirely of thick mud, called 'mudflows,' are akin to landslides.[73] Quoting from Longwell, Knopf, and Flint: [74]

"Another normal though infrequently operative process in arid regions is the mudflow. It occurs only where fine rock material becomes water-soaked on steep slopes after heavy rains, and moves downward as a slippery mass. . . . It advances in waves, stopping when it becomes too viscous to flow and damming the water behind it until it liquefies and again proceeds like an advancing flow of lava. Mudflows can carry boulders many feet in diameter. Observers have seen these great rocks bobbing 'like corks in a surf.' Successive mudflows play a part in the building of fans."

The transporting power of mudflows, their various peculiarities, and the resultant unsorted deposits have many characteristics much like those of glacial deposits and have frequently deceived engineers.

The distinguishing characters of glacial deposits are clearly summarized by Blackwelder,[75] who has made a special study of them:

"The deposits left by glaciers should be distinguished from those made by streams, lakes and other agencies.

"The ice tongue of a glacier leaves only one type of deposit called till. It is wholly unstratified and its components are quite unsorted—a jumbled orderless mass of clay, sand, and boulders. Blocks three to five feet in diameter are common and those 25 feet or more are not rare. In general, the size of such boulders depends upon the spacing of the joint-cracks in the rocks of the mountain sides. Usually till is an earthy mass well sprinkled with stones and boulders but in some cases the boulders predominate. This is particularly true of the deposits of small glaciers which have done little more than sweep the coarse talus from the valley slopes. The stones in till may be of any shape from well-rounded to angular but many have the corners and edges rounded. It is usual to find some that have been bevelled by being rasped along the bottom of the glacier. Hard rocks may thus be well polished. Such stones, like the bedrock, are covered with scratches which are easily recognized.

"It is often difficult to identify till, especially if it has been much decayed or eroded or is poorly exposed. It may then be confused with other bouldery deposits which are unstratified. From volcanic mudflow deposits, such as abound in the Miocene beds on the Sierra Nevada flanks, till may often be distinguished by its containing large quantities of nonvolcanic rocks. Even this criterion fails where glaciers occupied volcanic mountains such as Mts. Shasta and Rainier. Ordinary mudflow deposits are seldom as thick as glacial moraines and are generally interbedded with typical stream gravel and sand, as in the alluvial fans of the arid regions. Unless the surface topography is still preserved or unless one finds plenty of scratched stones, it may be almost impossible to distinguish till from landslide dumps. In many cases no one type of evidence can be relied on, but one must study all the facts and weigh the importance of each.

"The rivers which issue from glaciers deposit coarse gravel, then fine gravel, and finally sand as the current weakens near the edge of the mountains. These three grades of detritus are more or less interbedded, because variations in the river's power occur from time to time at a given place. Like river deposits in general, those of glacial streams are distinctly stratified, though usually cross-bedded. They are fairly well sorted into separate layers of sand and gravel of various sizes. The pebbles are normally well rounded and very rarely either faceted or scratched. Angular stones are rare. Although small boulders are carried by ice cakes and become stranded in the glacial river gravel, large boulders are generally absent.

---

[73] Blackwelder, Eliot, Landslide family and its relations (abstract): Pan-Am. Geologist, vol. 54, p. 73, 1930.
    Finch, R. H., Mud flow eruption of Lassen volcano: The Volcano Letter, no. 266, pp. 1-4, 1930.
    [74] Longwell, Knopf, and Flint, op. cit., pp. 77-78.
    [75] Blackwelder, Eliot, Glacial and associated stream deposits of the Sierra Nevada: California Div. Mines, Mining in California, State Mineralogist's Rept. 28, pp. 306-308, 1932.

"To distinguish the deposits of a glacial from a nonglacial river is difficult and often impossible, unless one can trace the gravel terraces into actual connection with a glacial moraine or can work out in detail the physiographic history of the district.

"The deposits made in glacial lakes are rather distinctive. On the bottom of the lake, clay and silt are laid down very evenly in thin sheets which are commonly banded as seen later in cross-section. This is due to the fact that the layer deposited in winter is finer and darker in color than the one laid down during the summer melting season. Unlike most lake deposits the glacial lake clays commonly contain scattered pebbles and even small boulders which have been dropped from cakes of ice floating over the lake. These laminated clays may be associated with beds of peat or chalky or diatomaceous earth formed by organisms that inhabited the clearer parts of the lake. Streams entering the lake form advancing deltas composed of gravel and sand in which the stratification is characteristic of deltas in general. In quantity the delta deposits often exceed the other lake deposits greatly, for the glacial rivers carry large quantities of coarse detritus all of which lodges in the deltas rather than upon the floor of the lake.

"Other deposits that may be formed in glacial valleys, such as landslides, talus, and alluvial fans, need not be described specifically. They are local and generally well known."

The deepening of canyons and the deposition of gravels by outwash streams which issue from the snouts of glaciers form a very important chapter in the robbing and destruction of earlier gold-bearing gravels. Much of the material carried by Pleistocene glacial outwash streams of the Sierra Nevada was dumped at the foot of the western slope of the range at the point where the major rivers enter the Great Valley. The extensive gold-dredging ground of California to a large extent owes its existence to these streams.

So far as the processes involved in stream action are concerned, much can be gained by the detailed study of ancient glacial stream channels found throughout the world, especially in its northern belt. The mass of literature published on this subject should contribute greatly to the building up of a systematic knowledge of stream habit.

Physiographic Terms Relating to Streams

The mere definition of some of the terms used in stream physiography gives a direct insight into the science.[76]

'Cycle of erosion' includes the series of changes from the initial cutting of a surface to the final reduction of a region to a baselevel. The surface of a region reduced to fairly low relief, but still undulating, is called a 'peneplain' (also spelled peneplane). It is a significant fact that the Sierra region in early Tertiary time was approaching the peneplain stage of erosion, when the area was covered with lava to form a more nearly plain-like surface, and later uplifted. The uplift caused deep dissection by streams.

Stages of stream development, from gulleys to completely worndown plains, consist of youth, maturity, and old age, with continuous transitional stages between. The early stages represent very rapid growth, which slows down gradually until at old age the changes may be extremely slow.

The genesis or origin of the stream takes into consideration the initial surface over which the stream first flowed. Several terms are used by geologists in relation to this subject. Consequent streams are those whose positions were determined by the initial slopes of the land surface. Subsequent streams are those which are established by growing headward along belts of weak rocks.

[76] Johnson, Douglas, Streams and their significance: Jour. Geology, vol. 40, p. 482, 1932.

Where the underlying structures of the rocks have affected the direction of the stream and its valley, the stream is said to have structural control. The terms fault, joint, strike, anticlinal, synclinal, and monoclinal are prefixed to the word 'valley'; thus, fault-valley, strike-valley, etc. It is especially significant that the richest gold-bearing channels have structural control—the streams originally ran on and along mineralized zones in bedrock.

Streams may start their courses over one sort of geological formation, but as time progresses they may cut through it and be let down on a lower and entirely different type of structure; such streams are said to be 'superimposed' (or simply superposed) on the underlying rock structure. When a certain stream pattern, originally developed because of previous topographic or structural conditions, is retained even after those conditions are removed, the stream is said to have inherited its peculiar features from much earlier periods of its life.

Streams may be intermittent or permanent. Depression of topography along a coast may cause the sea to invade the valleys, and the streams are drowned. Uprise along the coast may leave hanging valleys. Tributaries to a main stream which has been much faster-cutting may also be left 'hanging'; as, for example, in Yosemite Valley where the side streams reach the Merced River by way of beautiful waterfalls.

The longitudinal profile of a stream is taken from its source to its mouth, while its gradient represents its inclination at some particular part of its course. Tributaries are said to be accordant when they enter at about the same level as their main stream. A stream is said to be at grade when rate of degradation and rate of aggradation are about equal.

A stream which is able to maintain its course, even when a segment of the earth is gradually raised athwart that course, is called an antecedent stream. If, however, the uprise of the mountain causes the flow of the streams to be accelerated down its slope, so that they cut deeper gorges, they are said to be rejuvenated. Even stream meanders, developed on a plain, may be entrenched or incised deeply, to form a winding canyon by elevation of the plain.

Rejuvenation may be effected in other ways than mountain-making. Change of climate may make a decided change in stream cutting. Stream piracy is another important cause. This consists of the capture of one stream by another. The second lies at a lower elevation; its head cuts back until the first is tapped, or beheaded. Then the water from the first stream, from that point upward, is caused to flow into the capturing stream. In this manner the flow in the first is accelerated often to such an extent that a new gorge may be formed. Whole stream systems may thus be readjusted and repeatedly go through new life cycles. Piracy and stream adjustment were apparently very active in the Sierra Nevada during Tertiary time; this process partly accounts for many of the deep accumulations there of Tertiary gravel.

The pattern of an individual stream, or of the whole or any part of its system, develops in its own peculiar way because of certain controlling geological, topographical, and climatic features. The pattern, therefore, is a character significant enough to bear special study and to support many new descriptive terms. It is now best studied by means of aerial photographs, though detailed topographic and geologic maps once presented the only bases of accurate expression.

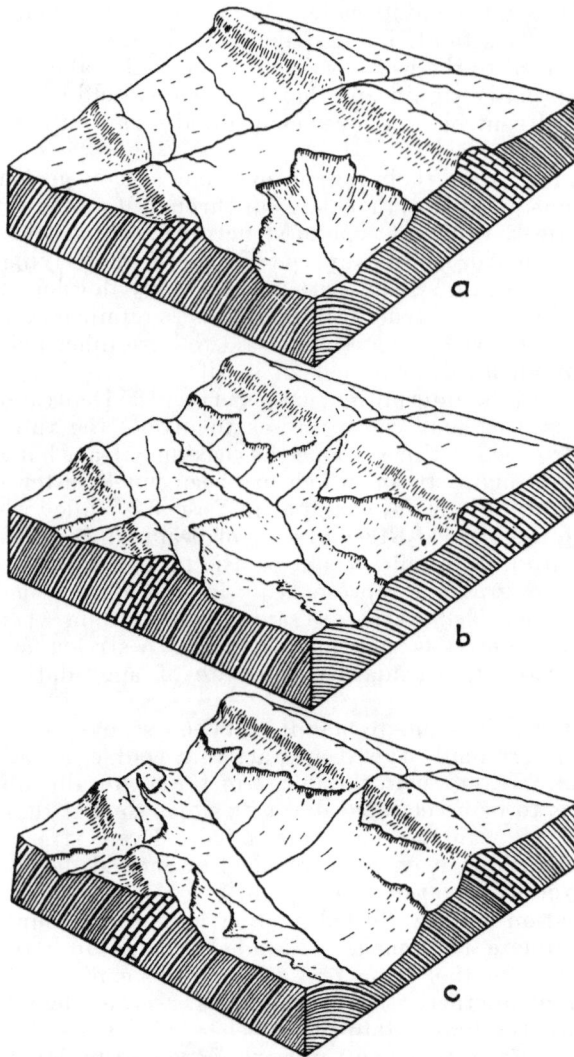

FIG. 72.   Diagrams to illustrate three successive
stages in stream piracy.   The stream cutting back at the
lower elevation beheads and captures the stream flowing at a higher level.   *After Blackwelder and Barrows,
American Book Co.*

Many clues as to the geologic structure and history of the underlying region are gained by the study of stream pattern, from either an intensive or regional point of view.   In the northern Sierra Nevada, the stream pattern developed prior to volcanic activity was structually controlled by bedrock; but during the later period of volcanism it suffered change by that widespread activity.   The major streams subsequent to volcanism followed the slope of the lava-covered, tilted, and uplifted fault-block-range.

An instructive outline of the subject of stream pattern is given by Zernitz,[77] who describes and illustrates by many actual examples such patterns as follows: dendritic, trellis, rectangular, annular, radial, and parallel.   He states that:

"The patterns which streams form are determined by inequalities of surface slope and inequalities of rock resistance.   This being true, it is evident that drainage patterns may reflect original slope and original structure or the successive episodes by which the surface has been modified, including uplift, depression, tilting, warping, folding, faulting, and jointing, as well as deposition by the sea, glaciers, volcanoes, winds, and rivers.   A single drainage pattern may be the result of one or of several of these factors."

The fact that lakes fill depressions, basins, and valleys along stream courses, and that their deposits are intimately associated with those of streams, makes their study interrelated with that of stream channels. The so-called 'pipe-clay' deposits, which nearly always immediately overlie the gold-bearing gravels, represent beds of silt and finely divided volcanic ash, which have settled in lakes and formed a series of thin layers.   They often contain impressions of leaves, showing the character of the forests that grew in that early period.   This feature indicates that before the volcanic flows came to cover them up, the stream valleys had been transformed into lakes, into which the volcanic ash settled.

In general, a lake is not as long-lived as a stream.   Streams tend to destroy lakes, either by gradually filling their basins with detritus or by cutting down their outlets to a point where their basins may be drained. Sometimes, however, lakes persist long enough so that the area whose drainage they receive is worn down to a local and temporary baselevel. Lakes are formed in a number of ways: by landslides across stream courses; by lava flows which dam up the drainage; by the down-faulting of segments of the earth which are then filled with water; by glacial action, either by scooping out rock basins or by damming with till; by peculiar action of rivers themselves, such as silting off oxbow loops in a meandering course.   An excellent paper written by the late Dr. W. M. Davis has recently been published, which not only discusses the present lakes of California,[78] but the origin of lakes in general.

Desert Processes

In the desert, the processes of erosion and of stream action differ very much from those in more humid areas.   Only quite recently have the geologic processes in the desert been given much consideration; so also has much serious thought only lately turned toward the possible development of desert placers on a larger scale than mere 'dry washing.'   The fact that adequate supplies of water may usually be derived

[77] Zernitz, Emile, Drainage patterns and their significance: Jour. Geology, vol. 15, p. 498, 1932.
[78] Davis, William Morris, The lakes of California: California Jour. Mines and Geology, vol. 29, pp. 175-236, 1933.

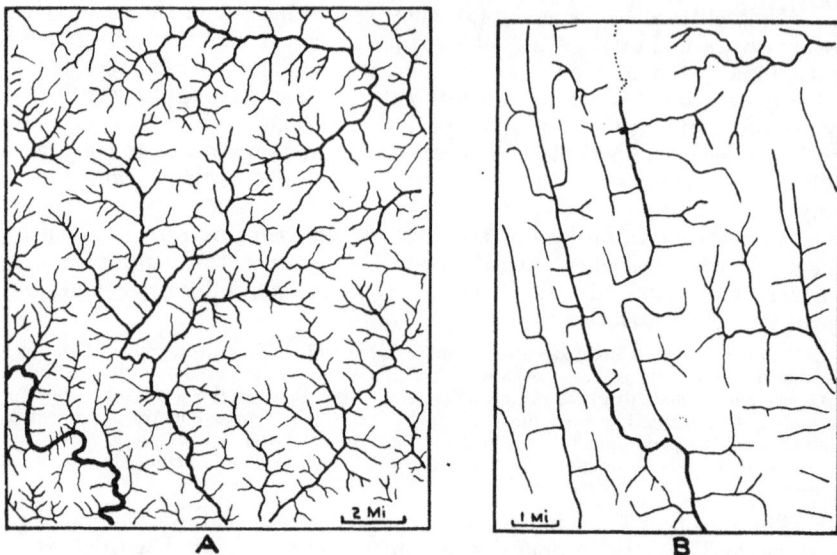

FIG. 73.   A, An example of dendritic drainage pattern, which develops where the underlying rocks lie in a horizontal position and offer uniform resistance to erosion. *After Zernitz, Jour. Geology, 1932.*   B, An example of trellis drainage pattern which develops in folded rocks differing in degrees of resistance to erosion.   The stream courses are therefore structurally controlled.   *After Zernitz, Jour. Geology, 1932.*

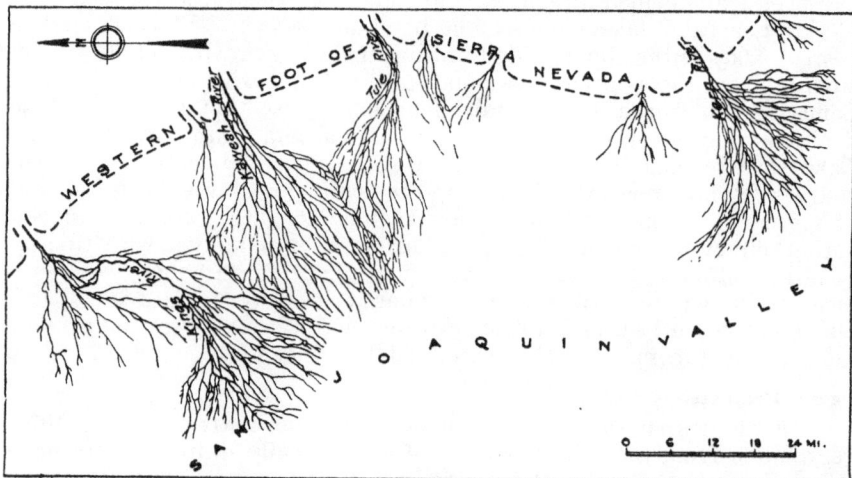

FIG. 74.   Map of southeastern margin of San Joaquin Valley showing fans built by streams which disappear after leaving their mountain canyons.   The coalescing alluvial fans form in this manner an extensive bajada.   *After Slichter, U. S. Geol. Survey Water-Supply Paper 67, 1902.*

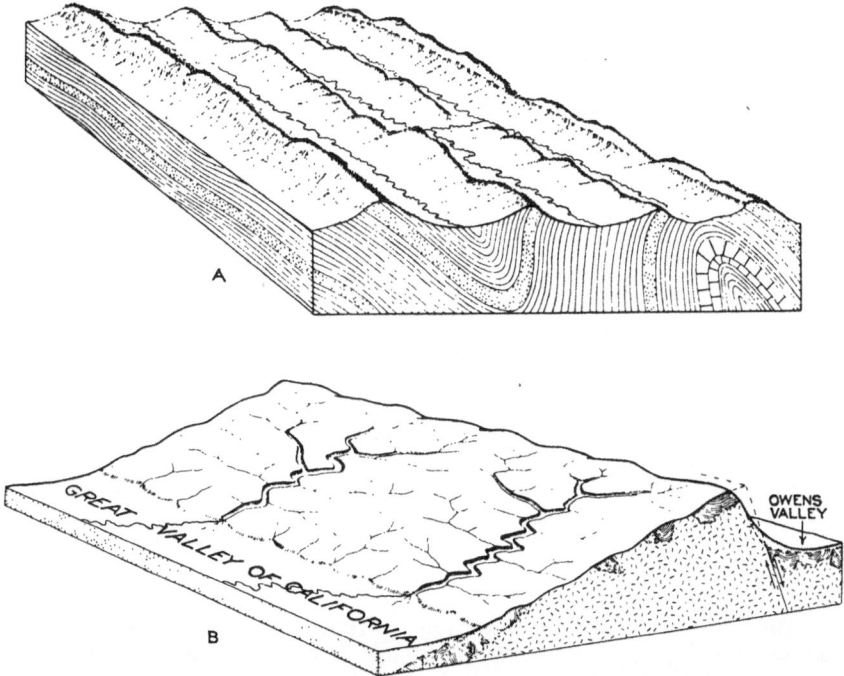

Fig. 75. A, Block diagram illustrating Cretaceous Sierra Nevada topography. The upturned edges of bedrock controlled the drainage pattern, which was later inherited by streams of the early Eocene period. *After Matthes, U. S. Geol. Survey, Prof. Paper 160, 1930.* B, Block diagram to show the tilting of the Sierra Nevada and its effect on stream cutting.' Erosion, prior to the tilting, planed down the surface and exposed the granite, leaving only occasional fragments of the intruded metamorphic rock bodies as roof pendants. The streams, at the point where they leave their mountain canyons and enter the Great Valley, form alluvial fans. *After Matthes, U. S. Geol. Survey, Prof. Paper 160, 1930.*

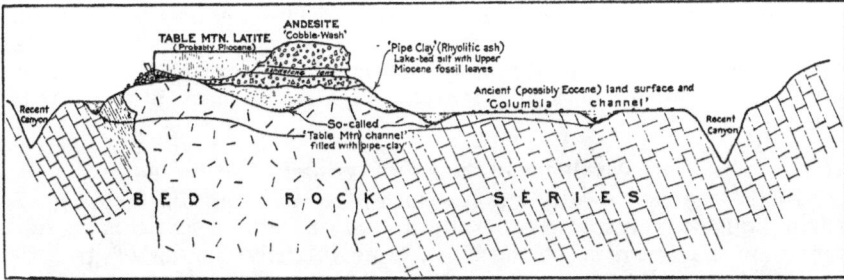

Fig. 76. Diagrammatic geologic cross-section of Table Mountain in the vicinity of Columbia, Tuolumne County. The following principal geologic events affecting the ancient channels of the district are graphically illustrated : (1) An early Tertiary surface, developed on bedrock over which flowed the old river system of the Columbia basin. (2) This surface, including its rich gold-bearing gravels, was later covered with lake sediments ('pipe-clay') consisting of fine silty volcanic ash. (3) Andesite 'cobble-wash' then covered the entire country. (4) A canyon was cut in this andesite mudflow. (5) A basic lava (latite) flowed down this canyon as a molten stream. (6) After this lava cooled and hardened, it became the bed of another river which deposited some gold-bearing gravels on its surface, washed from the surrounding eroded region. (7) Continued erosion during the Pleistocene resulted in the washing away of the less resistant volcanic materials, leaving the very resistant latite standing as Table Mountain. Furthermore, the earlier prevolcanic bedrock surface was uncovered, or resurrected, exposing the gold-bearing gravels originally deposited in the Columbia channel system. (8) Quaternary erosion has cut canyons of great depth, far below the level of the former Tertiary surfaces.

from underground sources in the desert, and the fact that these sources may be found through geological investigation and geophysical surveying, are gradually being accepted.

The most significant results of recent desert-process studies are summarized by Blackwelder [79] as he describes the five distinct types of desert plains:

1. Pediments (including those only thinly veneered with alluvial fans), which represent the desert slope, cut in bedrock, in contrast to the built-up thick alluvial fans or bajadas.
     "Pediments are essentially compound graded floodplains excavated by ephemeral streams * * * the pediment, not the bajada, is the normal and inevitable form developed in the arid regions under stable conditions."
2. Bajadas [80] (compounded alluvial fans), which are built up largely as a result of disturbed or interrupted development of graded slopes. The upward movement of a fault block causes renewed erosional activity, and thick gravel deposits are formed by the consequent torrents and mudflows when they are released from their canyons and enter a region of lesser gradient.
3. Dried-up lake bottoms of the desert, or playas, whose conditions indicate that once more permanent lakes filled the flats and were fed by streams that are now nonexistent.
4. Dip slopes, which are broad planes developed on a denuded, hard, flat-lying or gently tilted rock layer.
5. River floodplains, which are abnormal desert features, but which were apparently more widespread at an earlier time, when precipitation in a given region was much greater than it is today.

These earlier plains and stream courses have been covered by the bajadas of today, so that the older channels have become buried in the true sense of the word. Some of them may represent a large potential placer reserve, but they have not yet been well investigated.

Need for Establishment of Working Criteria

From the study of the complicated life history and habit of streams and the processes involved in their development, a series of working criteria should be developed by the exploration geologist for interpreting the conditions found in the deposits of all streams, including those of Tertiary age in the Sierra Nevada. There is no text available that completely covers all phases of streams—their life history, habits, deposits, etc., especially as this study is related to placers, but the subject offers an interesting field of research which would have a very broad economic application.

Geologic Conditions in the Gold Provinces of California

In the Sierra Nevada and Klamath Mountains of California gold-bearing quartz veins are generally found in metamorphic rocks not more than a few miles, at the most, from intrusive bodies of granitic rocks. The age of the metamorphics, which are made up of slates, schists, limestones, and meta-igneous bodies, is earlier than Cretaceous, namely pre-Paleozoic, Paleozoic, Triassic, and Jurassic. The time of intrusion of the granitic masses was late Jurassic.

The quartz veins were formed shortly after the igneous intrusion, during the last stages of the Jurassic. It would seem that the metamorphic rock masses were uplifted and intruded by molten magmas which then cooled and contracted, causing the surrounding and over-

---

[79] Blackwelder, Eliot, Desert plains: Journ. Geology, vol. 39, pp. 133-140, 1931.
[80] The name *bajada placer* as adopted somewhat erroneously by Webber and previously described in this paper, refers more to single alluvial fans rather than to these compound forms.

lying roof-rocks to crack along many planes of weakness and to form thousands of fissures. Into these openings the residual gaseous solutions, released from the crystallizing granites and composed largely of silica, were injected These, after solidifying, were crushed again, and solutions containing gold entered the complex mineralized zones, enriching especially the cross-fractures where openings were most abundant.

The gold-bearing veins thus formed deep beneath the surface of the earth, had then to be brought to the light of day by the erosion and removal of the covering layer of rocks, nearly two miles in depth. This gigantic work was accomplished during the Cretaceous period, and as a result thousands of layers of shales, sandstones, and conglomerates several miles in stratigraphic thickness were laid down in an adjoining marine basin. Some of these beds—especially the last ones to be deposited —are gold-bearing, showing that the last part of the Cretaceous erosion finally reached the hidden veins.

That part of the geologic history, however, which was most important so far as the making of rich placer deposits is concerned, was the Eocene period. The deep erosion which took place during the Cretaceous had worn the surface down to such an extent that the metamorphic rocks with their mineralized zones had been reached, so that the streams of the Eocene ran along and through them.

Structural control of the drainage, fully developed during the Cretaceous, was thus inherited by the Eocene streams. Ridges and valleys followed the north-south trending beds of hard and soft rock. The subtropical climate of the early Eocene together with other conditions particularly favorable to rock disintegration, such as a more prolonged time of stability in the earth's crust, made it possible for the gold in the surface rocks to be released from its matrix.

Then came the inception of the Tertiary Sierran uplift, which rejuvenated stream flow, causing the released gold in the disintegrated veins to be washed into the river channels, resulting in very rich concentrations. The streams were loaded with fine quartz sand and pebbles, together with clays derived from the decomposed feldspathic parts of the rocks. The finer particles were washed to the sea, and as a result the Eocene (Ione formation) today contains large deposits of commercial clay interbedded with quartz sands.

The westward tilting and resulting acceleration of stream flow interrupted the north-south drainage system inherited from the Cretaceous period. The readjustment of the streams resulted in their general direction of flow being finally changed from north and south trends to a westerly course, somewhat as it is today. The longest of these known streams even headed far to the east into what is now Nevada.

Hardly had the Eocene come to a close when much rhyolite ash, thrown into the air from volcanoes, settled over the region. By Oligocene time, rhyolite ash had covered much of the northern Sierra Nevada, damming rivers and forming lakes, the bottom sediments of which are now represented by thinly layered pipe-clay immediately overlying the richer gold-bearing gravel. The newly developed rivers, flowing directly down the western-tilted slope of the Sierra, over a volcanic cover, as consequent drainage, were repeatedly interrupted by continued ejections of lava and a further tilting of the region. Not until the late Pliocene or early Pleistocene did the constant out-pouring of lava cease. Then,

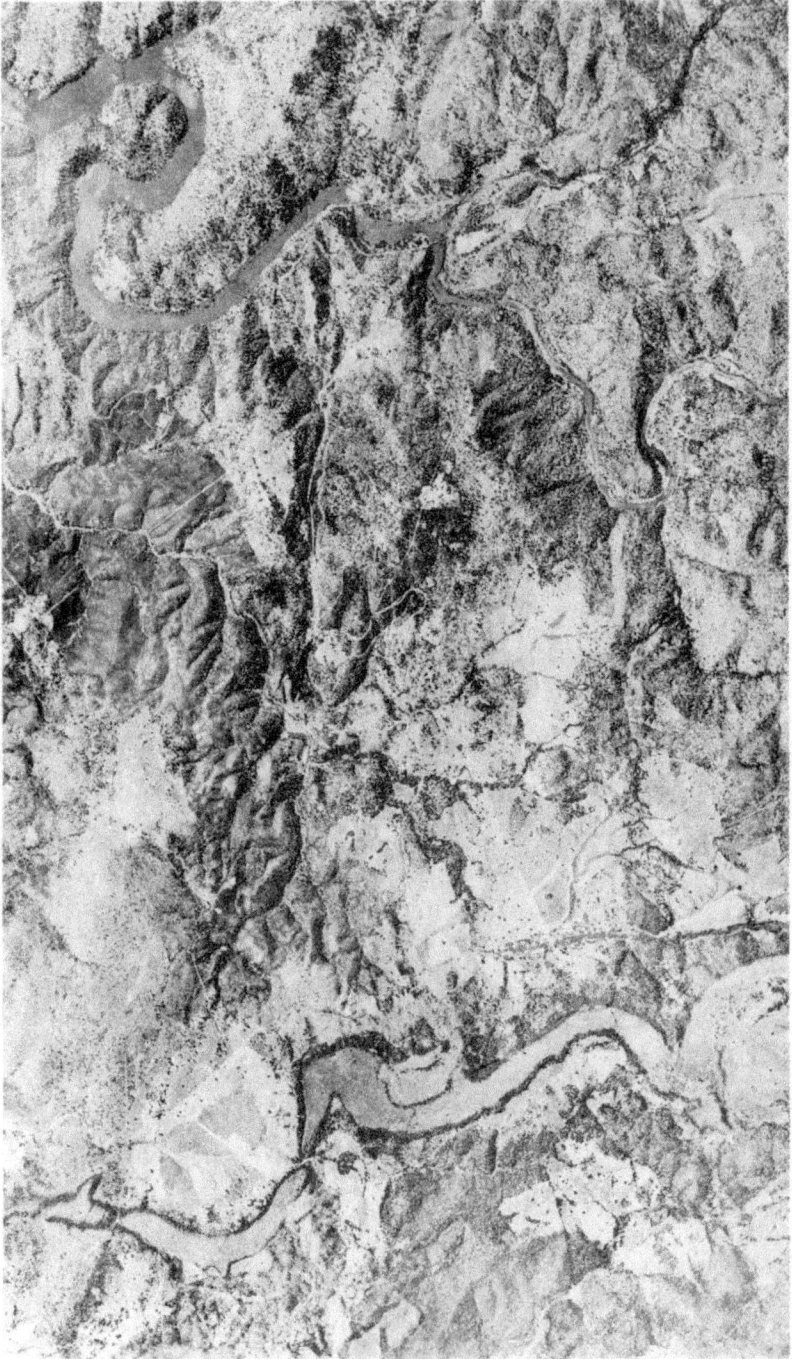

FIG. 77.   Aerial map (for explanation, see fig. 78 caption).
Photo by courtesy of Fairchild Aerial Surveys, Inc.

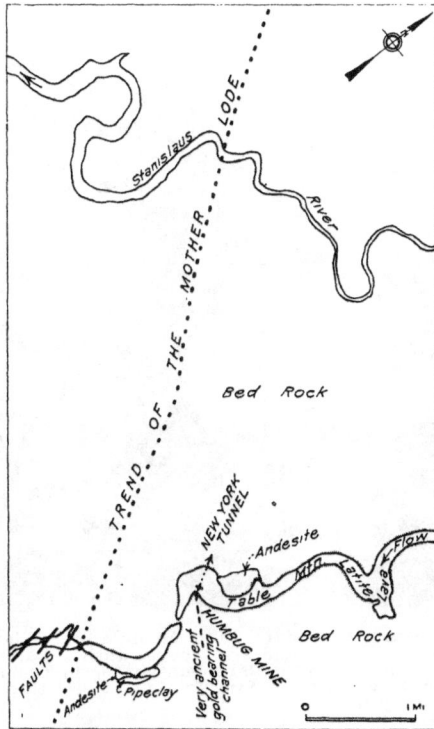

FIG. 78.   Index sketch (reduced) of the same area shown in figure 77. *Explanation of figure* 77: Mosaic made up of many overlapping vertical photographs taken from an airplane, elevation 10,000 feet; area 6½ by 4 miles, between Jamestown and Angels Camp on the Mother Lode (see fig. 78). The Stanislaus River winds through the upper half of the picture; in the lower, the Table Mountain latite flow, which occupies a late Tertiary canyon cut in andesite 'cobblewash', stands out in bold relief. The softer andesite rocks and the underlying 'pipe-clay' cut by this canyon have been stripped away by Pleistocene erosion, save for two or three small patches remaining in protecting curves of the harder lava flow. The surrounding country lies lower than Table Mountain in elevation, and is worn in bed-rock. The prevolcanic gold-bearing channels are represented only as gravel-filled fragments, which, however, show their northward trend, parallel to the strike of bedrock and the Mother Lode. These remnants can be recognized in the pictures only after field examination. One such mined-out channel passes under Table Mountain at right angles to it, near its central position in the picture. It was here that the Humbug drift mine on the south met the New York tunnel on the north, resulting in an underground fight, and later litigation.

FIG. 79. Surface of the resistant Table Mountain latite lava flow which once filled a canyon cut by a river in andesite 'cobble-wash'. Later the latite surface served as a stream bed; now it is the flat mountain top in Tuolumne and Calaveras Counties referred to in the writings of Mark Twain and Bret Harte. *Cut by courtesy of Engineering and Mining Journal.*

by a series of violent earth movements, the Sierra Nevada broke away from the region to the east along huge fault-scarps, which are formed at the foot of the present steep eastern slope, where displacements are now measured in thousands of feet. Within the Sierran slope, smaller adjustment faults also broke the continuity of the older buried Tertiary stream grades. Some of the ancient streams, the courses of which headed farther to the east, were virtually 'chopped' into many pieces; some were elevated and others depressed, and many were warped to various peculiar positions. Undoubtedly there are some segments of these old channels which now lie deeply buried beneath great thicknesses of alluvium in down-dropped fault-blocks east of the Sierran escarpment.

In the Great Basin and the Mojave Desert region of California and Nevada are remnants of Tertiary stream deposits, interbedded with or lying beneath lavas, all of which have suffered much by faulting and warping.

In these regions, however, the most important period of placer formation was in the early Pleistocene, rather than in the Tertiary. Two types of lode gold supplied the sources. One type was formed in much the same manner as the Sierra Nevada lodes. The other consisted of mineralized zones in rhyolite of early and middle Tertiary time. In the Pleistocene there were normal streams flowing through the desert, fed by melting glaciers of the higher mountains. Placers that were formed by these Pleistocene streams have since been largely covered by desert alluvial fans. Some have been elevated and are cut by more

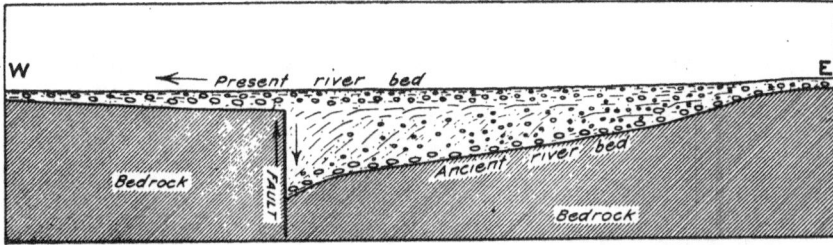

FIG. 80. Ideal cross-section of a river in the Sierra Nevada, the bed of which has suffered down-faulting on the upstream side, causing gravels, sand, and silt to accumulate in the pocket thus formed.

recent streams, when present in this arid region, so that recent concentrations from older river gravels provide one source for the desert dry placer.

In the Klamath Mountains, though the early geologic history was much like that of the Sierra Nevada, there were no lavas to fill the valleys in which the stream gravels were deposited. Uplifts, accompanied by renewed stream-cutting, caused terraces to be left on the valley sides, where the rich gravels have given up their gold to hydraulic mining. Some of the finer gold particles were washed by the rivers to the sea and have formed deposits known as beach placers along the northern shore of California.

Down-faulting of various degrees of magnitude in places caused accumulations of gravels to form, especially on the down-throw sides of faults. A number of such faults are located in both the Sierra Nevada and Klamath Mountains, and may hold a reserve of gold not yet entirely recovered. In the Sierra most of these minor displacements show that the east side of the fault-plane has been dropped down, so that where faults cross westward flowing rivers, accumulations of gravels have taken place in pockets thus formed, east of the fault-plane and upstream.

The whole Pleistocene period was one of great events for California. The eastern side of the Sierra Nevada was raised to very lofty heights. The westward-flowing streams, as a consequence, were so greatly accelerated that they cut deep and rugged canyons. The uprise, accompanied by faulting, caused such violent earthquakes that enormous masses of rock were shaken from the mountain sides, in many places to form local lakes which were later to be drained and destroyed by active erosion. Glaciers developed in the higher mountains and crept down the canyons, carving them wider and leaving them U-shaped in form. Their melting supplied much water to the streams. Some local volcanic cones were built up here and there near or over the fault planes.

The Tertiary stream gravels, which had long been buried deeply beneath lavas, were exposed by the Pleistocene canyon-cutting rivers. From the dissected portions of the old channels, gold was removed and washed into the newer streams, which concentrated it on their bedrock riffles. The remaining portions of the Tertiary deposits were left with their stubs exposed high up on the intervening ridges. In places, erosion merely stripped the covering of volcanic tuffs, sands, and gravels from the bedrock, leaving the channel with its rich gold deposits laid practically bare for the lucky early miner to win. Some of the finer particles of gold were swept clear out to the Great Valley where they were dropped on the edge of the plain. These areas are now the dredge grounds.

The general western tilt of the Sierra Nevada has been found to continue along the same slope (about two degrees) far beneath the alluvium and sediments of the Great Valley. Areas dredged for gold values in the gravels thrown down by the great canyon-cutting rivers of the range lie along the extreme western margin of the foothills near the place where bedrock passes beneath the alluvium, and aligned in a direction N. 20° W. The gravels dredged do not lie directly on bedrock but on tuffaceous clay layers, spread out above the detritus-covered down-warped Sierran surface. Beneath the 'false-bedrock' and cut in the true bedrock surface is a stream pattern with gold-bearing gravel-filled channels, now reached only in one or two places. This buried channel system undoubtedly holds in reserve a great wealth for future improved exploration and development. Excessive underground water is always encountered in these mines which are located beneath the level of the alluvial plain.

The great differences between the geology of the Coast Ranges and that of the Sierra and Klamath regions are fundamental in that the western area served frequently as a basin for deposition during the Tertiary and Cretaceous, while the latter represented land areas throughout that time. The Coast Ranges together with the Great Valley now contain enormous accumulations of marine Tertiary and Cretaceous sediments, while the Sierra Nevada and Klamath Mountains are not so covered. Cretaceous and Tertiary streams coursing down the mountain flanks brought gravel, sands, and clays into a marginal sea.

The very fact that streams are conveyors of materials, in contrast to the basins of deposition toward which they flow, accounts for the very different geologic conditions on the two sides of the Great Valley. Certain geologic time divisions of the Cretaceous and Tertiary of the Coast Ranges are represented by strata measured in many thousands of feet, while in the Sierra mere films of Tertiary gravels, or deposits of no greater thickness than a few hundred feet, trapped by volcanic coverings, represent some of these same later periods. Particles of gold, recurrently washed from the mineralized rocks of the mountain range, were dropped, by reason of their high specific gravity, and retained in the bedrock riffles of both the ancient and modern streams, while the lighter detritus was carried to the broad sea basins to form strata covering hundreds of square miles.

Conclusion

The depletion of the more accessible and more easily discovered gold placers, followed by losses due to poorly directed exploration, calls for a more effective technique to bring further success to placer mining. The technique is available; the next thing to do is to apply it.

First, there is aerial photography which may speedily and accurately give a wealth of valuable information as regards geology, and in addition, the finest sort of a map showing surface features in greatest detail.

Second, there is geophysical surveying which, when coordinated with geology, may greatly aid underground prospecting in making new discoveries and in reducing its cost by more intelligently directing its course of action.

Third, physiographic geology, advanced to a more systematic science than ever before, may be used in unravelling the history of the ancient streams and their corresponding topography. Contouring the pre-lava surface is found to be an excellent method of showing graphically this ancient topography, and especially the old valleys in which lay the early gold-bearing streams.

Fourth, a better understanding of desert processes in general and desert placers in particular should help to develop a gold reserve which has so far not received the attention it deserves.

Fifth, the technique recently developed in the examination of stratified sediments, their structure, texture, mineral-grain composition, etc., may be aptly applied to placers, to aid in tracing out their origin and the courses of the older drainage systems now extinct.

In taking stock of the possible reserves of placer gold in California, several sources would seem worth investigating. All of these require detailed exploration prior to any attempt at mining. For the most part, these reserves are buried or concealed in such a way that they have either been overlooked or considered too remote or too much of a speculation for a mining venture. Such factors as involved water-rights, litigation, difficulty in gaining title, laws unfavorably affecting hydraulic mining, lack of sufficient capital, and many other stumbling blocks now prevent good placer ground from being worked.

The possible reserves discussed in this report may be summarized as follows:

Pleistocene and Recent Placers

1. Deep river gravel deposits, over which the present larger rivers are now flowing. Recent and Pleistocene faulting caused gravel to be accumulated on the down-throw side of faults, while the rivers have continued to flow over the gravels without washing them completely out.

2. Pleistocene stream placers, buried beneath alluvial fans of the Great Basin and Mojave Desert.

3. Recent ephemeral stream deposits and alluvial fans or 'bajada placers' of the Great Basin and Mojave Desert regions.

4. Marine or beach placers along the coast, for the most part located in northern California.

5. Isolated high terraces or bench gravels, such as those which occur in the Klamath Mountains.

Tertiary Stream Placers

6. Gold-bearing channels cut in bedrock which lie beneath the false bedrock layers of the dredged areas along the western foot of the Sierra Nevada.

7. Buried Tertiary channels and associated covered benches located in the well-known gold-bearing districts of the state. Large areas still lie buried and unexplored in some of the older mining districts.

8. Buried Tertiary channels and benches in the lava-covered district which lies between the Sierra Nevada and Klamath Mountains. Most of this area is probably too deeply covered to be reached by mining, but the southern marginal area may have some possibilities.

9. Tertiary channels of the Great Basin and Mojave Desert areas, interbedded with volcanic rocks or lying beneath them.

10. Tertiary marine placers. Finely divided gold particles in the Ione formation at the point where the corresponding Eocene streams entered the Ione sea.

Cretaceous Marine Placers

11. Cretaceous marine placers, largely in the Chico conglomerate (Upper Cretaceous) beds of northern California. The Lower Cretaceous beds are also reported to contain some gold-bearing layers.

Largest Reserve

The largest of these possible reserves probably lie in the remaining buried Tertiary stream placers of the northern Sierra Nevada.

Bibliography of California Placers and Related Geological Subjects

·Allen, Victor—The Ione formation of California: Univ. Cal. Pub., Bull. Dept. Geol. Sci., Vol. 18, 1929, pp. 347–448
Alling, Mark N.—Ancient auriferous gravel channels of Sierra County, California: Amer. Inst. Min. Eng., Bull. 91, 1914, pp. 1709–1728; Trans. 49, 1915, pp. 238–257.
    Ancient river-bed deposits in California: Pacific Min. News, Eng. & Min. Jour. Press, vol. 1, 1922, pp. 134–140, 161–166.
    Geologic chart appertaining to the ancient river beds of California: Cal. State Min. Bur. Bull. 92, 1923, plate in pocket.
Anderson, Chas. A.—The Tuscan formation of Northern California with a discussion concerning the origin of volcanic breccias: Univ. Cal. Publ., Bull. Dept. Geol. Sci., vol. 23, 1933, pp. 215–276.
Anderson, Frank Marion—The physiographic features of the Klamath Mountains: Jour. Geol., vol. 10, 1902, pp. 144–159.
    Physiography and geology of the Siskiyou Range: (Abstract) Jour. Geol., vol. 11, 1903, p. 100.
    Jurassic and Cretaceous divisions in the Knoxville-Shasta succession of California: State Mineralogist's Rpt., XXVIII, 1932 (auriferous conglomerates), p. 326.
Anonymous—Mastodon tooth in Amador County: Amer. Jour. Sci., (2), vol. 34, 1862, pp. 135–140.
    A great gravel mining enterprise: Min. & Sci. Press, vol. 30, 1875, p. 70.
    The Inter-Yubas ridge: Min. & Sci. Press, vol. 32, 1876, p. 312.
    Placer County Mines: Min. & Sci. Press, vol. 33, 1876, pp. 33, 42, 65, 74.
    The mining interests of 1876: Min. & Sci. Press, vol. 34, 1877 (hydraulic mining), p. 41.
    Theory of auriferous gravel channels: Min. & Sci. Press, vol. 41, 1880, p. 226.
    Spring Valley mine: Min. & Sci. Press, vol. 42, 1881, pp. 441, 452.
    Pocket mining and nuggets: Min. & Sci. Press, vol. 44, 1882, p. 190.
    Dredging for gold: Min. & Sci. Press, vol. 45, 1882, p. 24.
    Gravel: Min. & Sci. Press, vol. 45, 1882, p. 328.
    Measurements of auriferous earth: Min. & Sci. Press, vol. 45, 1882, p. 328.
    The discovery of gold in California: Min. & Sci. Press, vol. 50, 1885, p. 37.
    Auriferous gravel in Placer County: Min. & Sci. Press, vol. 51, 1885, p. 165.
    New placer diggings: Min. & Sci. Press, vol. 55, 1887, p. 228.
    River system of California: Min. & Sci. Press, vol. 55, 1887, p. 228.
    Buried trees in gravel mines: Min. & Sci. Press, vol. 62, 1891, p. 339.
    Geology of Placer, El Dorado, and Amador Counties: Min. & Sci. Press, vol. 70, 1895, pp. 308–310.
    California's supposed great rivers of ancient times: Min. & Sci. Press, vol. 77, 1898, p. 401.
    Prehistoric rivers of California: Min. & Sci. Press, vol. 79, 1899, p. 544.
    Old channel placers: Min. & Sci. Press, vol. 82, 1901, pp. 175–176.
    Gold dredging in California: Min. & Sci. Press, vol. 91, 1905, pp. 125–126, 141–142, 160–161, 178–179.
    Hydraulic Mining in Humboldt County, California: Eng. & Min. Jour., vol. 79, 1905, p. 362.
    Hydraulicking in Trinity County, California: Min. & Sci. Press, vol. 101, 1910, p. 143.
    Hydraulicking at La Grange Mine, Trinity County: Eng. & Min. Jour., vol. 95, 1913, pp. 1005–1007.
    Map of black sand deposits: Min. & Sci. Press, vol. 118, 1919, p. 264.
    Unique placer development, River Placers, Ltd., engaged in unusual project on Middle Fork of the Yuba River: Min. Jour. (Ariz.), vol. 14, no. 23, 1931, p. 4.
Aubury, Lewis E.—Gold dredging in California: (Introductory) Calif. State Min. Bur. Bull. 57, 1910, pp. xiii–xiv.
Averill, Chas. Volney—Preliminary report on economic geology of the Shasta quadrangle: State Mineralogist's Rept. XXVII, 1931, pp. 56–60.
    Gold deposits of the Redding and Weaverville quadrangles: State Mineralogist's Rept. XXIX, 1933, pp. 2–73.
Ayres, William O.—The ancient man of Calaveras (California): Amer. Nat., vol. 16, 1882, pp. 845–854.
    Gravel deposits of the lower Klamath: Min. & Sci. Press, vol. 50, 1885, p. 394.

Becker, George F.—Structure of a portion of the Sierra Nevada of California: Bull. Geol. Soc. Amer., vol. 2, 1891, pp. 49–74.

Antiquities from under Tuolumne Table Mountain in California (with discussion by G. F. Wright) : Geol. Soc. Amer., Bull. 2, 1891, pp. 189–198.

The Witwatersrand Banket with notes on other gold-bearing pudding stones : U. S. Geol. Survey, 18th An. Rept., 1896 (California), pp. 32–33.

Auriferous conglomerates of the Transvaal: Amer. Jour. Sci, (4), vol. V, 1898, pp. 193–208 (including notes on California).

Becker, George F., Lindgren, W., and Turner, H. W.—Description of the gold belt: U. S. Geol. Survey, Sonora folio, 41, 1897 ; Bidwell Bar folio, 43, 1898 ; Downieville folio, 37, 1897.

Blackwelder, Eliot—Pleistocene glaciation in the Sierra Nevada and the Basin Ranges: Geol. Soc. Am. Bull., vol. 42, 1931, pp. 865–922.

Glacial and associated stream deposits of the Sierra Nevada : State Mineralogist's Rept. XXVIII, 1932, pp. 303–310.

Eastern slope of the Sierra Nevada : XVI Nat. Geol. Congress Guidebook 16, Ex. C-1, 1933, pp. 81–102.

Blake, Theodore A.—Remains of the Mammoth and Mastodon in California : Am. Jour. Sci. (2), vol. 19, 1855, p. 133.

Blake, William P.—On fossils in the auriferous rocks of California: Am. Jour. Sci. (2), vol. 43, 1867, pp. 270–271.

Note on a large lump of gold found on the middle fork of the American River : Calif. Acad. Nat. Sci., Proc. 3, 1868, p. 166.

1. New locality of fossils in the gold-bearing rocks of California : pp. 289–290. 2. Tooth of the extinct elephant, Placer County : p. 290. 3. Quarry of gold-bearing rock : pp. 290–291. Calif. Acad. Nat. Sci., Proc. 3, 1868, pp. 289–291.

On a fossil tooth from Table Mountain : Am. Jour. Sci. (2), vol. 50, 1870, pp. 262–263.

The various forms in which gold occurs : Rept., Director of the Mint, 1884, p. 573–597 (1885).

Bordeaux, Albert F. J.—Les anciens chenaux auriferes de Californie : An. Mines (10), vol. 2, 1902, pp. 217–258.

Boutwell, J. M.—The Calaveras skull (shown to be of recent origin) : U. S. Geol. Survey, Prof. Paper 73, 1911, pp. 54–55.

Bowie, A. J.—Hydraulic mining in California: Amer. Inst. Min. Eng., Trans. 6, 1879, pp. 27–100.

Mining debris in California rivers: Trans. Tech. Soc. Pac. Coast, vol. IV, Feb.-March, 1887.

A practical treatise on hydraulic mining in California: New York (Fifth Edition), 1893.

Bradley, Walter W.—Renewed activity in California gold mining: Min. & Met., vol. 13, 1932, pp. 385–390.

Itinerary, Yosemite to Mother Lode : XVI Int. Geol. Congress, Guidebook 16, Ex. C-1, 1933, pp. 62–65.

An echo of the days of '49 : Eng. & Min. Jour., vol. 135, 1934, pp. 494–496.

Brewer, William H.—Alleged discovery of an ancient human skull in California: Am. Jour. Sci. (2), vol. 42, 1866, p. 424.

Brooks, E. F.—Platinum in California : Min. & Sci. Press, vol. 114, 1917, p. 116.

Brown, C. J.—Hydraulic gold mining : Min. & Sci. Press, vol. 31, 1875, pp. 50, 114, 161, 178, 313, 316 ; vol. 32, 1876, pp. 50, 89, 121.

Browne, John Ross—Report on the mineral resources of the States and Territories west of the Rocky Mountains : U. S. Treasury Dept., 1868.

Resources of the Pacific slope : New York, 1869.

Browne, John Ross, and Taylor, James W.—Reports upon the mineral resources of the United States : U. S. Treasury Dept., 1867.

Browne, Ross E.—The ancient river beds of the Forest Hill divide : State Mineralogist's Rept. X, 1890, pp. 435–465. With maps, sections, and plates.

The channel system of the Harmony Ridge, Nevada County : State Mineralogist's Rept. XII, 1894, p. 202.

California placer gold : Eng. & Min. Jour., vol. 59, 1895, pp. 101–102.

Gold in ancient California river channels : Min. & Sci. Press, vol. 77, 1898, pp. 107–108 ; (Mysteries of the ancient rivers of the Forest Hill divide, Placer Co., Calif.) : vol. 78, pp. 285–290.

Bryan, Kirk—Geology and ground-water resources of Sacramento Valley, California: U. S. Geol. Survey Water-Supply Paper 495, 1923, 285 pp.

Bush, Edgar—Active placers in southern California : Min. Jour. (Ariz.), vol. 16, 1932, p. 7.

Casey, Thos. Lincoln—Mining debris, California : 51st Congress, House of Rep., Ex. Doc. 267, 1891.

Cassidy, Andrew—The California Blue Lead in Oregon and Washington Territory : Min. & Sci. Press, vol. 30, 1875, p. 82.

Chaney, Ralph W.—Age of the auriferous gravels: (Abstracts) Bull. Geol. Soc. Am., vol. 43, 1932, pp. 226–227 ; Pan. Am. Geol., vol. 55, 1931, p. 361 ; vol. 56, 1931, pp. 71–72.

Notes on occurrence and age of fossil plants found in the auriferous gravels of Sierra Nevada : State Mineralogist's Rept. XXVIII, 1932, pp. 299–302.

Further evidence regarding the age of the auriferous gravels : (Abstract) Bull. Geol. Soc. Am., vol. 44, 1933, p. 78.

Chase, A. W.—The auriferous gravel deposit of Gold Bluffs : Am. Jour. Sci. (3), vol. 7, 1874, p. 379.

Cleaveland, Newton—Giant gold dredges at Natoma, California : Min. & Sci. Press, vol. 103, 1911, pp. 446–448.

Crampton, F. A.—Auriferous gravel channels in Nevada County : Min. Jour. (Ariz.), vol. 16, no. 6, 1932, pp. 5–7.

Cranston, R. E.—Mechanical features of the California gold dredge : Min. & Sci. Press, vol. 104, 1912, pp. 303–307, 338–342, 372–375.

Cronise, Titus Fey—The natural wealth of California: H. H. Bancroft & Co., San Francisco, 1868, pp. 416–433, 531–547.

Crossman, J. H.—Auriferous gravels of San Gabriel Range, San Bernardino County: State Mineralogist's Rept. IX, 1888–1889, pp. 236–237.

Davidson, George—On the auriferous gravel deposits of California: Calif. Acad. Sci., Proc. 5, 1873, pp. 145–146.

Gravels in Placer County, California: Calif. Acad. Sci., Proc. 5, 1875, p. 41.

Davis, Horace—On auriferous gravels near Sonora, California: Calif. Acad. Nat. Sci., Proc. 1, 1855, p. 61; 2d edition, 1873, p. 62.

Day, David T.—Platinum: Min. & Sci. Press, vol. 78, 1899, p. 88.

Notes on the occurrence of platinum in North America: Trans. Am. Inst. Min. Eng., vol. 30, 1900, p. 702.

Day, David T., and Richards, R. H.—Useful minerals in the black sands of the Pacific slope: U. S. Geol. Survey, Min. Res. for 1905 (1906), pp. 1175–1258.

De Groot, Henry—Hydraulic and drift mining: State Mineralogist's Rept. II, 1882, pp. 133–190.

Recollection of California mining life: Min. & Sci. Press, vol. 47, 1883, pp. 292, 320–321, 330, 346, 382.

The San Francisco ocean placer—the auriferous beach sands: State Mineralogist's Rept. IX, 1890, pp. 545–547.

De La Bouglisse—Note sur les mines d'or de Golden River (California): 46 pp., Paris, 1885.

De Launay, L.—Geological description of the gold mines of the Transvaal (Witwatersrand, Heidelberg, and Kerksdorp districts): Trans. Inst. Min. Eng., vol. 11, 1896 (California, p. 434).

Alluvions auriferes de Californie: In L. De Launay, Traite de Metallogenie Gites Mineraux et Metalliferes, vol. 3, Paris et Liege, 1913, pp. 705–712.

Dickerson, Roy E.—The Ione formation of the Sierra Nevada foothills, a local facies of the upper Tejon-Eocene: Science (n. s.), vol. 40, 1914, pp. 67–70.

Diller, J. S.—Notes on the geology of northern California: U. S. Geol. Survey Bull. 33, 1886.

Geology of the Lassen Peak district: U. S. Geol. Survey, 8th An. Rept., 1889, pp. 395–432.

Geology of the Taylorsville region of California: Bull. Geol. Soc. Am., vol. 3, 1892, pp. 369–394.

Cretaceous and early Tertiary of northern California and Oregon: Bull. Geol. Soc. Am., vol. 4, 1893, pp. 205–224.

. . . on the auriferous gravel of lacustral origin in the region of Taylorsville, California: Am. Jour. Sci. (3), vol. 46, 1893, pp. 398–399.

Revolution in the topography of the Pacific Coast since the auriferous gravel period: Jour. Geol. vol. 2, 1894, pp. 32–54.

Tertiary revolution in the topography of the Pacific coast: U. S. Geol. Survey, 14th An. Rept., pt. 2, 1894, pp. 403–433.

Lassen Peak folio (No. 15): U. S. Geol. Survey, 1895.

Geomorphogeny of the Klamath Mountains: (Abstract) Geol. Soc. Am., Bull. 12, 1901, p. 461.

Topographic development of the Klamath Mountains: U. S. Geol. Survey Bull. 196, 1902.

Mineral resources of the Indian Valley region, California: U. S. Geol. Survey, Bull. 260, 1904, pp. 45–49.

Redding folio (No. 138): U. S. Geol. Survey, 1906.

Geology of the Taylorsville region, California: U. S. Geol. Survey, Bull. 353, 1908, p. 128.

The auriferous gravels of the Trinity River Basin, California: U. S. Geol. Survey, Bull. 470, 1911, pp. 11–29.

Trinity River gravels, California: Eng. & Min. Jour., vol. 92, 1911, pp. 495–497.

Auriferous gravels in the Weaverville quadrangle, California: U. S. Geol. Survey, Bull. 540, 1914, pp. 11–21.

Dolbear, S. H.—Dry placer mining in California: Eng. & Min. Jour., vol. 89, 1910, p. 359.

Donnelly, Mauriee—Geology and mineral deposits of the Julian district, San Diego County, California: State Mineralogist's Rept. XXX, 1934 (placer deposits), p. 369.

Doolittle, J. E.—Gold dredging in California: Cal. State Min. Bur., Bull. 36, 1905.

Dragaje de oro en California: Bol. Soc. Nac. Min. (3), vol. 18, 1907, pp. 409–428.

Dron, R. W.—Gold mining in the Sierra Nevada, California: Geol. Soc. Glasgow, Trans. 11, 1900, pp. 265–266.

Duling, John F.—Geophysics as an aid in gold placer drift mining: Min. Jour. (Ariz.), vol. 18, Mar. 30, 1935, pp. 5–6.

Dunn, R. L.—Drift mining in California: State Mineralogist's Rept. VIII, 1888, pp. 736–770.

River mining: State Mineralogist's Rept. IX, 1890, pp. 262–281.

Auriferous conglomerate in California (Siskiyou County): State Mineralogist's Rept. XII, 1893–1894, pp. 459–471.

Eddy, L. H.—Dredges on upper American River: Eng. & Min. Jour., vol. 93, 1912, pp. 997–1000.

Edman, J. A.—Notes on the gold bearing black sands of California: Min. & Sci. Press, vol. 69, 1894, pp. 294, 356, 372.

Platinum minerals of Plumas Co., California: Min. & Sci. Press, vol. 77, 1898, p. 401.

The auriferous black sands of California: Cal. State Min. Bur. Bull. 45, 1907, pp. 5–10.   Eng. & Min. Jour., vol. 83, 1907, pp. 1047–1048.

Effner, A. E.—Beach mining with surf washer: Min. & Sci. Press, vol. 86, 1903, p. 364.

Egleston, T.—The formation of gold nuggets and placer deposits: Trans., Amer. Inst. Min. Eng., vol. 9, 1881, pp. 633–646.

Working placer deposits in the United States: School Mines Quart., vol. 7, 1886, pp. 101–131.

Drift mining: School Mines Quart., vol. 8, 1887, pp. 204–220.

Ellsworth, Elmer W., Tracing buried-river channel deposits by geomagnetic methods: State Mineralogist's Rept. XXIX, 1933, pp. 244–250.
Endlich, F. M.—Mining in the Mojave Desert in California (Goler Dry Placers): Eng. & Min. Jour., vol. 62, 1896, pp. 197–198.
Eteson, A. C.—Suggestions on inland gold dredging: Min. & Sci. Press, vol. 81, 1900, p. 597; vol. 82, 1901, p. 36.
Fairbanks, Harold W.—Geology of the Mother Lode region: State Mineralogist's Rept. X, 1890, pp. 23–90, map.
        Red Rock, Goler, and Summit mining districts in Kern County: State Mineralogist's Rept. XII, 1893–1894, pp. 456–458.
        Auriferous conglomerate in California: Eng. & Min. Jour., vol. 59, 1895, pp. 389–390.
        Red Rock, Goler, and Summit mining districts in Kern County: Min. & Sci. Press, vol. 70, 1895, pp. 241, 245.
Ferguson, H. G., and Gannett, R. W.—Gold quartz veins of the Alleghany district, California: U. S. Geol. Survey Prof., Paper 172, 1932.
Foster, G. G.—The gold regions of California: New York, 1848.
Franke, Herbert A.—Selected bibliography on placer mining: State Mineralogist's Rept. XXVIII, 1932, pp. 219–224; XXX, 1934, pp. 283–289.
Fuchs, Edmond—Note sur les graviers auriferes de la Sierra Nevada de California: Soc. Geol., France, Bull. (3) 13, 1885, pp. 486–488.
Fuchs, Edmond, and de Launay, L.—Alluvious auriféres de California: In Traité des Gêtes Minéraux et Méialliféres, vol. 2, Paris, 1893, pp. 961–969.
Gardner, E. D., and Johnson, C. H.—Placer mining in western United States: U. S. Bur. Min., Inf. Cir. 6787, 1934, pp. 1–89.
Gibson, Arthur—Magnetometric determinations applied to placer mining: Eng. & Min. Jour. Press, vol. 114, 1922, pp. 1064–1069. (California.)
        Auriferous placer resources of California: Min. Cong. Jour., vol. 13, 1927, pp. 476–479.
Gilbert, G. K.—Stages of geologic history of Sierra Nevada: Wash. Phil. Soc., Bull. 9, 1887, p. 7.
        Terraces of the high Sierra, California: (Abstract) Science, (n.s.), vol. 21, 1905, p. 822.
        Transportation of detritus by Yuba River: (Abstract) Bull. Geol. Soc. Am., vol. 19, 1908, pp. 657–659.
        Quantity of mining debris: U. S. Geol. Survey, Prof. Paper 73, 1911, pp. 18–21.
        Hydraulic mining débris in the Sierra Nevada: U. S. Geol. Survey, Prof. Paper 105, 1917, 154 pp.
Gledhill, Edward—The analogy between the gold "cintas" of Colombia and the auriferous gravels of California: Trans., Inst. Min. Eng., vol. 20, 1902, pp. 391–400.
Goldstone, L. P.—Fresno, Sierra, and Tuolumne Counties: State Mineralogist's Rept. X, 1890, pp. 734–738.
Goldsmith, E.—The blue gravel of California: Acad. Nat. Sci., Phila., Proc. 1874, pp. 73–74.
Goodyear, W. A.—The auriferous gravels of California: Min. & Sci. Press, vol. 39, 1879, pp. 182–183.
Griffin, F. W.—The gold dredging industry: Min. & Sci. Press, vol. 88, 1904, pp. 260–261.
        Recent developments in gold dredging: Min. & Sci. Press, vol. 97, 1908, pp. 219–223.
Gruetter, F. W.—Platinum on the Pacific Coast: Min. & Sci. Press, vol. 113, 1916, pp. 20–21.
Haley, Charles S.—Dry placers of southern California: State Mineralogist's Rept. XVIII, 1922, pp. 321–324.
        Progress report on placer gold investigation: State Mineralogist's Rept. XVIII, 1922, pp. 373–374.
        Tertiary sluice robbers: State Mineralogist's Rept. XVIII, 1922, pp. 550–553.
        Gold placers of California: Cal. State Min. Bur., Bull. 92, 1923, 167 pp. including maps and charts.
        Primary and secondary gold concentrations: State Mineralogist's Rept. XIX, 1923, pp. 38–40.
Hall, C. L.—The gravel fields of Northern California: Min. & Sci. Press, vol. 74, 1897, p. 113.
Hall, Frank H.—Ancient gravels, Siskiyou County, California: Min. & Sci. Press, vol. 66, 1893, p. 85.
Hammond, John Hays—The auriferous gravels of California: In H. C. Burchard Report of the U. S. Director of the Mint upon the statistics of the production of the precious metals in the United States for the Calendar Year 1881, pp. 616–630 (1882).
        Auriferous gravels of California; geology of of their occurrence and methods of their exploitation: State Mineralogist's Rept. IX, 1888–1889, pp. 105–138.
Hanks, H. G.—Placer, hydraulic, and drift mining: State Mineralogist's Rept. II, 1882, pp. 28–192.
        Placer Gold: In H. C. Burchard, Report of the U. S. Director of the Mint upon the statistics of the production of the precious metals in the United States for the Calendar Year 1882, pp. 728–732 (1883).
        Gold in river channels: Min. & Sci. Press, vol. 58, 1889, p. 485.
        Deep gold placers of California: Min. & Sci. Press, vol. 60, 1890, pp. 231, 237, 249, 255, 264, 271, 280, 296, 314, 315, 330, 331, 337, 347, 353, 361, 362, 378, 384.
        The deep-lying auriferous gravels and table mountains of California: San Francisco, 1901, 15 pp.
Hanna, G. Dallas—Mammoth tusks found near Oroville, California: State Mineralogist's Rept. XXV, 1929, pp. 88–90.
Hay, Oliver P.—The geological age of the Tuolumne Table Mountain, California: Wash. Acad. Sci. Jour., vol. 16, 1926, pp. 358–361.
        The Pleistocene of the western region of North America and its vertebrated animals: Carnegie Inst. Wash., Publ. No. 322 B, 1927.

Heiland, C. A., and Courtier, W. H.—Magnetometric investigations of gold placer deposits near Golden, Colorado: Am. Inst. M. & M. Eng., Geophysical Prospecting, 1929, pp. 364–384. (Nothing on California.)

Hershey, O. H.—Neocene deposits of the Klamath region, California: Jour. Geol., vol. 10, 1902, pp. 377–392.

A supposed early Tertiary peneplain in the Klamath region, California: Science (n. s.), vol. 15, 1902, pp. 139–156.

The significance of certain Cretaceous outliers in the Klamath region, California: Am. Jour. Sci. (4), vol. 14, 1902, p. 33.

The relation between certain river terraces and the glacial series in northwestern California: Jour. Geol., vol. 11, 1903, pp. 431–458.

Certain river terraces of the Klamath river region, California: Amer. Jour. Sci. (4), vol. 16, 1903, pp. 240–250.

The river terraces of the Orleans Basin, California: Univ. Cal., Bull. Dept. Geol. Sci., vol. 3, 1904, pp. 423–475.

Hill, James M.—The mining districts of the western United States: Introduction by W. Lindgren: U. S. Geol. Survey, Bull. 507, 1912 (California), pp. 5–43.

The Los Burros district, Monterey County, California: U. S. Geol. Survey, Bull. 735, 1923, pp. 323–329.

California gold production, 1849–1923: Econ. Geol., vol. 21, 1926, pp. 172–179.

Historical summary of gold, silver, copper, lead, and zinc products in California, 1848–1926: U. S. Bur. Min. Ec., Paper 3, 1929.

Hinds, N. E. A.—Geologic formations of the Redding-Weaverville districts, Northern California: State Mineralogist's Rept. XXIX, 1933, pp. 114–122.

Late Cenozoic history of southern Klamath Mountains: (Abstract) Pan. Am. Geol., vol. 51, 1934, pp. 315–316.

Mesozoic and Cenozoic eruptive rocks of the southern Klamath Mountains, California: Univ. Cal. Publ., Bull. Dept. Geol. Sci., vol. 23, 1935 (Tertiary), pp. 367–377.

Hinds, N. E. A., and Russell, D. R.—Ione formation of the Redding district: (Summary statement) Pan. Am. Geol., vol. 51, 1929, p. 375.

Hitchcock, C. H.—The Calaveras skull: Eng. & Min. Jour., vol. 9, 1870, pp. 345–346.

Hittell, John S.—The dead rivers of California: Overland Monthly, vol. 1, 1868, pp. 430–435.

Dead rivers of California: In R. W. Raymond, Second Report on Mineral Resources of the States and Territories west of the Mississippi, 1870, pp. 63–67.

Hobson, J. B.—Placer County: State Mineralogist's Rept. X, 1890, pp. 410–434.

Hoffman, Charles F.—The Red Point drift gravel mine: Trans. Tech. Soc. Pacific Coast, vol. 10, No. 12, San Francisco, 1894, pp. 291–307.

Holmes, W. H.—Review of the evidence relating to auriferous gravel man in California: Amer. Anthropologist (n. s), vol. 1, 1899, pp. 107–121, 614–645; Smiths. Inst. An. Rept., 1899, pp. 419–472 (1901).

Holmes, W. H., and McGee, W. J.—Geology and archeology of the California gold belt: (Abstract) Am. Geol., vol. 23, 1899, pp. 96–99.

Horner, R. R.—Notes on the black sand deposits of southern Oregon and northern California: U. S. Bur. Min., Tech. Paper 196, 1918.

Hubbard, J. D.—Drift mining in California: Eng. & Min. Jour., vol. 104, 1917, pp. 863–866.

Hulin, Carlton D.—Geologic features of the dry placers of the northern Mojave Desert; State Mineralogist's Rept. XXX, 1934, pp. 417–426.

Hurst, G. L.—Resoiling after dredging in California: Min. & Sci. Press, vol. 107, 1913, pp. 719–720.

Hutchins, J. P.—The nomenclature of modern placer mining: Eng. & Min. Jour., vol. 84, 1907, pp. 293–296.

Irvine, C. D.—The beach placers of the south Pacific Coast: Min. World, vol. 29, 1908, pp. 321–322.

Jackson, C. F., and Knaebel, J. B.—Small-scale placer mining methods: U. S. Bur. Mines, Inf. Cir. 6611, 1932.

Jacobs, H. S.—Ancient river channels (California): Min. & Sci. Press, vol. 34, 1877, p. 264.

Jakosky, J. J.—Practical aspect of geophysical surveys: Min. Jour. (Ariz.), vol. 14, No. 16, 1931, pp. 7–8.

Use of geophysics in gold mining: Min. Jour. (Ariz.), vol. 15, no. 6, 1931, pp. 5–7, 37.

Jakosky, J. J., and Wilson, C. H.—Use of geophysics in placer mining: Min. Jour. (Ariz.), vol. 16, No. 14, 1929, pp. 3–4.

Geophysical studies in placer and water-supply problems: Am. Inst. M. & M. Eng., Tech. Publ. 515, 1934 (Trinity County placers), pp. 6–12.

Jamison, C. E.—Santa Clara Placers: (Los Angeles and Ventura Counties). Min. & Sci. Press, vol. 100, 1910, pp. 360–361.

Janin, Charles—Present day problems in California gold dredging: Min. & Sci. Press, vol. 103, 1911, pp. 474–478.

Present day problems in California gold dredging: Trans., America Inst. Min. Eng., 42, 1912, pp. 855–873.

Placer mining methods and operating costs: U. S. Bur. Min., Bull. 121, 1916, pp. 42–58.

Topography and geology of dredging areas: Min. & Sci. Press, vol. 117, 1918, pp. 763–764.

Gold dredging in the United States: U. S. Bur. Mines, Bull. 127, 1918.

Proposed regulation of gold-dredging: Min. & Sci. Press, vol. 106, 1931, pp. 881–384.

Janin, Charles, and Winston, W. B.—Gold dredging in California: Cal. State Min. Bur., Bull. 57, 1910.

Jenkins, Olaf P.—Report accompanying geologic map of northern Sierra Nevada: State Mineralogist's Rept. XXVIII, 1932, pp. 279–298.

Geologic map of northern Sierra Nevada showing Tertiary river channels and Mother Lode Belt: Cal. State Div. Mines, 1932; later reprinted, with modifications and report accompanying, printed on back of map.

Jenkins, Olaf P. (Cont.)
　　Use of geology in seeking gold: Min. Jour. (Ariz.), vol. 17, June 15, 1933, p. 3.
　　Middle California and Western Nevada: XVI Int. Geol. Congress, Guidebook 16, Ex. C-1, 1933.
　　Current notes: (with notes on geophysical prospecting of buried river channels by William Q. Wright, Jr.), State Mineralogist's Rept. XXX, 1934, pp. 3–4.
　　Resurrection of early surfaces in the Sierra Nevada, a fundamental geologic process involved in placer gold deposition: Cal. State Mineralogist's Rept. XXX, 1934, pp. 5–6.
　　Geologic map of Mother Lode Belt and vicinity (1932): In State Div. Mines Bull. 108, by C. A. Logan, 1934.
Jenkins, Olaf P., and Wright, W. Quinby—California's gold-bearing Tertiary channels: Eng. & Min. Jour., vol. 135, 1934, pp. 497–502.
Kellogg, A. E.—Origin of flour gold in black sands: Min. Jour. (Ariz.), vol. 14, no. 20, 1931, pp. 3–4.
Kemp, J. F.—Platinum and associated metals: U. S. Geol. Survey, Bull. 193, 1902, pp. 51–53.
Kimble, George W.—The Kimble Drift mine, El Dorado County, California: Min. & Sci. Press, vol. 85, 1902, p. 23.
　　Ancient river channels of California: Min. & Sci. Press, vol. 94, 1907, pp. 726–727.
Knopf, Adolph, The Mother Lode system of California: U. S. Geol. Survey, Prof. Paper 157, 1929.
　　Tertiary auriferous gravels: XVI Int. Geol. Congress, Guidebook 16, Ex. C-1, 1933, pp. 60–61.
Knowlton, F. H.—Flora of the auriferous gravels of California: U. S. Geol. Survey, Prof. Paper 73, 1911, pp. 57–58.
Knox, N. B.—Dredging and valuing dredging ground in Oroville, California: Canadian Min. Rev., vol. 22, 1903, pp. 211–213.
Koch, Felix J.—The Calaveras skull: Amer. Antiquarian, vol. 33, 1911, pp. 199–202.
Laizure, C. McK.—Elementary placer mining methods and gold-saving devices: State Mineralogist's Rept. XXVIII, 1932, pp. 112–204.
　　Elementary placer mining in California and notes on the milling of gold ores: State Mineralogist's Rept. XXX, 1934, pp. 121–289.
Lakes, Arthur—The deep leads of California: Colliery Eng., vol. 14, 1894, p. 170; Min. & Sci. Press, vol. 68, 1894, p. 136.
　　Fossilized big trees, California: Sci. Am. Supl. 39, 1895, p. 15862.
　　Placer mining in California: Mines and Minerals, vol. 19, 1899, pp. 297–298.
　　Gold bearing beach sands of California: Mines and Minerals, vol. 19, 1899, p. 369.
　　Calaveras County mines (California): Mines and Minerals, vol. 20, 1899, pp. 198–200.
Lang, Herbert—Black sand of the Pacific Coast: Min. & Sci. Press, vol. 113, 1916, pp. 811–813.
Laur, P.—Terrains auriferes de la Californie: Rv. des deux mondes 33 annee (second period), 1863, pp. 453–472.
Lawson, Andrew C.—The geology of Middle California: XVI Int. Geol. Congress, Guidebook 16, Ex. C-1, 1933, pp. 10–12.
Laylander, K. C.—Magnetometric surveying as an aid in exploring placer ground: Eng. & Min. Jour., vol. 121, 1926, pp. 325–328.
Le Conte, Joseph—Post-Tertiary elevation of the Sierra Nevada shown by the river beds: Am. Jour. Sci. (3), vol. 132, 1886, pp. 167–181.
　　The old river beds of California: Am. Jour. Sci. (3), vol. 132, 1886, pp. 176–190.
　　On the origin of normal faults and of the structure of the Basin region: Am. Jour. Sci. (3), vol. 138, 1889, pp. 257–263.
　　Critical periods in the history of the earth: Univ. Cal., Bull. Dept. Geol. Sci., vol. 1, 1895, pp. 313–336.
Leidy, Joseph—Remarks on vertebrate fossils from Table Mountain, Tuolumne Co., Cal.: Acad. Nat. Sci., Phila., Proc. 1870, pp. 125–227.
　　Notes on Pleistocene mammals in California: In Whitney, J. D., "Auriferous gravels of the Sierra Nevada of California." Harvard Coll., Mus. Comp. Zool., Mem. 6, 1880, pp. 256–258; (another edition) Whitney, J. D., "Contributions to American Geology," vol. 1, 1880, pp. 256–258.
Lesquereux, Leo—Reports on the fossil plants, etc.: Mem. Mus. Comp. Zool. Harvard Coll., vol. 6, No. 2, Cambridge, 1878.
　　The Cretaceous and Tertiary floras: Rept. U. S. Geol. Geog. Survey, Terr., (Hayden), vol. 8, 1883.
Lindgren, Waldemar—Two Neocene rivers of California: Bull. Geol. Soc. Am., vol 4, 1893, pp. 257–298.
　　An auriferous conglomerate of Jurassic age from the Sierra Nevada; Amer. Jour. Sci. (3), vol. 48, 1894, pp. 275–280.
　　Sacramento folio (No. 5); U. S. Geol. Survey, 1894.
　　The gold-quartz veins of Nevada City and Grass Valley districts, California: U. S. Geol. Survey, 17th Ann. Rept., Pt. 2, 1896, pp. 1–262.
　　Nevada City folio (No. 29): U. S. Geol. Survey, 1896.
　　Pyramid Peak folio (No. 31): U. S. Geol. Survey, 1896.
　　Truckee folio (No. 39): U. S. Geol. Survey, 1897.
　　Colfax folio (No. 66): U. S. Geol. Survey, 1900.
　　The gold production of North America, its geological derivation and probable future: Int. Min. Congress, 5th Proc., 1903, pp. 29–36.
　　The geological features of the gold production of North America: Trans. Am. Inst. M. Eng., vol. 33, 1903, pp. 790–845, 1077–1083.
　　Neocene rivers of the Sierra Nevada: U. S. Geol. Survey, Bull. 213, 1903, pp. 64–65.
　　The Tertiary gravels of the Sierra Nevada of California: U. S. Geol. Survey, Prof. Paper 73, 1911.

Lindgren, Waldemar, and Knowlton, F. H.—Age of the auriferous gravels of the Sierra Nevada, with a report on the flora of Independence Hill: Jour. Geol., vol. 4, 1896, pp. 881–906.
Lindgren, Waldemar, and Turner, H. W.—Placerville folio (No. 3): U. S. Geol. Survey, 1894.
    Marysville folio (No. 17): U. S. Geol. Survey, 1895.
    Smartsville folio (No. 18): U. S. Geol. Survey, 1895.
    Reprints from Placerville, Sacramento, and Jackson folios (Nos. 3, 5, 11): U. S. Geol. Survey, 1914.
Lindgren, Waldemar, and others—The production of gold and silver in 1904: U. S. Geol. Survey, 1905; (California by Charles G. Yale, pp. 41–66).
Locke, A.—Tuolumne Table Mountain (near Jamestown, California): Min. & Sci. Press, vol. 105, 1912, p. 85.
Logan, C. A.—Platinum and allied metals in California: Min. Bur., Bull. 85, 1919.
    Auburn field division: Amador, Nevada, Placer, El Dorado, Calaveras, Tuolumne Counties: State Mineralogist's Rept. XIX, 1923, pp. 13–21, 59–60, 94–97, 140–145.
    Sacramento field division: Amador, Nevada, Plumas, Shasta, Sierra, Tuolumne, Mono, Placer, Sacramento, Yuba, Calaveras, El Dorado Counties: State Mineralogist's Rept. XX, 1924, pp. 1–23, 73–84, 177–183, 355–367.
    Sacramento field division: Sacramento, Calaveras, Placer Siskiyou Counties: State Mineralogist's Rept. XXI, 1925, pp. 1–22, 135–172, 275–280, 413–498.
    Ancient channels of Duncan Canyon region, Placer County, California: State Mineralogist's Rept. XXI, 1925, pp. 275–280.
    Sacramento field division; El Dorado County, California: State Mineralogist's Rept. XXII, 1926, pp. 397–452.
    Sacramento field division; Trinity, Shasta, El Dorado Counties: State Mineralogist's Rept. XXII, 1926, pp. 1–67, 121–216, 313–452.
    Sacramento field division; Amador, Placer Counties, California: State Mineralogist's Rept. XXIII, 1927, pp. 131–202, 235–286. 373.
    Sacramento field division; Tuolumne County, California: State Mineralogist's Rept. XXIV, 1928, pp. 3–53.
    Sacramento field division; Sierra County, California: State Mineralogist's Rept. XXV, 1929, pp. 151–212.
    Sacramento field division; Butte County, California: State Mineralogist's Rept. XXIV, 1928, pp. 191–210; XXVI, 1930, pp. 383–407.
    Sacramento field division; Nevada County, California: State Mineralogist's Rept. XXVI, 1930, pp. 90–137.
    Mother Lode gold belt of California: Cal. State Div. Mines, Bull. 108, 1934, including 6 geologic maps.
Louderback, George D.—Morphologic features of the Basin Range displacements in the Great Basin: Univ. Cal., Bull. Dept. Geol. vol. 16, 1826, pp. 1–42.
    Notes on the geologic section near Columbia, California; with special reference to the occurrence of fossils in the auriferous gravels; Carnegie Inst. Wash., Publ. Paleo., vol. 440, 1934, p. 440.
MacDonald, D. F.—The Weaverville-Trinity Centre gold gravels, Trinity County, California: U. S. Geol. Survey, Bull. 430, 1910, pp. 48–59.
MacGinitie, Harry—Ecological aspects of the floras of the auriferous gravels: (Abstract) Bull. Geol. Soc. Am., Proc. 1, 1934, p. 356; (Abstract) Pan. Am. Geol., vol. 42, 1934, pp. 75–76.
Maclaren, J. Malcolm—Gold: Its geological occurrence and geographical distribution: London, 1908 (California, pp. 503–514).
Marcou, Jules—Sur le gisement de l'or en Californie: Arch. Sci. Phys. Nat., vol. 28, 1855, pp. 124–135. Also in his Geology of North America, Zurich, 1858, pp. 81–84.
Matthes, F. E.—Geologic history of the Yosemite Valley: U. S. Geol. Survey, Prof. Paper 160, 1930.
    Geography and geology of the Sierra Nevada: XVI Int. Geol. Congress, Guidebook 16, Ex. C-1, 1933, pp. 26–40.
    Up the western slope of the Sierra Nevada by way of the Yosemite Valley: XVI Int. Geol. Congress, Guidebook 16, Ex. C-1, 1933, pp. 67–81.
Maxson, John H.—Economic geology of portions of Del Norte and Siskiyou Counties, Northwesternmost California: State Mineralogist's Rept. XXIX, 1933, pp. 123–160.
McGillivray, J. J.—The old river beds of the Sierra Nevada of California: In H. C. Burchard, Report of the U. S. Director of the Mint upon the statistics of the production of the precious metals of the United States for the Calendar Year 1881, pp. 630–644 (1882).
Mendell, J. H.—Report upon a project to protect the navigable waters of California from the effects of hydraulic mining: In Repts. Chief Eng. U. S. Army; also House Ex. Doc. No. 98, 47th Cong., 1st sess., 1882.
Merriam, John C., and Stock, Chester—Tertiary mammals from the auriferous gravels near Columbia, California: Carnegie Inst., Wash., Publ. Paleo., vol. 440, 1934, pp. 1–6.
Miscellaneous—Statistical reports by the California State Mining Bureau, later called the Division of Mines.
    California Miners' Association Annuals.
    Mineral Industries, McGraw-Hill.
    Mineral Resources of the United States: U. S. Geol. Survey and later by U. S. Bureau of Mines.
    Débris Commission reports.
Nason, F. L.—The Golder gold diggings: Eng. & Min. Jour., vol. 59, 1890, p. 223.
Oldham, C.—Placer gold discovery in Greenhorn Gulch: Min. Jour. (Ariz.), vol. 15, no. 13, 1931, pp. 3–4.
Pardee, J. T.—Placer deposits of the western United States. In Ore deposits of the Western States: Am. Inst. M. & M. Eng. Publ., 1933, pp. 419–450.
    Beach placers on the Oregon Coast: U. S. Geol. Survey, Cir. 8, 1934, pp. 4–41.
Patmon, C. G.—Methods and costs of dredging auriferous gravels at Lancha Plana, Amador County, California: U. S. Bur. Min. Inf. Cir. 6659, 1932.
Phillips, J. Arthur—A treatise on ore deposits: London, 2d edition, 1896 (California placers, pp. 19–27).

Radford, G. K.—Mining for gold in the auriferous gravels of California: Inst. Min. Eng., Trans. 17, 1900, pp. 452–481.
Ransome, F. L.—Some lava flows of the western slope of the Sierra Nevada, California: U. S. Geol. Survey, Bull. 89, 1898, 74 pp.
       Mother Lode district folio (No. 63): U. S. Geol. Survey, 1900.
Ransome, F. L., and Turner, H. W.—Sonora folio (No. 41): U. S. Geol. Survey, 1897.
       Big Trees folio (No. 51): U. S. Geol. Survey, 1898.
Raymond, Rossiter W.—Mineral resources of the States and Territories west of the Rocky Mountains: U. S. Treasury Dept., 1869.
       Statistics of mines and mining in the States and Territories west of the Rocky Mountains:.U. S. Treasury Dept., 1870–1876.
Reid, J. A.—A Tertiary river channel near Carsön City, Nevada: Min. & Sci. Press, vol. 96, 1908, pp. 522–526.
Rensch, H. E. & E. G., and Hoover, Mildred B.—Historic spots in California: Stanford Press, 1933.
Ricketts, A. H.—Manner of locating and holding mineral claims in California: Cal. State Div. Mines, Bull, 106, 1931 (Placer locations, pp. 9–10).
Richards, R. H., and Day, D. T.—Investigation of the black sands from placer mines: U. S. Geol. Survey, Bull. 285, 1906, pp. 150–164.
Robinson, F. W.—Notes on hydraulic mining: State Mineralogist's Rept. II, 1882, pp. 119–129.
Robinson, L. L.—Impounding reservoir for mining and other debris. House Committee on water rights and drainage, Assembly Bill 451, Feb. 17, 1887.
Romanowitz, C. M., and Young, George J.—Gold dredging: Eng. & Min. Jour., vol. 135, 1934, pp. 486–490.
Rose, T. Kirke—The metallurgy of gold: 5th edition. London, 1906, pp. 45–92.
Russell, Israel C.—Rivers of North America: New York, 1898 (California, pp. 275–278).
Sampson, R. J.—Placers of southern California: State Mineralogist's Rept. XXVIII (Supplement), 1932, pp. 245–255.
Sauvage, Ed.—De l'exploitation hydraulique de l'or en Californie: An Mines (7), vol. 9, 1876, pp. 1–77.
Scupham, J. R.—The buried rivers of California as a source of gold: Mines and Minerals, vol. 19, 1898, pp. 150–152.
Shimmer, G. L.—Ancient river channels of California: Eng. & Min. Jour., vol.·114, 1922, p. 857.
Silliman, B. Jr.—On the deep placers of the South and Middle Yuba, Nevada County, California: Am. Jour. Sci. (2), vol. 40, 1865, pp. 1–19.
       On the existence of the mastodon in the deep-lying gold placers of California: Am. Jour. Sci. (2), vol. 45, 1868, pp. 378–381.
       Notice of the peculiar mode of the occurrence of gold and silver in the foothills of the Sierra Nevada, and especially at Whiskey Hill, in Placer County, and Quail Hill, in Calaveras County, California: The Acad. Nat. Sci., Proc. 3, 1867, pp. 349–351; Am. Jour. Sci. (2), vol. 45, 1869, pp. 92–95.
Simpson, Edward C.—Geology and mineral deposits of the Elizabeth Lake quadrangle, California: State Mineralogist's Rept. XXX, 1934, (placer deposits, p. 409).
Sinclair, W. J.—Recent investigation bearing on the question of the occurrence of Neocene man in the auriferous gravels of the Sierra Nevada: Cal. Univ. Publ. Am. Archaeology and Ethnology, No. 7, 1908, pp. 107–131.
Skidmore, W. A.—Deep placer mining in California: In R. W. Raymond, Third Report on Mineral Resources of the States and Territories west of the Mississippi, 1870, pp. 52–90 (1872).
       Hydraulic mining at Gold Run—The Blue Lead ancient river channel: Min. & Sci. Press, vol. 30, 1875, pp. 168–169.
       Gravel channels of ancient rivers: Min. & Sci. Press, vol. 51, 1885, p. 6.
Smith, A. M., and Vanderburg, Wm. O.—Placer mining in Nevada: Univ. Nev., Bull., vol. 26, no. 8, 1932.
Smith, J. P.—The geologic formations of California, with reconnaissance geologic map: Cal. State Min. Bur., Bull. 72, 1916.
Smyth, H. L.—Origin and classification of placers: Eng. & Min. Jour., vol. 79, 1905, pp. 1045–1046, 1179–1180, 1228–1230.
Sperry, E. A.—Investigation of Feather River black sands: Min. & Sci. Press, vol. 105, 1912, pp. 624–626.
Steffa, Don—Gold mining and milling methods and costs at the Vallecito Western Drift Mine, Angels Camp, California: U. S. Bur. Mines, Inf. Cir. 6612, 1932.
Stewart, J. D.—Hydraulic mining again interesting to capital: Eng. & Min. Jour., vol. 135, 1934, pp. 491–493.
Storms, W. H.—Ancient channel system of Calaveras County: State Mineralogist's Rept. XII, 1894, pp. 482–492.
       Peculiar effect of subterranean corrosion of rocks (Tuolumne Co., Cal.): Min. & Sci. Press, vol. 79, 1899, p. 5.
       Mother Lode region of California: Cal. State Min. Bur., Bull. 18, 1900.
       Ancient gravel channels of Calaveras County, California: Min. & Sci. Press, vol. 91, 1905, pp. 170–171, 192–193.
Taylor, G. F.—The Brandy City hydraulic mine, Sierra County (California): Eng. & Min. Jour., vol. 89, 1910, pp. 1152–1153.
Theller, J. H.—Hydraulicking on Klamath River: Min. & Sci. Press, vol. 108, 1914, pp. 523–526.
Thurman, C. H.—Possibilities of dredging in the Oroville district, California: Min. & Sci. Press, vol. 118, 1919, pp. 257–258.
Toleman, P.—Auriferous gravels of the counties: The Nevada City Nugget, Special Mining Number, May 10, 1929, pp. 19–21.
Trask, John B.—Report of First State Geological Survey, Sacramento, 1853.
       Report on the geology of the Coast Mountains and part of the Sierra Nevada: Calif. Assembly Doc. No. 9, Session of 1854.
       Report on the geology of northern and southern California: Calif. Assembly Doc. 14, Session of 1856.

Tucker, W. B.—Notes on mining activity in Inyo and Mono Counties in July, 1931: State Mineralogist's Rept. XXVII, 1931, pp. 543–546.
    South of the Tehachapi gold mining makes new gain: Eng. & Min. Jour., vol. 135, 1935, pp. 517–521.
Tucker, W. B., and Sampson, R. J.—Los Angeles Field Division; San Bernardino County: State Mineralogist's Rept. XXVI, 1930, pp. 221–260; XXVII, 1931 (Gold), pp. 280–333.
    Los Angeles Field Division; Ventura County (Gold): State Mineralogist's Rept. XXVIII, 1922, pp. 253–257.
    Gold resources of Kern County, California: State Mineralogist's Rept. XXIX, 1933, pp. 271–339.
Turner, H. W.—Mohawk lake beds: Bull. Wash. Phil. Soc., vol. 11, 1891, pp. 385–410.
    Jackson folio (No. 11): U. S. Geol. Survey, 1894.
    The rocks of the Sierra Nevada: U. S. Geol. Survey, 14th An. Rept., pt. 2, 1894, pp. 441–495.
    Auriferous gravels of the Sierra Nevada: Am. Geol., vol. 15, 1895, pp. 371–379.
    The age and succession of the igneous rocks of the Sierra Nevada: Jour. Geol., vol. 3, 1895, pp. 385–414.
    Further contributions to the geology of the Sierra Nevada: U. S. Geol. Survey. 17th An. Report, pt. 1, 1896, pp. 521–762.
    Bidwell Bar folio (No. 43): U. S. Geol. Survey, 1898.
    The occurrence and origin of diamonds in California: Am. Geol., vol. 23, 1899, pp. 182, 191.
    Post-Tertiary elevation of the Sierra Nevada: Bull. Geol. Soc. Am., vol. 13, 1903, pp. 540–541.
    Cretaceous auriferous conglomerate of the Cottonwood Mining district, Siskiyou County, California: Eng. & Min. Jour., vol. 76, 1903, pp. 653–654.
Turner, H. W., and Ransome, F. L.—Sonora folio (No. 41): U. S. Geol. Survey, 1897.
    Big Trees folio (No. 51): U. S. Geol. Survey, 1898.
von Bernewitz, M. W.—Dredging and resoiling: Min. & Sci. Press, vol. 118, 1919, p. 471.
Weatherbee, D'Arcy—A hydraulic mine in California: Min. & Sci. Press, vol. 93, 1906, pp. 296–298.
    Dredging and horticulture: Min. & Sci. Press, vol. 94, 1907, p. 151.
Whitney, J. D.—Geological Survey of California, Geology: vol. 1, Philadelphia, 1865.
    Notice of a human skull recently taken from a shaft near Angels, Calaveras County, California: Calif. Acad. Nat. Sci., Proc. 3, 1867, pp. 277–278.
    Sur les amas detritique de la Californie: Soc. Geol., France, Bull. (2) 24, 1867, pp. 624–625.
    The auriferous gravels of the Sierra Nevada: Mem. Mus. Comp. Zool. Harvard Coll., vol. 6, 1879, p. 659. Includes chapters by W. A. Goodyear and W. H. Pettee.
    The climatic changes of later geological times: Mem. Mus. Comp. Zool. Harvard Coll., vol. 7, pt. 2, 1884, p. 394.
Wiel, S. C.—The ancient channel of Gibsonville, California: Min. & Sci. Press, vol. 91, 1905, p. 73.
Williams, Howel—Geology of the Marysville Buttes, California: Univ. Cal. Dept. Geol. Sci. Bull., vol. 18, 1929, pp. 103–220.
Wilson, C. H.—Economic value of geophysics in mining: Min. Jour. (Ariz.), vol. 15, no. 21, 1932, pp. 3–5.
Winslow, C. F.—On human remains along with those of *Mastodon* in the drift of California: Boston Soc. Nat. His., Proc. 6, 1857, pp. 278–279.
Wright, G. F.—The lava beds of California and Idaho, and their relation to the antiquity of man: (Abstract) British Asso. Rept. 61, 1892, p. 651.
    Evidence of Glacial man in America: (Abstract) Amer. Geol., vol. 12, 1893, pp. 173–174.
    Recent volcanic eruption in California: Amer. Nat., vol. 27, 1893, pp. 813–816.
    Recent date of lava flows in California: Records of the Past, vol. 4, 1905, pp. 195–198.
    The latest concerning pre-historic man in California: Records of the Past, vol. 7, 1908, pp. 183–187.
Yale, C. G.—Auriferous gravel—a theory of its formation: *In* H. C. Burchard, Report of the U. S. Director of the Mint upon the statistics of the production of the precious metals in the United States for the Calendar Year 1880, pp. 376–379 (1881).
    Mining débris legislation: Cal. Mines Assoc., Mines and Minerals, 1889, pp. 255–262.
    Beach mining in California: U. S. Geol. Survey, Min. Res. for 1912, pp. 253–254 (1913).
    Dry placers of California: U. S. Geol. Survey, Min. Res., pt. 1, 1913, pp. 262–263.
Yeatman, J. A.—The pump in placer mining: Min. & Sci. Press, vol. 88, 1904, p. 226.
Young, George J.—California's gold dredge in operation over twenty years: Eng. & Min. Jour., vol. 121, 1927, pp. 1042–1046.
    Drift mining at Vallecito, Calaveras County, California: Eng. & Min. Jour., vol. 128, 1929, pp. 394–397.
Zadach, Stanley—Placer mining in San Gabriel Canyon: Min. Journ. (Ariz.), vol. 17, no. 3, 1933, p. 3.

SECTION III

PROSPECTING AND SAMPLING PLACER DEPOSITS

Nothing is more important to the success of a placer mine than to determine in advance that the gravel in question contains enough gold and possibly metals of the platinum group, to return a profit. Money received for the metals sold must be sufficient to pay for the cost of operation plus the cost of all equipment plus royalties or cost of land plus interest on the investment plus a reasonable profit. Of course, if the equipment is to be used on more than one deposit, amortization of its cost may be distributed accordingly. A large proportion of all placer operations has failed because the gold in the gravel was insufficient to repay the cost of even the most efficient mining, not to mention the return of money invested or interest thereon.

Many methods of sampling are available, including the simple panning of gravel from natural exposures, drifting, test-pitting or trenching, shaft-sinking, and churn-drilling. Actual mining on a small scale often is done as a method of sampling prior to investing considerable money in development or equipment. Panning, rocking, and small-scale machines have been described in Section I, under *Small-Scale Methods*. All of these are useful in prospecting and sampling. Small machines driven by gasoline engines have been much used in recent sampling jobs to save labor.

Weight of Placer Gravel

Placer gravels vary greatly in weight per cubic yard in place and percentage increase in volume on being loosened. Yet in making estimates of yardage and value it often is necessary to use some factor to convert volume in place into loose volume or into tonnage. In common placer terminology "heavy" gravel indicates coarse rather than weighty material. The weight is greater in tight or cemented than in loose ground, and it increases with the proportion of large boulders and heavy rock material such as diorite, greenstone, or hornblende schist. The amount of moisture present likewise affects the weight.

Three contiguous samples taken by Gardner and Johnson [1] from the same bed of tight, fine clayey gravel overlying a pay streak in the Greaterville district, Arizona, indicated weights of 3,450 pounds, 3,540 pounds, and 3,000 pounds to the cubic yard, respectively, or an average of 3,300 pounds. A sample of clean gravel with 30 to 40 percent cobbles over 2½ inches in diameter, from another gulch in the same district, weighed 3,600 pounds to the cubic yard in place. The samples contained 5 to 8 percent of moisture. The expansion of the first three samples was 50, 54, and 33 percent, respectively, an average of 46 percent. The last sample expanded 17 percent.

The gold-bearing river gravel at the pit of the Grant Rock-Service Company, Fresno, California, weighs 2,850 pounds per cubic yard. Some engineers in calculating tonnage allow 3,000 pounds to the cubic yard, bank measure. Handbooks give weights per cubic yard ranging from 2,600 to 3,650 pounds. An average weight probably is between 3,000 and 3,300 pounds to the cubic yard.

---

[1] Gardner, E. D., and Johnson, C. H., Placer mining in the western United States, Part 1, General information, hand-shoveling, and ground-sluicing: U.S. Bur. Mines. Inf. Circ. 6786, p. 27, 1934.

Sampling Natural Exposures

Whenever a gold pan is used skillfully the result not only proves or disproves the presence of gold but also usually shows accurately the amount of gold contained in the sample chosen. Panning along a creek bed thus is the most elementary method of sampling a placer deposit. However, the samples commonly are taken so as to render the quantitative results worthless.

A gravel deposit of much size seldom can be sampled directly from its natural exposures; but a few creek banks, steep-sided gulches, or old excavations such as hydraulic pits may be available, in which case certain precautions should be taken to get true samples. First, the vertical extent of gravel to include in a given sample should be determined. If hydraulicking is to be done, the whole depth of gravel ordinarily is included in one sample, except when it is planned to pipe off the barren overburden to waste, in which event it is desirable to know the depth of barren material and samples may be taken of each distinguishable stratum. If drifting is planned, only the lower, economically minable gravel need be sampled. After the location and extent of a sample cut have been decided, care must be taken to have equal quantities of gravel from all points along its length. The best way to do this is to cut a channel or groove of uniform shape and size from top to bottom of the sample distance. Enough such samples must be taken to prove the continuity of the "pay streak."

Mechanically the procedure of sampling a gravel face resembles closely that pursued in lode mines. A pan and a pick are the requisite tools. The bank should be trimmed well and cleaned along the sample line to eliminate effects of surface weathering. The pan may be used to catch the loosened material, or a canvas may be spread on the ground. If conditions favor it, a measurable channel or groove should be cut so that the volume taken can be measured. Otherwise, the only alternatives are to use a factor for pans per cubic yard or to measure the loose gravel in a box which has been calibrated in terms of bank measure.

Drifting

Drifting is a common method of prospecting a deep placer deposit when conditions are favorable. The cost of driving a small drift in placer gravel ordinarily ranges from $2 to $6 per foot. Difficult ground conditions or excessive water may increase the cost to $10 or $15 per foot.

For sampling purposes the gravel usually is taken to the surface and concentrated in sluice boxes. If an old drift is being sampled various methods may be followed. If the ground will permit, the most satisfactory method is to slab 1 or 2 feet from the side of the drift and wash the gravel thus broken. If not, vertical channels may be cut on one or both sides of the drift at intervals of about 5 feet. If the latter is done, the volume of the sample cut may be measured, which is preferable under most conditions, or a factor may be used for reducing loose measures of gravel to solid measure, which facilitates taking the sample but introduces some uncertainty and leads to carelesness in sampling. If values are to be expressed in cents per ton it is still necessary to decide what conversion factor to use in making estimates of tonnage.

Test-Pitting

Test-pitting and trenching are applicable only to gravels so shallow that a man can throw out the dirt by hand. The best procedure is to mark out the area of the pit on the surface, making it rectangular, as small as convenient, and preferably in dimensions of even feet such as 2 feet wide and 3 or 4 feet long. Then it should be excavated to bedrock with smooth, vertical walls. Sometimes a cleared space is prepared and the dirt thrown on the bare ground, but in view of the greater difficulty and possible small error involved in rehandling the dirt it is better to shovel it onto a canvas or board platform or into a receptacle. If a large boulder projects into the pit no correction of the theoretical volume of gravel should be made, regardless of whether or not the boulder is removed or allowed to remain in place, as obviously it can not be ignored in mining operations and is the equivalent of so much barren gravel. The gravel taken from the pit may be thrown into one or more piles, depending on whether or not information is desired regarding one or more individual strata. Sometimes alternate third, fourth, or tenth shovelfuls are used for the sample to reduce the volume to be panned or otherwise concentrated. This should be avoided whenever practical because of the possible error introduced.

The cost of sinking test pits or running trenches in earth and gravel has been noted often enough for fair generalizations to be established. Gardner [2] states that opencast work in placer ground costs from $0.40 to $1 per cubic yard, depending on the nature of the ground and on wages. Wages for such work then ranged from $3 to $4 per 8 hours. Furthermore, a man should be able to pick and shovel about 8 cubic yards of fairly loose gravel in 8 hours. In test pits the worker's efficiency would be lowered somewhat by the cramped quarters and by the care necessary to square out the corners and trim the sides to vertical planes.

The spacing of test pits depends on the nature of the deposit. If the pay gravel occurs in narrow channels the best plan is to space the pits in lines across the channels, as is done with churn-drill holes when sampling dredge ground. The holes must be close enough to yield an average value which represents the average value of the channel at that point. In practice the spacing ranges from a few to 50 or 75 feet, depending on the uniformity of results. The transverse lines of pits theoretically should be placed close enough to show either fairly uniform values from line to line or a reasonable upward or downward trend of gold content along the channel. Unfortunately, this is seldom possible, and the usual practice of spacing the lines from 100 to several hundred feet apart is a compromise in the engineer's mind between the cost of sampling and the need for accurate results.

Shaft-Sinking

Shaft-sinking is a common method of testing placer ground. Prospect and sampling shafts, unless intended for later use in drifting or mining and unless exceptionally deep (75 feet or more), are sunk as small as practicable. The usual section is rectangular and 3 by 4 to 4 by 6 feet in size. Round timber 4 to 6 inches in diameter, which is available in most districts, is commonly used to crib shafts in loose gravel. In gravel tight enough to stand safely without lagging the only timber

---

[2] Gardner, E. D., Cost of mine openings: Eng. and Min. Jour., vol. 100, pp. 791-794, Nov. 13, 1915.

necessary is stulls set to hold the ladder.   A hand windlass is the usual means of hoisting, using a light steel, 2-cu.ft. bucket and a ¾- or 1-inch-diameter manila rope, as ordinary wire rope is unsuitable for a windlass; 75 to 100 feet is the maximum depth at which such equipment can be used most efficiently.   For greater depths a power hoist of some kind should be installed.   Large boulders can be lifted with the ordinary hand windlass if it is provided with long cranks; as much as 800 pounds can be raised by two men.   Such feats, however, are considered dangerous because of the general absence of safe brakes or catches on windlasses, and the possibility of killing or injuring the operator if he loses control of the crank handle.

All of the gravel removed from such a shaft is often trucked to a stationary washing plant to recover the gold.   The gravel is dumped into a hopper from which it is fed to a revolving screen or trommel equipped with a water-spray, and undersize is washed in sluices.   If these are placed in steps, so that the gravel drops from one to the next below, the necessity for tight cross-joints is avoided.   Slopes of the boxes can be adjusted to give good recoveries for different types of gravel. In one actual case 4 boxes were used, each 10 feet long, 10 inches wide, and 6 inches deep.   Grade of the first three boxes was set at 1½ inches per foot, and of the fourth at ½-inch per foot.   In shafts that stand without timber, the gravel may be sampled by cutting a vertical channel. A common size is 1 foot wide and 1 foot deep.   Care should be taken with bedrock, which should be removed for at least a few inches in depth, and then brushed clean.   Recovered gold should be weighed on an accurate balance, and an occasional assay should be made to determine the fineness of the gold.

Where gravel is loose, sometimes steel tubes about 4½ feet in diameter, called caissons, are sunk as the shaft is excavated, to hold the walls. An example of this method in which telescoping caissons were used is given in the table at the end of this section.   If the shafts are wet, and pumps driven by engines or motors are needed, this method becomes very expensive, and drilling may be used instead; although drilling is a less accurate method of sampling.

Drilling

Placer deposits, particularly those being considered for dredging, are often sampled by drilling, which could be called drive-pipe sampling with more accuracy.   The casing, usually a 6-inch pipe with a special shoe, is driven a foot at a time into virgin gravel, then the core of gravel that rises in relation to this pipe is cut by the bit of the churn-drill and bailed out with a sand pump.   Water is added to all holes.   The cuttings from each foot are panned by an expert panner, and the weight of the gold in each pan is estimated.   The colors of gold from the pans from a single hole are combined and amalgamated with a drop of quicksilver.   This gold is later recovered from the amalgam by the engineer and actually weighed.

The drill in common use is the Keystone, made by Keystone Driller Company, Beaver Falls, Pennsylvania.   California agents are Harron, Rickard and McCone Company, 2070 Bryant Street, San Francisco, or 3850 Santa Fe Avenue, Los Angeles.   Late designs include model 70 mounted on a truck, and model 71 on crawler tracks.   According to the manufacturer, model 71 will climb grades of 50 percent when loaded with necessary tools and equipment.

The object of this type of drilling is not to make rapid progress with the hole but to get accurate samples.   Drilling would proceed at a much greater speed if the bit of the churn-drill were kept ahead of the casing, but this is never done in sampling unless a boulder must be broken to allow driving of the casing, because the volume of gravel removed from the hole would not be known.   When the casing is driven a foot at a time, volume may be calculated as that of a cylinder a foot long and with diameter equal to the outside diameter of the cutting shoe, usually $7\frac{1}{2}$ inches.   Water is added to dry holes so that the cuttings may be withdrawn by means of a suction sand-pump.   Water is added to wet holes to keep the water-level in the hole higher than the natural ground-water level.   This helps to prevent excess material from entering the casing. When bedrock is reached, the casing can no longer be driven, but drilling is continued for a few feet to recover gold on bedrock and to make sure that the casing is not resting on a boulder.   Sometimes the drill strikes a fissure in bedrock and an abnormal amount of gold is recovered.   In such a case the panner should keep this gold separate from gold recovered from the gravel, and hand it to the engineer as a separate button of amalgam.

The engineer dissolves the mercury from the amalgam with nitric acid of specific gravity 1.42 diluted with an equal volume of distilled water, washes the gold well with boiling water, then dries and anneals the gold in a small porcelain cup.   Sometimes a little alcohol is added to the wash-water to prevent sputtering when drying.   The dried gold dust is weighed on a balance sensitive to a milligram or less.   Fineness of the gold must be determined by an occasional fire-assay.   If platinum is present, concentrates from panning must be saved and weighed, then sent for chemical analysis.   The amount of gold in cents per cubic yard for the gravel removed from the hole is calculated as follows:

$$V = \frac{W \times M \times 27}{D \times 0.3068}$$

in which V=value of the gravel in cents per cubic yard
　　　　W=weight of gold recovered from the hole in milligrams
　　　　M=price in cents of a milligram of gold
　　　　27=number of cu. ft. in 1 cu. yd.
　　　　D=depth of hole in feet
　　0.3068=area in square feet of shoe-circle $7\frac{1}{2}$ inches in diameter.

The value of M depends on the price of gold and its fineness.   When gold is $35.00 per ounce and the gold recovered is 900/1000 fine, it is calculated as follows:

$$M = \frac{3500 \times .9}{31103} = .101 \text{ cents per milligram.}$$

When a long narrow deposit is being sampled, such as bars along a river, rows of holes are often drilled across the bars at distances of 1000 feet along the river.   Holes in the rows are often about 150 feet apart. Whenever erratic occurence of gold in the bars is found, the spacing should be changed to suit local conditions.   When the spacing of holes is irregular, the average value per cubic yard of the whole bar is found by weighting the value found in each hole by the area represented.   The area between adjacent holes is multiplied by the average value in cents

Table 1. Representative placer prospecting costs; prospecting carried to conclusion and followed by dredging except in two cases

	(1)	(2)	(3)	(4)	(5)	(6)
Style of drill	Keystone No. 3	Keystone No. 3	Keystone No. 3	Shafting	Shafting	Shafting
Power	Gasoline-driven	Gasoline-driven	Steam	Hand; no timber	Caissons hand-driven	Frozen; sunk by hand
Year	1931	1931	1908, 1910	1906	1931-32	1907
Location	Folsom, California	Waldo, Oregon	Sumpter, Oregon	Salmon, Idaho	Snelling, California	Dawson, Yukon Territory
Months	March	May-June	Aug.-Nov.; July-Nov.	June-September	November-May	June-October
Formation	Gravel with much clay	Deep gravel with much clay	Loose gravel; 16 per cent soil	Shallow gravel	Loose gravel; 20 per cent soil	71 per cent frozen muck; 19 per cent frozen gravel
Operating conditions	Favorable	Favorable	Favorable	Favorable; water pumped by hand	Unfavorable; much water; power pumps	Very favorable; steam thawing
Wage rate per day:						
Foreman	$9.60	$9.60	$7.67	$7.50	$8.25	$8.00
Driller	8.25	7.25	4.50		7.25	
Panner	7.25	6.25	4.50	4.17	6.45	5.65
Helper	4.40	4.00	3.00	3.00	6.45	5.65
Assistant engineer			2.67	5.67	6.00	4.65
Rent of team per day	8.00		5.50			
Rent of truck per day	5.00	5.00	1.50			
Rent of drill per day	12.50	10.00	5.00	6.00	10.00	
Board per month	35.00	37.50	30.00	20.00	37.50	50.00
Number of holes		16	226	51	46	64
Total depth .... feet	747.5	1,386	4,400.5	697	778	1,553.5
Number of crew-shifts	37.5	66	320			
Number of man-shifts				354.5	294.5	490
Number of days	22	48	160	75	194	98
Advance per shift ...... feet	20.4	21	12.2	1.97	2.64	3.17
Advance per day .... feet	34	29	24.4	15.5	4.00	15.85
Average depth per hole .... feet	68	86.7	19.47	13.67	16.9	²24.27
Costs:						
Labor	$1,142.55	$2,671.34		$1,853.00		
Repairs	92.15	82.13				
Freight	70.00	352.65				
Travel expenses	162.67	141.14				
Drill rental	325.00	480.00	800			
Gasoline and oil	70.56	40.66				
Engineering	¹60.75	1,061.32	562.50			
Teams	8.00					
Rent of truck	200.00	250.00				
Totals	$2,231.62	$5,079.24	$10,169.43	$2,415.50	$7,666.69	$5,435.47
Cost per foot	2.985	3.665	2.31	³3.46	⁴9.85	⁵3.50

¹ Wagon.　² 17 muck; 7 gravel and bedrock.　³ $2.01 sinking; $1.45 sampling.　⁴ $7.48 sinking; $2.37 sampling.　⁵ Sinking and sampling.

per cubic yard for the two holes. This is continued across the line of holes and the two columns of figures representing areas and areas times cents per cubic yard are added. Total of the last column divided by total area gives the average value per cubic yard of the line of holes. Allowance may be made for small areas to be worked beyond the last holes at either end of the row. If the gold value of the holes is decreasing rapidly toward the ends of the row, this small area will be assigned a suitable fraction of the value per cubic yard of the last hole in the row. Allowance must be made for rising bedrock at the point where the bottom of the dredge would strike it; also for a foot of barren badrock to be excavated by the dredge.

For the constant 0.3068 in the denominator of the above formula, some engineers have substituted the quantity 0.27, because the sampling done by drills usually gives a figure that is too low for the value per cubic yard of the gravel. Other engineers wish to discount their results, and substitute 0.3333 for 0.3068. Such substitution is bad practice because it causes confusion. If the engineer feels that his figures should be increased or decreased for any reason, he should state the reason and the percentage of increase or decrease. In all cases he should report the actual number of milligrams of gold recovered from the hole.

A log-sheet for such a drilling job contains 11 columns, and an entry is made in each column every time the casing is driven a foot. Column no. 1 is an entry of the total depth to which casing has been driven in feet and tenths of a foot; column no. 2 is the length of core after driving (length of casing minus length of tools and cable below top of casing with tools resting on top of core); column no. 3 is length of core in casing after drilling (about 0.3 feet); column no. 4 is length of core after pumping; column no. 5 is actual depth of hole to top of core; column no. 6 is an estimate of the number of milligrams of gold and is divided into three sub-columns; (a) milligrams of gold in particles over 7 milligrams in weight, (b) milligrams of gold in particles 2 milligrams to 7 milligrams in weight, (c) milligrams of gold in particles less than 2 milligrams in weight; column no. 7 is an estimate of amount of gravel caught in pan expressed in units of 'panfulls' (normally 0.75); column no. 8 is time of day at which pumping was finished; column no. 9 is a description of the firmness of the gravel (loose, firm, or very firm); column no. 10 is type of material (soil, clay, cemented gravel, fine gravel, medium gravel, coarse gravel, or large boulders); column no. 11 is remarks (occurrence of some unusual mineral, rusty gold, change in character of gold, causes of delays, or anything else important or unusual).

Gardner and Johnson [3] have published a table of drilling results compared with actual recovery by dredging on 40 different tracts of land. On 23 of the tracts recovery was more than 100 percent of the gold content indicated by the drill-sampling, on 16 tracts recovery was less than 100 percent, and on one tract, recovery was practically 100 percent. When recovery is less than 100 percent, the dredge may have left some of the gold behind. This is practically unavoidable where bedrock is very hard, and contains cracks and fissures into which the gold can settle beyond reach. No satisfactory method has yet been devised for accurately calculating percentage recovery by a dredge. However, when

[3] Op. cit., pp. 42-45.

15

recovery is more than 100 percent, it can mean only that sampling results were too low. The majority of cases mentioned above fall into this class. Shafts that have been sunk around drill holes have indicated the same thing; recovery of gold per cubic yard from the shaft has been greater than that from the drill-hole.

Results of drill-sampling are likely to be low in loose gravel such as tailing, because of poor core-recovery. Driving the casing will push some of the gravel aside, and the bit of the drill will push more of it aside. The sand-pump is used as much as possible after the casing is driven and before drilling is started. However, some of the gravel is usually too large to pick up with the pump, and the bit must be used to break it up.

Table 1 and the following quotation are reproduced from U. S. Bureau of Mines Information Circular 6786 by Gardner and Johnson.[4]

"The prospecting was carried to a conclusion in each instance. Dredging followed the prospecting in four of the six cases.

"The large difference in cost of engineering between jobs no. 1 and no. 2 was due to the much larger overhead at no. 2 on account of the nature of the deposit. The distance from headquarters at San Francisco also affected the size of the engineering force necessary on the ground. Job no. 3 was in effect two jobs. The work was interrupted for a full year.

"Job no. 4 was favored by an ample force of experienced gravel miners, a low water level, and not a difficult quantity to handle. Diaphragm bilge pumps were used and in some shafts too deep for suction, a second shaft sunk 10 feet deep and adjacent to the first was used for increasing the effective depth, by means of two lifts.

"Job no. 5 was a very difficult undertaking and required the use of steel telescoping caissons, especially designed for the job. Gasoline-driven pumps of the jackhead deep-well cylinder type proved very awkward but most effective.

"In both jobs no. 4 and no. 5 the total contents of the shafts were washed in a long-tom device by hand and in job no. 4 a check sample was cut from the side of the shaft and washed in a rocker.

"Job no. 6 is a typical case of shaft prospecting in frozen ground where the gravel deposit is unusually thin or shallow. Here the conditions for shaft work were very favorable, but the high cost of living was reflected in the unit cost. Where experienced men are available, as in this case, and equipment developed by the miners for inaccessible places is at hand for their use, a very low unit cost is obtained.

"Drilling in frozen ground is also very economical, owing to the speed with which the work is accomplished and the absence of casing costs. The volume of sample is quickly and accurately obtained by water measurement after completion of the hole.

"The unit cost or cost per foot for placer prospecting is usually uncertain, since it depends upon the total footage. The number of holes or cross sections as the case requires may prove to be very few, and the total cost of starting and clearing up the job falls upon a small total footage. Two cases can be cited as follows: One in Colombia, S. A., cost about $25,000 for about 1,500 feet of drilling where the equipment was left behind and never salvaged. Another in central Alaska cost about the same for less than 500 feet of drilling where the equipment was not salvaged on account of cost. In such cases and in many others that constantly arise costs can be reduced to a very low figure by a preliminary examination made by an experienced and reliable placer-mining engineer and are usually represented by the engineer's fee and expenses. The number and distribution of prospect holes needed to provide the essential information can be readily determined by the preliminary survey."

4 Op. cit., p. 35.

Geophysical methods have been used to a limited extent for tracing buried river channels containing gravel deposits. Such methods must be used in connection with a careful study of the local geology; otherwise interpretation of the results will be impossible. Geophysical methods may give information about the course and the depth of the channel, but they tell nothing whatever about the gold-content of the gravel. This must be determined by one of the sampling methods described above.

Magnetic methods may be used where the gravel has either a higher or a lower magnetic permeability than the underlying bedrock. However, if a lava capping exists over the gravel magnetic conditions in this may interfere so much that the method cannot be applied. If the gravel contains concentrations of magnetite-sand, a magnetic survey will reveal magnetic 'highs'. On the other hand, if the bedrock is more magnetic than the gravel, the channel may be outlined by observing the magnetic 'lows' above it. Ellsworth[1] describes such a survey made of a channel, of which the position had already been determined by mine-workings, at Forest Hill Divide, California. The instrument used was the Hotchkiss Superdip magnetometer.

The U. S. Bureau of Mines was developing a method of tracing buried river channels by resistivity measurements before the war. This method shows promise of considerable success, but those in charge of the work have not yet been able to put the results in report-form because of the search for strategic minerals after the war started.

A channel located in California has been outlined by a large geophysical engineering company with seismic work by the refraction method. The results were later confirmed in part by drilling. Although officials of the company which did this work feel that the method possesses merit, they do not want to report on it because results obtained by them have not yet been checked by actual mining.

Some results of geophysical surveys at the Roscoe Placer of Humphreys Gold Corporation in Colorado have been described by Dart Wantland.[2] Both resistivity and magnetic methods were used.

---

[1] Ellsworth, E. W., Tracing buried river channel deposits by geomagnetic methods: California Jour. Mines and Geology, vol. 29, pp. 244-250, 1933.
[2] Wantland, Dart, A comparison of geophysical surveys and the results of operations at the Roscoe Placer of the Humphreys Gold Corporation, Jefferson County, Colorado: Colorado School of Mines Quart., vol. 32, no. 1, January 1937.

SECTION IV

PLACER MINES BY COUNTIES

In the following pages are listed properties which have been operated recently, or which are thought to contain reserves of placer-gravel. They are to be regarded as examples only; more complete lists will be found in reports on individual counties contained in our California Journal of Mines and Geology.

The California State Division of Mines does not sample placer mines. This involves a campaign of drilling or one of the other methods described herein in Section III, *Prospecting and Sampling Placer Deposits*, and the Division of Mines is not equipped to do this work. Any statements contained herein regarding the amount of gold in placer gravels have been obtained from persons not connected with the Division of Mines, but only figures that are believed to be reliable have been used. Readers should not invest money in equipping or purchasing placer mines on the basis of these figures. The value of deposits should be checked by a competent engineer using the methods described in Section III before an investment is made.

Figures on the production of certain operations are given below. Such statistics are collected by the U. S. Bureau of Mines and have been compiled by Merrill.[1]

All but a few of California's gold mines were shut down late in 1942 by Limitation Order L-208 of the War Production Board, which remained in effect until July 1, 1945.

---

[1] Merrill, C. W., and Gaylord, H. M., Gold, silver, copper, lead, and zinc in California (Mine Report) : U. S. Bur. Mines Minerals Yearbooks, 1940 to 1943.

FIG. 81. Wallace dredge. This 3-cu. ft. dredge owned by Wallace Dredging Company was built by Yuba Manufacturing Company in 1935 and has been used on two properties. It is one of the earliest portable pontoon-type dredges built—a method which can be used to advantage for placer-dredge construction if dredges are to be used on more than one property. Photo by courtesy of Yuba Manufacturing Company.

AMADOR COUNTY

Allen Ranch. Henry and Weaver operated a dragline dredge on the Allen Ranch property on Sutter Creek Gulch during part of 1941. The dragline excavator was equipped with a ¾-cubic yard bucket.

Amador Dredging Company, Ione, operated a dragline dredge in the Ione district during 1941.

Arroyo Seco Gold Dredging Company, 351 California Street, San Francisco, operated an electric connected-bucket dredge, equipped with eighty-six 6-cu.ft. buckets from January 1 to May 15, 1941. The company operated throughout 1940 also at a property 3 miles west of Ione.

Detert Estate. Mountain Gold Dredging Company of Plymouth and W. D. Ingram operated dragline dredges on this property near Plymouth during 1941.

Ingram operated also on Indian Creek 4 miles west of Plymouth with two dragline excavators equipped with 2½-cu.yd. and 1-cu.yd. buckets. Land was restored for agricultural use by leveling tailing and replacing overburden.

Elephant Hydraulic Mine. This mine, near Volcano, was operated in 1932 by the K. D. Winship estate. One and one-half to 3 feet of gravel on decomposed slate bedrock are overlain by 40 to 45 feet of volcanic ash. Water amounting to 135 miner's inches was delivered to one no. 2 giant through 18-inch and 8-inch pipe-lines under 115 feet of head. Gold was recovered in a sluice 18 inches wide, 16 inches deep and 32 feet long, equipped with Hungarian riffles. Two men whose wages were $3.50 per 9-hour shift washed 3600 cubic yards during a 90-day season at a total cost of $0.20 per cubic yard for labor and supplies. The volcanic ash was drilled with hand augers, blasted and piped through the sluice. Gravel was tight but was cut by the giant. Bywash-water amounting to 40 miner's inches raised the total through the sluice to 175 inches.

Garibaldi Mine. Garibaldi Bros., of Volcano operated a nonfloating washing plant, to which gravel was delivered by mechanical means, at their mine on Pioneer Creek half a mile east of Volcano, intermittently during 1941. The yield from 33.200 cubic yards of gravel was 229 ounces of gold and 35 ounces of silver. This mine was worked during parts of 1940 also. Garibaldi or Boardman property was operated by dragline and dry-land dredging in 1942. The dragline dredge recovered 775 ounces of gold and 102 ounces of silver from 172,200 cubic yards of gravel, and the dry-land dredge recovered 157 ounces of gold and 23 ounces of silver from 35,500 cubic yards of gravel.

Horseshoe Dredging Company, Ione, operated a dragline dredge in the Ione district from May 2 until July 26, 1940.

Horton Mine. This mine is in Jackson Valley 5 miles south of Ione and was operated by H. G. Kreth of Ione by hydraulicking from January to June and from October to December, 1941.

Independence Gold Mines. This company treated 12,100 cubic yards of gravel in the Camanche district at a stationary washing plant between July 30 and October 12, 1941, recovering 187 ounces of gold and 19 ounces of silver.

Irish Hill Mine. McQueen and Downing, 1040 38th Street, Sacramento, operated a dragline dredge at this mine from March 28 to June 25, 1941.

Kent Property. During 1940, E. A. Kent, 351 California Street, San Francisco, operated two dragline dredges on Sutter Creek between the towns of Sutter Creek and Volcano. One was equipped with a 1¾-cu.yd. bucket, the other with a 2½-cu.yd. bucket.

Lancha Plana Gold Dredging Company, La Lomita Rancho, Lockeford, operated an electric connected-bucket dredge, equipped with sixty-five 4½-cu.ft. buckets on Jackson Creek near Buena Vista throughout 1940 and from January 1 to May 4, 1941, when it was dismantled and moved to Butte County.

Lorentz Property. Lorentz and Swingle and *Long Bar Gold Dredging Company* operated dragline dredges on this property on the Cosumnes River in 1942. The company recovered 1140 ounces of gold and 138 ounces of silver from 200,000 cubic yards of gravel. Lorentz and Swingle also operated on Cosumnes River, 7 miles northwest of Plymouth in 1941.

Matulich Property. *Mountain Gold Dredging Company* of Plymouth operated a dragline dredge at this property near Drytown intermittently during 1941.

McCulloh Property. *Pacific Placers Engineering Company*, 3400 H Street, Sacramento, operated a dragline dredge on this property in the Ione district during parts of 1940 and 1941. Yield from 350,000 cubic yards of gravel was 2749 ounces of gold and 258 ounces of silver in 1941.

Pension Mine. *Long Bar Gold Dredging Company*, 935 Forum Building, Sacramento, operated a dragline dredge at this mine from September 13 to 28, 1942. A volume of 14,000 cubic yards of gravel yielded 87 ounces of gold and 12 ounces of silver.

Placeritas Mining Company, 245 North Gramercy Place, Los Angeles, operated a dragline dredge on six different properties in 1940, all within a radius of 4 miles from Plymouth.

Rim Cam Gold Dredging Company operated a dragline dredge on two properties near Drytown and on the Cosumnes River between Plymouth and Nashville in 1940. (See also *Yager Ranch.*)

River Pine Mining Company. This company, address of which was Plymouth, operated a dragline dredge near Aukum, which was equipped with a 1¾-cu.yd. bucket, from January 1 to June 12, 1941, when it was moved to El Dorado County. The yield from 300,000 cubic yards of gravel was 1380 ounces of gold and 192 ounces of silver in 1941. This was the most productive dragline dredge in the county in 1940.

Rupley Ranch. This property is on Willow Creek 5 miles west of Drytown. J. C. Pantle operated a dry-land dredge here from January 1 to October 15, 1942. He recovered 1290 ounces of gold and 183 ounces of silver from 230,000 cubic yards of gravel. A dragline excavator equipped with a 1½-cu.yd. bucket delivered gravel to the washing plant. The yield in 1941 was 1850 ounces of gold and 254 ounces of silver from 360,000 cubic yards of gravel. In 1940, Pantle's recovery from 265,000 cubic yards of gravel was 1154 ounces of gold and 139 ounces of silver.

San Andreas Gold Dredging Company, 960 Russ Building, San Francisco, operated a dragline dredge with 1½-cu.yd. bucket on the Arroyo Seco Ranch during part of 1940.

Treble Clef Mine.   E. L. Lilly, 706 California Building, Stockton, operated a dragline dredge with 2¼-cu.yd. bucket during parts of 1941. The outfit was operated throughout 1940 on two different properties.

Yager Ranch.   Rim Cam Gold Dredging Company operated a dragline dredge on this property in the Ione district from February 4 to May 26, 1941.

BUTTE COUNTY

A number of drift mines in Butte County such as the Emma, Indian Springs, and Perschbaker were very productive years ago, but drift mining has not been active in the county recently.   Segments of the underground channels remain unworked, but prospecting, development-work, and sampling are needed to determine whether they can be worked at a profit.   Further details are given by Logan[1] and Lindgren.[2]   Recent production of the county has been mostly from dredging.   The principal producers of placer gold from 1940 to 1943 are mentioned below.

Amo mine in Oroville district was operated in 1942 by F. C. Peterson, Box 550, Oroville, who used a non-floating washing plant and recovered 379 ounces of gold and 12 ounces of silver from 60,000 cubic yards of gravel.

Baker and McCowan, Palermo, operated a dragline dredge on the Farnan Ranch in the Oroville district in 1940.

Butte Operating Company, Oroville, operated a dragline dredge in the Oroville district throughout 1940.

Cory and Strong Placer is 2.8 miles by road north of Stirling City and contains 320 acres in the E½ sec. 16, T. 24 N., R. 4 E., M. D., on the west side of West Branch of Feather River.   A large lava-covered trough, locally known as the Mammoth Channel, crosses the property. Bedrock rims are 1500 to 2000 feet apart.   Years ago, an adit was driven at a point 400 feet south of the north property-line to a length of 700 feet.   It is in lava-wash supposed to be higher than the main channel but George E. Strong of Dixon, California, states that it produced enough gold to pay for the work.

Strong says that a shaft, which was kept unwatered by a 3-inch pump was sunk on the bank of the river, and that a drill-hole was put down 454 feet from the bottom of the shaft in sand and clay to reach hard bedrock.   He is planning some more drilling in an effort to locate gravel containing gold in paying amounts.

Gold Hill Dredging Company, 311 California Street, San Francisco, operated a connected-bucket dredge with seventy-four 9-cu.ft. buckets on the Wilton Kister property, on the east side of Feather River 7 miles south of Oroville during 1940, 1941, 1942.

Golden Feather Dredging Company, 817 25th Street, Sacramento, operated a dragline dredge, using a 5-cu.yd. bucket, on the Feather River opposite the town of Oroville from February 1940 until late in 1944.   This operation was given a special permit by the War Production

[1] Logan, C. A., Butte County: California Div. Mines, Mining in California, State Mineralogist's Rept. 26, pp. 383-406, 1930.
   Logan, C. A., Butte County: California Jour. Mines and Geology, vol. 31, p. 6, 1935.
[2] Lindgren, Waldemar, The Tertiary gravels of the Sierra Nevada of California: U. S. Geol. Survey Prof. Paper 73, pp. 84-93, 1911.

Board because gravel was cleared from the Feather River channel and stacked as a levee to protect the town of Oroville. Further details have been published by Wiltsee.[3]

Humphreys Gold Corporation, 910 First National Bank Building, Denver, Colorado, operated a dry-land dredge on the L. I. Kister property, Oroville district, during part of 1940.

Interstate Mines, Inc., Chico, moved a dragline dredge from Trinity County to the Gianella Ranch, Oroville district, and operated it during part of 1940 and 1941.

Kaufield and Danison, Oroville, worked the Ford property during part of 1941. The excavator used a 1-cu.yd. bucket.

Lancha Plana Gold Dredging Company, La Lomita Rancho, Lockeford, moved its connected-bucket dredge from Amador County to the Butte Creek district, resumed operations in October 1941, and worked until October 12, 1942. The dredge had sixty-five 4½-cu.ft. buckets.

Lemroh Mining Company, 2401 Bayshore Boulevard, San Francisco, operated a dragline dredge in the Magalia district during part of 1940. In 1941, this company operated a dragline dredge using a 2½-cu.-yd. bucket; 504,848 cubic yards of gravel yielded 2739 ounces of gold and 195 ounces of silver.

Lord and Bishop (Lobicasa Company after January 1, 1941), Box 812, Sacramento, washed 66,300 cubic yards of gravel by dragline dredging in the Oroville district during part of 1940 and recovered 336 ounces of gold and 11 ounces of silver. The dragline excavator had a 1½-cu.yd. bucket. In 1941, Lobicasa Company operated on the Peters Ranch.

Morris Ravine Mining Company, Oroville, operated a drift mine in the Oroville district in 1942 and recovered 658 ounces of gold and 65 ounces of silver from 850 cubic yards of gravel. This mine was worked in 1943 also.

Oroville Gold Dredging Company, 2052 Bird Street, Oroville, operated a connected-bucket dredge with seventy-two 8½-cu.ft. buckets on the Hazelbusch and T. M. Rogers tracts on Feather River 9 miles southwest of Oroville during 1941 and part of 1943.

Piedmont Dredging Company operated a Becker-Hopkins type dredge on Butte Creek during part of 1941. Recovery from 29,592 cubic yards of gravel was 121 ounces of gold and 10 ounces of silver.

Piombo Bros. & Company, 1517 Turk Street, San Francisco, operated a dragline dredge, using a 1½-cu.yd. bucket, on French Creek throughout 1940 and 1941.

Placer Development Company, 2401 Bayshore Boulevard, San Francisco, operated a dragline dredge in the Oroville district during part of 1940, and at Meadows 3 miles south of Oroville in 1941. The dragline excavator used a bucket of 2½-cubic yards.

Placer Exploration Company, Box 498, Chico, operated two dragline dredges in the Oroville district in 1941. One was equipped with a 5-cubic yard bucket, the other with a 2½-cubic yard bucket. Several properties were worked including the following: Dagorret, California Lands, Inc., and Innis. In 1941, this company worked on the Gianella

[3] Wiltsee, E. A., Operations of Golden Feather Dredging Company: Min. Cong. Jour., pp. 21-24, 35, August 1944.

Ranch; also during part of 1942. The Innis Ranch was worked in 1942 also.

William Richter & Sons, Oroville, operated a dragline dredge on the Douglas Jacob, Mary Harrin, V. Gamble, and John Bilkli properties in the Oroville district in 1940. They worked on the following properties in Oroville district in 1941: Belkriet, Bilkli, Freidel, Helen Whittier, Hume and Coleman, John Alm, Lorrie, Ray Angle, Rottinger, and Wyandotte.

Sunmar Dredging Company, Box 228, Oroville, operated a dragline dredge in Weymans Ravine 4 miles from Oroville during part of 1940. The dragline excavator used a 2-cubic yard bucket. In 1941 this company used equipment with 1½-cubic yard bucket on the following properties: Clark, Cratt, Schwartz, Crowder and Binney, Darby, Darby and Crowder, Leal, and Schwartz and Pedrazzini properties. This company worked on the Gianella and Peter properties in 1942.

Yuba Consolidated Gold Fields, 351 California Street, San Francisco, operated four electric connected-bucket dredges in the Oroville district during the 4-year period under consideration until October 15, 1942. Bucket-lines were as follows: eighty-four 9-cubic foot buckets, eighty-nine 9-cubic foot buckets, eighty-seven 9-cubic foot buckets, and seventy-one 6-cubic foot buckets.

CALAVERAS COUNTY

The drift mines of Calaveras County are of outstanding interest although all other types of placer mines have been operated in the county, and dredges have been very productive. For further information on such operations the reader is referred to State Mineralogist's Report XXXII [1], which contains descriptions of many mines and a long table of references to descriptions of older operations. Maps showing the ancient channel system at Mokelumne Hill, San Andreas, Angels Camp, and Vallecito are included in this report, and are sold separately by the Division of Mines. Considerable detailed information on these channels is contained in Bulletin 413 of the U. S. Bureau of Mines.[2] Descriptions of the Vallecito-Western and Calaveras Central mines are reprinted herein because they show recent trends in the mechanization of drift mines. The need for extensive preliminary work to prepare such a mine for the production of a large tonnage per day is clearly indicated. The principal channels in this county are at a lower elevation than the present surface and are reached by shafts.

Calaveras Central Mine

The Calaveras Central [3] mine is of especial interest because it is the largest drift mine in California, and its management has pioneered in the application of more recent and more effective engineering methods to produce large tonnage at low cost. The results so far achieved, with

[1] Logan, C. A., and Franke, H., Mines and mineral resources of Calaveras County: California Jour. Mines and Geology, vol. 32, pp. 324-364, 1936.

[2] Julihn, C. E., and Horton, F. W., Mineral industries survey of the United States; California, Calaveras County, Mother Lode District (south), Mines of the Southern Mother Lode Region, Part I, Calaveras County: U. S. Bur. Mines Bull. 413, pp. 21-94, 1938.

[3] Julihn, C. E. and Horton, F. W., op. cit.

much yet remaining to be done in the development of improved practice, at least point the way toward the only course likely to revitalize drift mining, which has so long been moribund. They demonstrate that reasonably large, steady production attained through increased mechanization, similar to that already in use in many lode mines, is highly effective in reducing the costs of drift mining as well.

The mine is about 1 mile north of Angels Camp. It is operated by the Calaveras Central Gold Mining Co., Ltd., 705 Hobart Building, San Francisco. The data in regard to these operations were made available for study by Harry Sears, president and general manager, from records of his company.

The company is said to control, by long-term leases, 876 contiguous acres, extending about 3½ miles along the Central Hill Channel of the main Tertiary Calaveras River, in secs. 21, 22, 23, 26, 27, and 28, T. 3 N., R. 13 E. This includes mining rights in areas formerly controlled by the Victor, McElroy, Pierano, and Reiner mines, the E. W. Johnson ranch, the Slab Ranch, and other mines of the Calmo Mining & Milling Co.

History and Production

The first recorded development on this property occurred before 1866, when the McElroy shaft was sunk on an intervolcanic channel called by the same name. This shaft, perhaps one of the first in California to be sunk in an attempt to mine deep gravels by such means, was about 200 feet deep. It was equipped with a hoist operated by a primitive overshot water wheel. It penetrated rich gravel, but after about $100,000 had been produced from a short distance along the channel, operations were stopped by flooding.

About 1866 the Mattison shaft was partly sunk on Bald Hill, the site of the present hoist and compressor buildings, and in the bottom gravels of that shaft the much publicized Calaveras skull, that proved to be a hoax, was alleged to have been found. Other desultory attempts to reach and work the gravels of the area continued but accomplished little, and production amounted to only about 2000 ounces of gold from 1900 to June 1931, when the present company began operations from a previously existing three-compartment shaft 350 feet deep. Since then it is said to have produced 20,000 ounces of gold, bringing the total production during the present century to 22,000 ounces.

The Gravels and Channels

The Tertiary system of gravels is well-represented by the various channels within the boundaries of the property, ranging from a system of at least three prevolcanic channels in the bedrock to later superimposed intervolcanic channels, and so up to recent gravels, which are present at many places on the surface. Most of these surface gravels contain a little gold, and the richer ones were worked in early years. They have, of course, no relation to the deep pay gravels of the ancient channels, which are buried beneath 250 to 350 feet of rhyolitic and andesitic tuffs, intercalated with beds of gravel, all of which are said to carry some gold.

Just east of Slab Ranch the bedrock rims outlining the Tertiary valley in which the ancient river flowed are less than 500 feet apart, but the valley widens rapidly downstream, and a mile below this narrow neck it is more than one-half mile wide and forms a considerable basin.

At Slab Ranch the old river flowed due west, but on entering the basin it coursed northwesterly and in turning cut several distinct channels separated by bedrock ridges or islands. Three bedrock channels in the basin have already been partly explored, but there is at least one other channel that lies somewhere in the deeper ground to the northeast that has not yet been reached. The southwest or Aetna Channel is unquestionably the youngest of the three, as it has cut through both the others. Similarly, Central Hill Channel, on which the main shaft is situated, is younger than No. 5 Channel, which it has cut through both to the southeast and northwest. Although the local relationships of these three channels are known, more exploratory work must be done, particularly on the channels to the east, before they can be definitely correlated with each other and with the main channel of the ancient river. There is evidence that Aetna Channel is not one of the main-stream channels but that it entered the basin through a gap in the rimrock to the south. Its gravels differ considerably from those in the other channels and contain a greater proportion of quartz sand and boulders. Its gold is relatively small and for the most part little worn, numerous pieces with quartz matrix adhering to them indicating that the gold has not traveled far from its source. On the contrary, in Central Hill Channel the gold is large and well worn. Nuggets weighing 1 pennyweight to 1 or 2 ounces occur frequently and the gold is identical in character with that from the Vallecito-Western and Golden River mines upstream. There is little question, therefore, that this channel was formed by the Tertiary Calaveras River, but there is evidence that it may not have been its original or main channel. In No. 5 Channel the gold is medium coarse and flaky and there are very few large nuggets. Further, the gravel in this channel is relatively shallow and the overlying rhyolite tuff in some places reaches the bedrock on the benches, practically all of the gravel having been swept away. In most of these cases, however, enough gold was left in the creviced bedrock to make it worth mining. In the Aetna and Central Hill Channels the gravel is very thick, and nowhere have the workings reached the overlying volcanic ash. In one instance the gravel of the Central Hill Channel overlaps a thin layer of volcanic ash, which caps No. 5 Channel. This not only confirms the greater age of No. 5 Channel but suggests that the undiscovered channel in the deeper ground to the northeast may have been the original channel of the main river through this basin. If this is true, it has an important economic bearing, as it is reasonable to suppose that the original channel contains the richest concentration of gold.

Superimposed above this system of bedrock channels in the basin are intervolcanic channels in rhyolite, of which little is known and of which only the McElroy has ever been worked. This channel lies above the lowest stratum of rhyolitic tuff and enters the basin from the south along the eastern edge of the Mother Lode. Presumably it derived much of its gold from the erosion of quartz veins and stringers of the Mother Lode formation, which passes through the southwestern end of the property. This section is near such famous mines as the Utica, Lightner, Angels Quartz, and Sultana, and numerous quartz stringers have been cut in running bedrock drifts, which invariably have shown values in gold; in one instance a 4-foot vein found in the Aetna workings yielded a cut sample which assayed 0.20 ounce of gold.

No molten lavas invaded this basin, but there were two distinct periods in which heavy blankets of rhyolitic tuffs were laid down, and a third and later period during which several hundred feet of andesitic tuffs were deposited. In some places the andesitic tuffs have been completely or nearly removed by erosion, while in others 100 feet or more of them still remain.

The Calaveras Central shaft passes through the following sequence of strata: 75 feet of andesitic tuff, 85 feet of brown gravel, 90 feet of rhyolitic tuff, 20 feet of gray gravel, 15 feet of rhyolitic tuff, and 60 feet of bluish gray gravel, which is more or less cemented by calcium carbonate. Normally, the lower 3 or 4 feet of gravel within the confines of the channels comprises the richer pay. The vertical limits of the pay gravel, however, vary widely, and in one place on No. 5 Channel good ground was mined for 21 feet above bedrock. This is an exceptional instance and in most cases only the lower 4 to 6 feet of gravel and 1 to 3 feet of the bedrock are mined. The width of the pay gravel varies greatly. In No. 5 Channel it is 50 to 70 feet and in Central Hill and Aetna Channels 150 to 200 feet. The bedrock throughout the workings is either Calaveras slate or schist. The original grade of the bedrock channels has been altered by a gradual uplift of the country toward the northwest, probably caused by faulting movements. A gradual elevation of the bedrock downstream has resulted.

The gravel itself is composed largely of well-rounded porphyry, granite, granodiorite, and considerable quartz. Many large boulders are found. The gold is coarse, most of it being retained on a 10-mesh screen, and it is about 885 fine, the impurity being largely silver. It is associated with considerable pyrite and a little black sand.

Development

When the property was taken over by the present company there were three shafts on the tract. The three-compartment Reiner shaft is 350 feet deep; the Aetna shaft, about 900 feet southeast of it, is 240 feet deep; the McElroy shaft, about 1200 feet southwest of the Reiner shaft, is 200 feet deep. The Reiner shaft was reconditioned to become the main Calaveras Central working shaft, the Aetna shaft being retimbered and connected underground with the main workings so that it affords the emergency exit required by California law and assists materially in ventilating the mine. Although the McElroy shaft was sunk about 80 years ago, it has not caved and may later be retimbered and used in working the McElroy channel. Connecting it with the present workings would aid greatly in their ventilation. The position of these shafts with relation to the underground workings and the channels that have been developed so far are shown in figures 13 and 14.

The first work done by the present company, starting in 1931, was to run a crosscut from the station of the main shaft 800 feet northeast, with the object of reaching the original channel, which lies in this direction; but 520 feet from the shaft No. 5 Channel was intersected, carrying excellent values in fine and medium-coarse gold. Because of the necessity of producing promptly, most of the developing and mining done during the next 2 years was confined to this channel, which was explored for 1600 feet upstream, where it was cut off by a deeper channel, apparently the one on which the main shaft is situated. Excellent pay gravel was found in this deeper ground, but no breasting was done, as new, low-level, bedrock haulage tunnels must be driven before this area can be mined economically.

From 1934 to 1936, inclusive, most mining and development were on the Central Hill Channel, which was followed downstream 2700 feet from the shaft. About 1500 feet from the shaft this channel was cut through by the Aetna Channel, the discovery of which at this intersection has assisted materially in establishing its position and bedrock grade and proves that it can be reached by crosscuts to the southwest from the workings on the Central Hill Channel, thus opening it for mining at many different places and over a considerable distance. Development has been confined to the three channels mentioned, which have been opened for a maximum length of over 4000 feet; but, owing to the large number of crosscuts and parallel drifts, total development approximates 30,000 feet. The total length of the three channels opened is estimated in a recent report to comprise only about one-tenth of the entire length of channels in the property, and only a small part of the gravel in the developed portion has been mined, with the previously stated yield of 20,000 ounces of gold.

The company has not yet started operations on the area obtained from the Calmo Mining & Milling Co., later called Mound City Gold Mines, Inc., but two shafts were sunk previously. One of them, the Slab Ranch shaft, was sunk in a narrow gorge where the rimrocks were only about 500 feet apart at the surface. The negative results should have been foreseen, as retention of gold there should not have been expected.

The original Calmo shaft was sunk farther downstream near the north rim, and a small amount of gold, valued at about $25,000, was produced. It included a nugget weighing 20 ounces, but this gold came from rim gravels, as the main channel appears to lie farther south in deeper ground not reached as yet.

The plans of the present company contemplate exploration of that area by proceeding upstream on the channels that may be followed from present workings in the normal course of operations. The main crosscut is to be extended to the northeast until it encounters the main channel, which has not been reached as yet. This channel likewise will be sought at a higher level by a crosscut from the southeast end of the present workings. The Aetna Channel will also be crosscut in several places and opened for mining, a number of existing drifts being connected to serve partly developed sections of the mine.

Reserves and Values of Gravel

A recent report estimates that in present workings there are 328,000 tons of high-grade gravel, averaging $5.32 a ton, and 117,000 tons of low-grade gravel, averaging $1.19 a ton, shown by present development work. These values are based upon past recovery from adjacent gravels. It can be increased in future operations by reducing the tailings losses, which have been excessive. The low-grade gravel is so situated that it can be mined by drag scraping at little cost, which is expected to permit some profit. These reserves, claimed as now assured, represent but a small part of the total length of the channels in the property. Much of the unexplored ground is close to and will be accessible from present workings.

The company is planning to increase its hoisting and mill capacity to 500 tons a day, after which mining will be expanded gradually by addition of units of underground equipment to provide that tonnage. The company records show that since 1931 it has mined 185,000 tons of material, of which about 75 percent has been auriferous gravel and bedrock that have been milled, the balance being waste bedrock.

Three channels so far developed in the Calaveras Central property—the Aetna, Central, and No. 5—are said to have yielded $240, $255, and $530, respectively, per linear foot. The average recovery during 1935 was $5.17 per ton, and from the beginning of operations in June 1931 to 1936, inclusive, the average was $4.94 a ton. These averages include recoveries from much low-grade gravel extracted during exploration work, which might be excluded in breasting developed reserves. The range of values found varied considerably with the character and grade of the bedrock and the position of the pay in relation to the channels. Rich concentrations were occasionally found yielding an average as high as $72 per ton for a single day and $18.71 per ton for an entire month. The rich pay streaks provided many specimens of consolidated gravel in which numerous small nuggets are plainly visible. In one speciment the nuggets are all flat and lie, with reference to each other, precisely like shingles on a roof. Such rich concentrations illustrate the effectiveness of bedrock as a riffle under some circumstances.

The yields obtained do not represent the total gold content of the gravel, as considerable tailings losses occurred. About 42,000 tons of fine tailings have been produced since mining began in 1931. Of these, 12,000 tons were re-treated and yielded $1.35 per ton with incomplete recovery. It is thought that the remaining 30,000 tons contain about $2 per ton. This would give a gold content of $84,000 for the fine tailings, to be distributed over the total of 140,000 tons milled, giving 60 cents a ton. Added to the average of $4.94 per ton recovered, $5.54 per ton is indicated as the average gold content of all gravel mined by the present company, exclusive, of course, of any lost in the coarse tailings.

Mining and Milling

The room-and-pillar method of mining is used. Bedrock drifts, generally under the rims or in the troughs of the channels, are used as haulageways.

The gravel face and 1 to 3 feet of the bedrock are shot down with light loads of 40-percent dynamite. In past operations the product has been loaded into 2-ton cars largely by Eimco-Finley loaders. Light drifters were used for drilling. The loaded muck was hauled to the shaft by storage-battery locomotives, generally in trains of four to six cars, and dumped into a gravel pocket, from which it was drawn into $2\frac{1}{2}$-ton skips and hoisted. The skips were dumped automatically into a 70-ton mill bin adjoining the head frame. Waste was delivered to a 20-ton bin, from which it was discharged by a 200-foot stacker with a 24-inch belt.

In the milling system used for most of the past production the material in the bin was fed to a 21- by 5-foot cylindrical washer with a 10-foot punched-plate screen section having 1-inch holes. The oversize from the washer was piled by a 275-foot stacker having a 24-inch belt. The undersize passed through a specially designed sluice and gold trap 24 inches wide and then through a second sluice 12 inches wide and 24 feet long, lined with Hungarian riffles. The discharge from the sluice was dewatered by a drag classifier and conveyed to a Leahy vibrating screen, which made three products—plus $\frac{1}{2}$-inch material, which was piled by a 150-foot stacker having a 10-inch belt; minus $\frac{1}{2}$-inch but plus $\frac{1}{4}$-inch; and minus $\frac{1}{4}$-inch. The two smaller sizes were treated in a Huelsdonk concentrator, the discharge from which was raised by a bucket elevator to a launder, which conveyed it to the fine-tailings dump.

The washing plant described is now being dismantled, as the company plans to replace it with a plant of 500 to 600 tons capacity.

The present double-drum hoist has a capacity of 1,000 tons a day. Two Sullivan angle-compound compressors of 400 and 800 cubic feet capacity, respectively, supply air for drilling and the operation of compressed-air machinery. The mine makes about 150,000 gallons of water daily, which is pumped from the shaft sump by a 50-horsepower, Sterling vertical turbine pump into a 100,000-gallon tank on the hillside above the mill. This tank supplies ample water for milling. The property has a large change room, a blacksmith and machine shop, a first-aid building, office, and other structures. The number of men employed has ranged from 30 to 70, but this number will be doubled under the contemplated expansion program.

Costs

The following table shows the average daily mining and milling costs during 1933 and 1934, a period selected because of the wide range in the average daily tonnage treated. The costs given cover the mining and milling of gravel and auriferous bedrock and the mining of waste bedrock, the latter usually amounting to 25 or 30 percent of the total. The daily tonnages range from a high of 209 tons in May 1933 to a low of 41 tons in May 1934, the corresponding working costs being $1.40 and $3.04 per ton, respectively. The adverse effect of the inverse ratio is at once apparent.

The costs given include labor, compensation insurance, power, explosives, lubricants, and salaries of the general manager and superintendent. Depreciation, depletion, and other overhead costs are not included.

Mining and milling costs at the Calaveras Central mine from January 1933, to December 1934, inclusive

1933	Monthly gross tonnage	Average daily tonnage	Mining and milling costs per ton	1934	Monthly gross tonnage	Average daily tonnage	Mining and milling costs per ton
January	3,399	109	$2.27	January	4,363	140	$1.64
February	3,719	132	2.11	February	3,056	109	1.88
March	5,433	175	1.73	March	2,234	72	2.30
April	5,472	182	1.61	April	1,264	42	2.73
May	6,507	209	1.40	May	1,271	41	3.04
June	5,753	191	1.54	June	1,645	54	2.74
July	4,313	143	1.73	July	2,440	81	1.85
August	3,315	106	2.02	August	3,114	100	1.91
September	3,137	104	2.03	September	3,350	111	1.84
October	2,526	81	2.52	October	4,203	135	1.54
November	2,471	82	2.60	November	4,087	136	1.63
December	3,491	116	1.84	December	2,856	95	2.31
				Average, 1933 and 1934	3,476	114	1.89

The very irregular rate of production indicates that development or equipment was inadequate to maintain steady production necessary for holding costs at a low level. During the 2-year period 83,419 tons of material was mined, an average of 114 tons a day, at a total cost of $157,699 and an average cost of $1.89 per ton.

Based upon an analysis of cost details, the management has estimated that if the output had averaged 200 tons daily, the corresponding cost would have been $1.40 a ton, which was actually attained in May 1933; that at 300 tons a day the corresponding cost should be about $1.25 a ton; and that at 500 tons a day about $1 might be attained. The tenor of these conclusions is obviously correct. There remains, however, the question whether the larger rates of production can be maintained steadily. If so, it would be accompanied by some lowering of average grade, as low costs would naturally induce the mining of greater widths, including the low-grade gravels of the channel margins, and perhaps abandonment of all selection within the channels proper. This would exactly parallel the usual trend toward nonselective mining that proves inevitable in lode mines.

Even now the costs attained compare favorably with those current when drift mining was at the height of its prosperity, when much cheap labor was available. Then, under favorable conditions for mining and milling, cemented gravel was worked at a cost of from $1.75 to $3.50 a cubic yard, equivalent to $1.16 to $2.32 a ton.[4]

Character of Production

The gross tonnages do not, of course, consist entirely of pay gravel or income-producing product. A certain proportion must be bedrock waste, taken out in running approach and haulage tunnels, and the proportion of such material to the pay gravel has an important effect.

In mining 185,000 tons of material, 75 percent was gravel and 25 percent waste. As the gross tonnage increases, the proportion of waste decreases, because in a small-tonnage operation the gravel is mined more selectively to offset high cost. To maintain this high average grade there must be constant additions to the main haulage system in order to pass through or around low- or even medium-grade gravel to reach the grade being mined; but in an operation where the daily tonnage is large and the cost lower, the gravel need not be mined so selectively, and its bulk in proportion to waste becomes greater.

The general effect of increased tonnage in reducing the percentage of waste is thought by the management to be somewhat as indicated below. For 100 and 200 tons the figures are based upon records; for the higher tonnages they are estimates as to effects that might be reasonably expected.

Daily production	Waste	Gravel	Pay material
	Tons	Tons	Percent
100 tons	30	70	70
200 tons	40	160	80
300 tons	50	250	83
400 tons	60	340	85
500 tons	70	430	86

[4] Hammond, John Hays, The auriferous gravels of California: California State Min. Bur., State Mineralogist's Rept. 9, p. 119, 1889.

The use of mechanical loaders in this mine has been an important means of reducing costs.   To obtain from them the benefit of continuous operation, empty cars must always be at the face for loading. Switch spurs are therefore driven at frequent intervals as the headings are extended, to aid in placing cars.   Usually these spurs develop later into crosscuts or other haulageways.   Sometimes, however, cars are lifted bodily and transferred to the end of the train by means of a device, developed at this mine, that may be called a car derrick.   It consists of a 6-inch I-beam, about 9 feet long, with a 1-ton, geared chain hoist on a carriage that rolls on the lower flanges of the beam.   The beam is wedged securely at each end into niches cut in the walls near the roof, or it may be supported partly by stulls.   A steel spreader with hooks at each end depends from the hook of the chain hoist.   The hooks are swung under either the ends or sides of the car and it is lifted a few inches off the track and pushed into a shallow recess in the wall, far enough to clear the train.   When the locomotive has pulled the train by, the empty car is returned to the track.   Use of this derrick is said to save a great deal of time and expense that would be required to prepare switching spurs.   It is especially useful in driving long bedrock drifts.

The great savings effected by mechanical loading underground have led the management to devise a system of mining that seeks to utilize it to the utmost.   The following notes concerning this and other means of reducing costs and increasing efficiency are condensed from data furnished by Harry Sears, general manager.

Technology of Drift Mining

The gravel is tightly cemented; bedrock drifts usually hold firm without timbering, hence both gravel and bedrock are treated like hard rock of a lode mine.   The mine is being laid out with the object of opening it in a number of sections with separate equipment provided for their operation.   Enough of them are to be provided to assure production of the mine as a whole at a steady rate and to permit the sections to remain idle after blasting long enough for the smoke to be dispelled.

The workings are planned so as to permit good circulation of air, and many crosscuts and connecting drifts have been run for this purpose, which, incidentally, have served as exploration drifts.   In most cases the efficiency gained through the better air conditions has largely paid for the exploration drifts.

Headings generally are 7 by 7 feet.   Drifts are run on a grade of 1 percent to assure drainage away from the faces, except in very wet gravel with considerable sand, where the grade sometimes is increased to 2 percent, as otherwise the sand would settle on the tracks and too much time would be lost in keeping them clear.   Normally there is 1 to 2 feet of bedrock in each gravel face, the rest of the 7 feet being gravel.

The whole of the face is drilled, usually with 8 to 10 holes 6 to 7 feet deep.   Most of the headings are loaded out with mucking machines and, before shooting, the rails with slide rails in place are brought practically up to the face.   The holes are loaded and shot so as to deliver the muck, combined gravel, and bedrock in a compact pile close to the face.   It is therefore a simple matter to bring a loader into the heading, clean up the track, and quickly reach the main muck pile.   The slide rails are advanced as needed, the lip of the loader scoop being used to push them under the muck when required.

Many areas have been worked by driving roughly parallel drifts with long walls between them, these walls later being shot through at regular intervals to form pillars. This was done with the tracks in both drifts in place, and nearly all of the muck was handled with the loaders. For this work the track is laid on flat steel ties, and it can be slid or barred over from its original position when required. Considering each of these working areas as a room containing a number of pillars or walls, if it is desired to mine it out completely and retreat, this can be done by pointing the track at individual pillars as they are finally shot out, and taking up and shortening the track while retreating to the entrance of the room. Stulls with headboards or caps may be required to support the ground while the final pillars are being shot and loaded out, but often they are not.

The loaders have also been used on longwall work, by drilling the wall and sliding the track over almost against it so that the muck will be thrown on the track. In all of these methods a moderate amount of hand-shovel work is necessary to clean up completely spots beyond the radius of the loader scoop. The best method of handling each mining area must, of course, be determined somewhat by the character of the ground.

All of the muck is loaded, and no boulders are separated or piled underground. The old practice of piling boulders underground requires much hand labor and is far more expensive than bringing them to the surface. Of course, very large boulders must be bulldozed or broken with sledges before they can be loaded out, and occasionally some of these larger boulders are left underground, but never piled up as walls. The bedrock is never scraped, but is shot and treated as gravel, being taken to sufficient depth to get all the gold in its crevices.

Work with a multiple-track system not only provides several headings for loading in a comparatively small area, but makes it possible to maintain switches close to the faces, so that empty cars may be readily available and the loaders can be kept in steady operation.

In driving long, single-track, bedrock haulage drifts by means of a car derrick, several thousand feet of drifts have been driven at the rate of 18 feet per day, with two men in each heading, three shifts per day. Each shift drilled, shot, and loaded out the muck for a 6-foot advance and did the necessary track and pipe work. Each crew of two men, however, was assisted by a motorman who brought in the empties and took out the loaded cars. Two-ton cars hauled by storage-battery locomotives were used. The motorman's services were only required during an approximate 1-hour period while mucking out. Gravel loading does not generally proceed as fast as this, particularly with the room-and-pillar work described, for considerable time is required to break large boulders, place stulls, and do track and hand-shovel work. For most rapid work, the loaders must be kept on straight tracks for they are slowed materially when on curves. The foregoing remarks apply to the use of Eimco-Finlay loaders, which operate within a fixed radius along or from the end of the track, and which throw the muck back into a car that is coupled to the loader and moves with it.

A Nordberg-Butler shovel on a caterpillar tread also has been used. This shovel is very efficient in room work where a train of several cars can be delivered on the track, and the shovel can load them all without spotting or shifting them. With its ability to move around on both sides of the cars and its consequent wide loading radius, this shovel has distinct advantages while working in a prescribed area, but much more time is required to move it from place to place in the mine than with the Eimco-Finlay loaders.

In opening various channels and preparing multiple sections of each for production, the management endeavors to maintain independent haulage drifts, preferably beneath the lowest point of the channel troughs; or, if this cannot be done, then in the bedrock rim alongside the channel. If it is necessary to pass through the gravel itself, the drifts are kept as much as possible in low-grade gravel. With this system, entrance may be had to the channels from any required point in the haulageway and as much or as little gravel extracted as desired without impairing the efficiency or adding to the upkeep of the main haulage system. Another point of advantage is that good gravel need not be tied up indefinitely to keep the haulage drifts open.

Where the main haulageway is deep enough below the gravel to permit it, entrance to the gravel is made through a raise and a bin is built, which can be loaded as required from the upper level and drawn when convenient by the train crew on the haulage level. Where the difference in levels is only 5 or 6 feet, transfer platforms are built to accommodate at least three cars, and these are loaded directly, either by center or side dumping, from the cars above. Car loading with drag scrapers is invariably done by dragging the load onto a platform beneath which a car is spotted, the load being delivered directly into the car through a slot in the platform.

Drag Scraping. A particularly interesting phase of mining at this property has been the work done with drag scrapers—probably their first use in drift mining. Though only a small tonnage has been handled by this method, it has been sufficient to establish facts on procedure and costs and its use is to be extended. The following is intended to be merely suggestive of possibilities foreseen.

In prepared ground, ninety 2-ton cars were loaded with a small scraper by three men during an 8-hour shift. It is estimated that four man-shifts were required to prepare the ground by drilling and shooting, that one man-shift was utilized for electric tramming, one-half man-shift for drawing into the skips, and one-half man-shift for hoisting, a total of nine man-shifts for drilling, shooting, loading, tramming, and hoisting 180 tons, a production of 20 tons per man-shift.

Assuming a labor cost of $5 per man-shift, including compensation insurance, the cost delivered in the mill bin was $45. Adding one man-shift for milling and sluicing gives a total direct labor cost of $50. With an allowance of $15 for explosives, power, and incidentals, the mining and milling cost for the 180 tons handled is estimated at approximately $65, or 36 cents per ton.

It is thought that with a heavier scraper and larger cars this production could have been increased 50 percent with additional expense of two man-shifts for drilling, one-half man-shift for tramming, one-half man-shift for drawing into skips and hoisting, and one-half man-shift for milling, making an additional three and one-half man-shifts, costing $17.50. With $7.50 added for explosives, power, and incidentals, a cost of $90 for mining and milling 270 tons of gravel, or 30 cents per ton, is indicated. To the foregoing costs must be added the proportion of expense properly chargeable to the ground, including approach and installation work. In handling caved gravel, drilling and shooting costs are eliminated.

Unfortunately, however, the application of this method has limits, experience gained at this mine indicating that drag scraping in drift mines is suitable only for areas presenting certain favorable conditions of bedrock and of the gravel body itself where a large prepared tonnage can be made available. It is expected that this method will find its principal use in handling large tonnages of low-grade gravel, but that in many cases it can also be used for mining the high-grade gravel remaining after all the ground that can be mined safely by other methods has been taken out.

The preparation of ground for scraping entails considerable dead work. The gravel should first be opened and thoroughly drained, as scraping can be done best in dry ground and water on the bedrock causes a loss of gold in scraping. In dry ground more gravel is broken per shot.

For safety, the tail pulley and the tugger hoist should be placed where there is no danger from caving, and they should be accessible through drifts or raises independent of the scraping area. Large scrapers and powerful tugger hoists are desirable, as the essence of success in scraping is quantity, and with heavy equipment the same number of men can load a much larger tonnage than they can with light equipment. There should be an ample supply of both cars and locomotives, for when loading is done by this method empty cars, preferably of large capacity, must be constantly available.

Hoisting and milling capacities must be suitable for handling the large tonnage necessary for operating on low-grade gravel. Otherwise, they must be used for gravel of higher value. In future operations it is thought that a tonnage balance will be worked out between high- and low-grade gravels.

Under some roof conditions drag scraping can best be used after the gravel is partly or largely worked out by drifting, and has then been caved or has been allowed to cave of itself to a limited height before scraping is attempted.

Reserve scrapers and cables must always be kept on hand, as they may be buried by caving under conditions that will prevent men from safely entering the room to recover them. Occasional losses of this equipment should be taken as a part of the cost of using the method.

Project for Improvements

During forthcoming construction of a new plant, the management plans a number of fundamental changes underground as well as on the surface.

The shaft is to be entirely retimbered, new water, air, and power lines installed, and the shaft deepened to create a larger sump and provide additional depth for drawing from gravel and waste pockets, each to be enlarged to 100 tons capacity. A separate gravel pocket will be provided for tonnage for sampling and testing. The bins will be arranged so that the cars on an incoming loaded train may be dumped into their respective pockets without disturbing the make-up of the train.

A new, 50-horsepower, Sterling, vertical turbine mine pump is to be installed and mounted on a roller carriage, so that when motor or pump repairs are necessary the complete unit may be disconnected, brought to the center of one of the hoisting compartments of the shaft, hooked onto the bottom of the skip, and brought to the surface.

Heavier rail will be laid throughout the main-haulage drifts and the track gage will be widened to permit the use of low-slung, 3-ton, drop-bottom cars. Trolley locomotives will be used for main haulage, and storage-battery locomotives in newly opened ground and for gathering and train make-up. New 3½-ton skips will be provided in the shaft for hoisting both waste and gravel, with screened cages above them for raising and lowering the men.

A new steel headframe 100 feet high will replace the present 65-foot wooden one. For surface storage, a 400-ton bin will be divided into three compartments—250 tons for the regular production of gravel, 75 tons for gravel to be sampled or tested, and 75 tons for waste.

The gravel will be delivered from the bins by magnetic vibrating feeders, which will assure the regular feed essential for proper milling. In order that the cemented gravel may be completely disintegrated and the gold liberated from it, the gravel must be subjected to much attrition and scrubbing. In the old washing plant the trommel did not retain enough water to effect complete scrubbing of the gravel, so a watertight cylindrical washer will be installed, which will carry a much larger load than the old trommel. In this the bulk of the material will be immersed in water, and the grinding and scrubbing action of the gravel and boulders is expected to produce a much cleaner product. Breaking up the larger cemented pieces does not present as much of a problem as does material between one-sixteenth and three-eighths inch in size. The greatest loss of gold in the tailings in the past has been in this size of material in which light flakes of gold adhere to bits of gravel or light sulphides or are encased therein. It is thought that this loss may be eliminated by the increased grinding and scouring action of boulders in the new scrubber; but if it is not, this fine material will be separated by screening, to be returned for re-treatment or to be ground separately.

The coarse gold, constituting the principal part of the clean-up, will be recovered as in the past in sluices immediately following the scrubber. The sands and fines will be fed in a thin flow to Pan American jigs, which will make a rough concentrate in their hutches; this will be cleaned by another jig to a final concentrate consisting of fine gold and black sand. This concentrate may be shipped for final recovery of its values to a plant treating black sands from dredges. The coarse material discharged from the mill will be piled by stacker belts, the fines being delivered by launders to the tailings pond.

A pilot mill will duplicate the processes of the main washing plant. Its function will be to make batch runs of 10 to 30 tons of gravel, so that any face in the mine may be separately sampled. Such large-lot sampling will determine how far mining may be profitably carried into areas of questionable value, and will serve as a constant check upon the average gold content of the mill tonnage.

Vallecito Western Drift Mine

The Vallecito Mining Company, Inc., Murphys, worked the Vallecito-Western drift mine throughout 1940 and recovered 221 ounces of gold and 25 ounces of silver from 685 cubic yards of gravel. The following description of this mine is reprinted from U. S. Bureau of Mines Bulletin 413.[5]

The Vallecito-Western mine is 3 miles east of Angels Camp in Sec. 24, T. 3 N., R. 13 E. It is on the same channel as the Calaveras Central mine, about 2 miles downstream; and the Golden River mine adjoins it upstream. This is the Central Hill or main channel of the Tertiary Calaveras River. The property is owned by Thomas B. Bishop Co., 166 Geary Street, San Francisco, but is under lease to the Vallecito Mining Company, of which Don Steffa, of Murphys, California, is general manager. However, from December 1932 to October 1936 the mine was operated under sublease by the Tonopah-Belmont Development Company under the direction of Frederick Bradshaw, of San Francisco, and the operating data given herewith cover most of this period.

Since the mine was opened in 1925 it has produced over $400,000 in gold, reckoned at its present price. It is developed by a 167-foot vertical shaft with a 4- by 4½-foot skipway and 2½- by 4-foot manway. This shaft was completed in May 1924, its site having been determined by a line of seven churn-drill holes sunk across the channel. The shaft passed through about 50 feet of white rhyolite cobble and ash and then entered a bed of well-rounded, bluish gravel made up largely of porphyritic material and containing frequent beds of sand and volcanic ash. Near the bedrock the porphyry gravel disappears, and the pay gravel is largely made up of granite with a small proportion of quartz, quartzite, jasper, and slate. Pieces of chocolate-brown semipetrified wood are sometimes found in the pay gravels.

At a depth of 153 feet, a tunnel on bedrock runs upstream from the shaft station for 300 feet and then, after a vertical raise of 5 feet to surmount an abrupt rise in the bedrock, it continues on a 1¼-percent grade for 4,000 feet. This tunnel serves as a haulageway and drain. At the shaft the bedrock is slate, but it changes upstream first to granodiorite and then to granite cut by bands of granodiorite. As the grade of the bedrock is slightly greater than the tunnel grade, particularly in its upstream end, the tunnel gradually passes into the bedrock, and access to the gravel above is gained by short raises used as chutes. The spacing of the raises depends on the volume of the gravel that is to be extracted through them and on the contours of the bedrock. Where the gravel is mined for widths of 100 to 150 feet the raises may be only 50 feet apart, but in narrow workings the distance between them may be as much as 125 feet.

The gravel is mined by breast stoping across the full width of the paystreak, which ranges from 40 to 150 feet and is extracted for an average height of 6 feet. Two rows of 6-foot holes, spaced 4 feet apart in the row, are drilled in the gravel faces, the lower one 1½ feet and the upper one 3½ to 4 feet above bedrock. These holes, loaded with three or four sticks of 40-percent dynamite, break to a height of 5 feet. If a greater height is mined, to include an occasional paystreak well above bedrock, an additional row of holes is drilled. Generally 75 percent of the gold recovered is on the bedrock or within a foot of it. When the bedrock is soft or badly creviced, 1 to 2 feet of it are mined. Such a condition occurs where the bedrock is slate or granodiorite, but if the pay-lead actually rests on bedrock even a granite bedrock is taken up.

---

[5] Julihn, C. E., and Horton, F. W., op. cit.

The gravel is well-rounded and heavy and contains many large boulders, which are piled underground in mined-out areas. Huge boulders 6 to 10 feet in maximum diameter occur frequently, and these are mined around; but if they are too much in the way they are drilled and broken. Normally the gravel is not cemented but stands well. However, in some places it must be held by square-sets, though usually posts 10 to 12 feet apart, with headboards, furnish ample support.

The broken gravel is loaded at the face into wheelbarrows and dumped into the nearest chute, from which it is drawn into 1-ton cars that are hauled, in trains of four, by a storage-battery locomotive to a car dump at a point 300 feet from the shaft, where the bedrock rises abruptly. Here it is transferred to other cars trammed by hand to the shaft, where they are discharged directly into a 1½-ton skip.

The shaft has a 76-foot head frame with a built-in, 30-ton ore bin. The skip is operated by a 50-horsepower electric hoist. An Ingersoll-Rand compressor, belt-driven by a 60-horsepower motor rated at 355 cubic feet per minute, supplies air for drilling. About 100,000 gallons of water per day is pumped from the shaft sump by a Sterling electric pump with a capacity of 135 gallons a minute. Pumping 12 to 16 hours a day handles the water.

The mine is well-ventilated. A 12-inch churn-drill hole from the surface intersects the haulageway about 2,400 feet from the shaft and has a down draft providing fresh air. A blower sends air from this point through 12-inch galvanized iron pipe to the tunnel face and the active mining areas.

The gold at the Vallecito-Western property is coarse and averages 895 fine. The balance is almost entirely silver. Ninety percent of the gold by weight stays on a 20-mesh screen. Nuggets having a value of $1 to $5 are common, and many worth from $10 to $20 have been found. The largest nugget yet discovered weighed a little less than 10 ounces.

Washing Plant. The washing plant adjoins the shaft. Gravel from the head-frame bin, into which the skip discharges, falls by gravity into the upper end of a sluice 12 feet long and 2 feet wide, where it is washed by only sufficient water to move rocks 4 to 5 inches in diameter. Larger rocks are picked from the sluice by hand, thrown into a car on a lower floor, and trammed to the dump. Fifty to sixty percent of the gold is recovered in this short sluice. The finer gravel flows into a revolving washer 36 inches in diameter and 9 feet long, making 35 revolutions per minute. This washer has a 4-percent slope and is equipped with lifting plates, which elevate the gravel and then let it fall, thus breaking lumps and scouring off adhering clay. A 4-foot double trommel of punched steel plate is attached to the discharge end of the washer. The inner screen is 20 inches in diameter and has 1½-inch round holes. The outer screen has one-half inch holes and is 36 inches in diameter. Plus 1½-inch gravel is sluiced to the dump, but the two finer sizes are washed in separate sluices.

These sluices are 1 foot wide inside and have a 4-percent grade. They are lined with riffles of a novel design, which are very effective in preventing packing and also in retaining the gold. These riffles are of bar iron 15½ inches long, 2 inches wide, and a quarter of an inch thick, bent at right angles 2 inches from each end and the bent portions beveled and drilled. The drill holes receive rivets that fasten the riffle to the sluice liners. The bar sits at an angle of 45° against the current and its upper edge forms a lip that produces a distinct boil in the pocket beneath it. The lower edge of the riffle face does not contact the bottom of the sluice, but the intervening space fills in with small pebbles, which keep the gold from sliding along the bottom of the box.

Some loss of gold occurs in the washing plant when pieces of cemented gravel are discarded by the trommel or go through the sluices. However, relatively little cemented ground is found in the workings, and this is segregated and allowed to air-slake before it is washed.

The recovered gold averages about 0.17 ounce, or $5.95 per ton of gravel washed, but it varies, of course, within wide limits. In August 1936, 80 tons of gravel were mined and washed per 24 hours. Twenty-five men were employed on three shifts.

The following tables show the production and costs at the mine for a period of 45 months ended September 1, 1936. In interpreting the costs, it must be remembered that they are based on the tonnage hoisted rather than on tonnage mined. The waste hoisted was equal to 44.2 percent of the gravel milled. Most of this waste was derived from bedrock cuts, and the rest consisted of boulders which it was necessary to hoist wherever new breasts were started, until enough room was cleared to pile them back of the face. About 30 percent of the ground mined consisted of boulders, which were left underground, so the total waste mined approximately equaled the tonnage of gravel milled.

Costs and production of the Vallecito-Western mine, 45 months, to September 1, 1936

	Tons
Gravel mined and milled	40,967
Waste hoisted	18,102
Total gravel and waste	59,069

	Value		
		Per ton of—	
	Total	Gravel	Gravel and waste
Production:			
Gold, 7,222.5 ounces	$232,268.00	$5.6696	$3.9321
Silver, 814.9 ounces	492.00	.0120	.0083
Total recovered	$232,760.00	$5.6816	$3.9404
Loss in tailings	11,713.00	.2860	.1984
Gross value in gravel	$244,473.00	$5.9676	$4.1388
Costs:			
Mining and milling during—			
Development	$62,747.05	$1.5316	$1.0623
Breasting	84,111.17	2.0531	1.4239
Maintenance and repairs	14,302.64	.3491	.2421
Water supply	13,344.31	.3257	.2259
Compensation insurance	13,064.07	.3190	.2212
Total of direct mining and milling	$187,569.24	$4.5785	$3.1754
Indirect mining and milling	19,852.06	.4846	.3361
Marketing (mint charges, etc.)	1,171.73	.0286	.0197
Total operating cost	$208,593.03	$5.0917	$3.5312
Profit before payment of royalty	$24,166.97	$.5899	$.4092
Royalties	24,946.00	.6089	.4223
Loss	779.03	.0190	.0130

The marginal character of drift mining, when conducted as described, is well-illustrated by these data of costs and production.   With an average gold content of $4.14 per ton of gravel and waste hoisted, which becomes $5.97 per ton for gravel alone, the actual recovery is $5.68 per ton of gravel.   However, this very substantial recovery exceeds the operating cost of $5.09 per ton of gravel by only 59 cents per ton, while royalties amount to nearly 61 cents per ton, so that there is a final loss of nearly 2 cents per ton of gravel.   The net result is merely the maintenance through 4 years of the labor employed and payment of the royalties required by the owner of the ground, while no excess is left to reward the operating company that carried the burden of responsibilities.

The rate of production is very low, however, averaging a trifle less than 30 tons of gravel per day for the whole period.   It is immediately obvious that a marginal operation at that low rate of production would probably become profitable at a higher rate of production.   That, in turn, would require an additional capital investment and probably a preliminary determination of the underground contours of the channels in order to plan with certainty production adequate for profit.

Such determinations have been virtually impossible in the past because of the lack of any known method for their accomplishment except at prohibitive cost.   If, however, it should prove that the buried channels of the ancient placers can be mapped completely by geophysical methods, the mining of such properties as this should no longer prove marginal but should yield substantial profits.

A portion of the gold mined during the period under consideration was marketed before its present price was attained, and the average received was $32.15 per fine ounce.

The early history and operation of this mine under the management of the Vallecito Mining Company has been described in an information circular of the Bureau of Mines.[6]

Other Mines

Bacon, E. A., 303 Delmar Way, San Mateo, recovered 286 ounces of gold and 14 ounces of silver from 10,000 cubic yards of gravel near Wallace in 1942.   Hydraulicking and a mechanical excavator were used.

Burson Mining Company, 2054 University Avenue, Berkeley, operated a dry-land dredge on the Foster Ranch in the Camanche district intermittently in 1941.   Recovery from 55,200 cubic yards of gravel was 495 ounces of gold and 37 ounces of silver.   (See Cat Camp Placers also.)

Calaveras Central Mine.   See page 235.

Cat Camp hydraulic mine was operated by J. E. Biallas of Valley Springs in 1940.   Operation of a nonfloating washing plant, for 6 months of 1941, to which gravel was delivered by a carry-all, yielded 605 ounces of gold and 34 ounces of silver from 100,000 cubic yards.   In 1942, Cat Camp Placers and Burson Mining Company produced from this mine 483 ounces of gold and 32 ounces of silver from 71,818 cubic yards of gravel with nonfloating washing plants.

---

[6] Steffa, Don, Gold mining and milling methods and costs at the Vallecito-Western drift mine, Angels Camp, Calif.: U.S. Bur. Mines, Inf. Circ. 6612, 14 pp., 1932.

Church Union mine (Kraemer) in sec. 26, T. 5 N., R. 11 E., M. D., is owned by J. J. McSorley and Thos. E. McSorley, Mokelumne Hill. It includes 80 acres of mineral rights and 30 acres surface rights on the same ground, containing the Coffee-Mill channel, which runs in an east and west direction. According to J. J. McSorley, 20 acres of this ground is virgin, and should yield about an ounce of gold per cubic yard for a 6-foot cut on bedrock. A tunnel 1000 feet long has been driven in bedrock beneath the channel.

Clark, W., subleased the Val Ranch in 1942 and operated a stationary washing plant, to which gravel was delivered by power shovel and trucks. From May 7 to October 19, the yield was 69 ounces of gold and 7 ounces of silver from 6245 cubic yards of gravel.

Deep Lead placer mine is in sec. 13, T. 5 N., R. 11 E., M. D., and is owned by Miss Theresa Rooney, 1106—3rd Street, Corpus Christi, Texas, who holds 90 acres on the Chili Gulch Blue Lead channel. According to J. J. McSorley, Mokelumne Hill, about 20 acres of this ground remains unworked, and will yield about an ounce of gold per cubic yard for a 6-foot cut on bedrock. A shaft from 90 to 100 feet deep will be required to open the mine.

Glo-Bar Mines, 370 East 37th Street, Long Beach, operated the Glo-Bar drift mine in the Campo Seco district throughout 1940. Recovery from 4464 cubic yards of gravel was 321 ounces of gold and 44 ounces of silver. Operations continued in 1941.

Gold Hill Dredging Company, 311 California Street, San Francisco, operated an electric connected-bucket dredge on the Arlington and Osterman properties along the Mokelumne River during 5 months of 1941.

Golden River Mining Company, Roland Rich Wooley, president, 915 Transamerica Building, Los Angeles, has operated a drift mine near Vallicita in secs. 21, 29, 30, T. 3 N., R. 14 E., M. D., on the same main channel of the Tertiary Calaveras River on which are the Vallecito-Western and Calaveras Central mines. This company has also done geophysical work and drilling on the adjoining Kentucky placer 2 miles northeast of Angels Camp in secs. 23, 26, T. 3 N., R. 13 E. The Kentucky apparently contains a mile of virgin channel. The company was preparing to sink a new shaft on this property at the time (Oct. 8, 1942) that gold mining was shut down by Limitation Order L-208 of the War Production Board.

Gruwell, C. E., Angels Camp, operated a dragline dredge on the Hogate ranch, 3½ miles north of Angels Camp in 1940.

Henry, J. H., 740 West Willow Street, Stockton, washed 17,200 cubic yards of gravel by dragline dredging on the E. A. Marsh property 4 miles southeast of Valley Springs between May 30 and July 27, 1940, and recovered 137 ounces of gold and 13 ounces of silver. The same equipment washed 45,600 cubic yards of gravel on the Genochio property on the North Fork of the Calaveras River 1½ miles north of San Andreas between September and the end of the year; 362 ounces of gold and 40 ounces of silver were recovered.

Horseshoe Dredging Company, Mokelumne Hill, operated a dragline dredge on the Calaveras River 2 miles southwest of Jenny Lind during 1941; also on the Beers, Gertzen, and Osborn ranches.

Lancha Plana Gold Dredging Company of Camanche, operated its dredge No. 2 on Mokelumne River in Calaveras County during part of 1940. The dredge has 84 buckets of 6-cu. ft. capacity.

Lord and Bishop (Lobicasa Co.), Box 812, Sacramento, operated three dragline dredges on the Stockton Reservoir property on the Calaveras River 3 miles from Valley Springs in 1940. The dragline excavators used 3-, 1½-, and 1¾-cu. yd. buckets respectively. Operation of the 3-cu. yd. outfit was continued from July 1 to December 23, 1941, when the ground was worked out.

Mehrten Bros. operated a nonfloating washing plant, to which gravel was delivered by carry-all in the Camanche district from January 1 to July 22, 1941. From 16,200 cubic yards of gravel, 146 ounces of gold and 15 ounces of silver were recovered.

Midas Placer Company, of Camanche washed 50,000 cubic yards of gravel in a dry-land plant between April 14, 1940, and the end of the year. Recovery amounted to 659 ounces of gold and 52 ounces of silver. During a few months of 1941, high-channel gravel at the Penn gold-copper-zinc lode property was worked.

Quartz Hill Placers and *A. W. Ellis,* Box 116, Angels Camp, operated a stationary washing plant, to which gravel was delivered by a power-shovel, on the Quartz Hill property in 1941. Recovery from 11,270 cubic yards of gravel was 298 ounces of gold and 30 ounces of silver.

R. and M. Mining Company of La Porte operated a dragline dredge on Coyote Creek from January 1 to April 15, 1940; the excavator had a 1¼-cu. yd. bucket.

Ralford Mining Company, operated a dragline dredge with a ⅞-cu. yd. bucket on the Wm. P. Hiatt ranch in the Campo Seco district, from February 1 to July 10, 1941. Recovery from 25,000 cubic yards of gravel was 199 ounces of gold and 14 ounces of silver.

San Andreas Gold Dredging Company, 960 Russ Building, San Francisco, operated two dragline dredges each equipped with a 1½-cu.yd. bucket in 1940. Gravel was washed at the Airola-Costa, Albert Guttinger, John Guttinger, Batten, Bishop (Bowling Green), Calaveras Cement Company, Reed, Canepa, Byers, Fisher, Nuland, Nuner, Tanner, Bishop (Lot 29), and Solari properties. This company, which was sold to Thurman and Wright, 960 Russ Building, San Francisco, on March 7, 1941, operated its two dredges on the Fisher, Hageman-Huberty, Hageman, Lombardi, and Nuner properties in 1941.

Stagan Mining Company, 1440 North Hunter Street, Stockton, operated a dragline dredge on the Hunt and Robie ranches in the Jenny Lind district in 1940. At the Hunt ranch 86,000 cubic yards of gravel yielded 615 ounces of gold and 37 ounces of silver. At the Robie ranch, 414,000 cubic yards of gravel yielded 2287 ounces of gold and 145 ounces of silver. In addition, a dry-land dredge washed a small quantity of gravel on the Robie ranch.

Thompson, W. C., Box 77, Linden, operated a dragline dredge on the Calaveras River throughout 1940. In 1941 the dredge, equipped with 2½-cu. yd. bucket, was operated for 4 months on the Gregory, Sinclair, and Dickhaut ranches 1½ miles southwest of Jenny Lind. Later in 1941, the outfit was moved to Siskiyou County and operated by Shasta Dredging Company.

Thurman and Wright,[7] Russ Building, San Francisco, operated dragline dredges in Calaveras County in 1941. (See under San Andreas Gold Dredging Company.)

Tomboy Gold Mines, c/o W. H. Wise, Redondo Beach, California, holds 300 acres in sec. 19, T. 5 N., R. 12 E., M. D. According to J. J. McSorley, Mokelumne Hill, this property contains 3000 feet of channel blocked out but not worked on the Happy Valley Blue Lead. The channel is opened by 3000 feet of drifting from a 400-foot incline. The channel has been crosscut each 200 feet and was found to be 250 to 350 feet wide. Gravel is said to run roughly $2 to $3 per cubic yard.

Vallecito-Western Drift Mine. See page 247.

Vanciel, C. F., Oakdale, operated a dragline dredge using a 1½-cu.yd. bucket at the Halter mine 5 miles west of Jenny Lind on Calaveras River from February 20 to April 26, 1941. From 87,848 cubic yards of gravel, 552 ounces of gold and 27 ounces of silver were recovered.

What Cheer mine, of 140 acres in sec. 24, T. 5 N., R. 11 E., M.D., is owned by Mokelumne Placers, Ltd., Box 156, Plymouth, and contains a segment of the Chili Gulch Deep Blue Lead gravel channel. According to J. J. McSorley, Mokelumne Hill, about 20 to 30 acres on the north and 20 acres on the south have not been worked. The channel is the same as that in the Deep Lead placer mine, which see. Mildred S. Barker, and three others, c/o Herbert E. Barker, 641 Boulevard Way, Oakland, own a tract of 170 acres adjoining that contains this channel also.

Wolholl Dredging Corporation of Natoma operated a dragline dredge on San Domingo Creek during part of 1940. The excavator had a 2-cu.yd. bucket.

Young & Son Co., Ltd., Mokelumne Hill, moved 40,000 cubic yards of gravel at the Yale and Allyn property to a stationary washing plant with tractors and carry-alls between April 9 and August 28, 1940; 420 ounces of gold and 37 ounces of silver were recovered.

---

[7] *See also* Thurman, C. H., Costs in dragline gold dredging: Am. Inst. Min. Met. Eng. Tech Paper 1900, Mining Technology July 1945, 6 pages.

EL DORADO COUNTY

Below are notes on recent (1940-43) placer mining operations in El Dorado County. Further details of mineral resources of this county are contained in State Mineralogist's Report XXXIV for 1938, including lists of gold mines both quartz and placer, by Logan.[1]

Barker Corporation, Box 696, Patterson, operated a dragline dredge on the Explorers property in 1942. The yield from 95,000 cubic yards of gravel was 754 ounces of gold and 94 ounces of silver.

Big Canyon Dredging Company, Box 656, Fresno, operated a dragline dredge with a 3-cu.yd. bucket during part of 1940. During 11 months of 1941, operations on Deer Creek yielded 3,160 ounces of gold and 321 ounces of silver from 540,000 cubic yards of gravel. During the first 7 months of 1942, operations yielded 1,269 ounces of gold and 137 ounces of silver from 250,000 cubic yards of gravel.

Duffy Property. George L. Duffy, Foresthill, holds placer mining claims in sec. 24, T. 13 N., R. 9 E., sec. 3, T. 13 N., R. 11 E., M. D., and adjoining sections, to cover a dredging project on Middle Fork American River. The river is a line between El Dorado County and Placer County, and additional details are contained in the chapter on Placer County.

El Dorado Dredging Corporation, Greenwood, operated a dragline dredge using a 1½-cu.yd. bucket on Greenwood Creek and on Coloma Creek during 1940. In 1941, operations from January 1 to March 6 on Coloma Creek yielded 833 ounces of gold and 124 ounces of silver from 106,078 cubic yards of gravel. Equipment was moved to the Hughes property on Rock Canyon Creek, where 338,940 cubic yards of gravel yielded 2,630 ounces of gold and 281 ounces of silver. At the end of the year the equipment was on Irish Creek.

General Dredging Corporation, Natoma, operated a dragline dredge in 1940 and part of 1941 on the South Fork of American River near the point where James Marshall discovered gold in 1848. The dragline excavator had a 1½-cu.yd. bucket. A second dredge with a 2-cu.yd. bucket was operated in this same area near Coloma in 1941. The smaller dredge was operated on a site near Shingle Springs during the last quarter of 1941.

Good Luck mine yielded 164 ounces of gold and 22 ounces of silver from 23,000 cubic yards of gravel in 1942.

Greenhorn Dredging Company, Youngs, operated a dragline dredge on the Middle Fork of Cosumnes River near Youngs during part of 1940 and 1941. The excavator had a 2-cu.yd. bucket.

Hook and Ladder mine at Smiths Flat, sec. 10, T. 10 N., R. 11 E., is held by Charles Fossetti, Smiths Flat. Lindgren[2] has published a map showing the underground channels of this region and states that they are inter-rhyolitic but have yielded several million dollars by drifting. Bert Bryan of Smiths Flat worked a channel on this property known as the Gray Lead channel until 1932. An old shaft 114 feet deep was available at a point 350 feet south of the State highway. He sank an additional 38 feet, ran 330 feet south in bedrock, and raised to the channel, which he worked for a length of 1,150 feet. He states that gravel

[1] Logan, C. A., Mineral resources of El Dorado County: California Jour. Mines and Geology, vol. 34, 206-280, 363-365, map, 1938.
[2] Lindgren, op. cit., U.S. Geol. Survey Prof. Paper 73, p. 173.

was 70 feet wide and $1\frac{1}{2}$ to $3\frac{1}{2}$ feet deep at this point, and that 10 feet back from the face a sample ran $19.50 per cubic yard from a 2-foot depth of gravel with gold at $20.67 per ounce. He was working down-stream (southward), and a drop of 11 feet in 50 feet of channel made further work too difficult. He thinks that several hundred feet of this channel remain in place, but that it is cut off by the Deep Blue Lead channel. Workings have been surveyed and mapped by Andrew Nesbit of Hammon Engineering Company.

Horseshoe Bar. W. D. Ingram, Gridley, operated a dragline dredge on this bar on Middle Fork American River just below Michigan Bluff during part of 1941. Dredging was done in Placer County also, as the river passes through the property, and the center of the river is the county line.

Horseshoe Dredging Company, Youngs, operated a dragline dredge on the Frank Kipp property during part of 1940.

W. D. Ingram, Foresthill, operated dragline-dredging equipment on the following properties in 1941: Craig Osborne, Craig Royce, Craig Salt Water, Emma J. Hodgkin, and Red Raven. (See Horseshoe Bar also.)

Irish Creek Mining Company, Georgetown, operated a nonfloating washing plant on the Morgan property during part of 1940.

Lemroh Mining Company, 2401 Bayshore Boulevard, San Francisco, operated a dragline dredge during part of 1940.

Max and Junction mine yielded 110 ounces of gold and 15 ounces of silver from 14,000 cubic yards of gravel in 1942.

McQueen and Downing of Weaverville operated a dragline dredge on Carson Creek during part of 1940.

Orlomo Company, Box 548, Placerville, operated a dry-land dredge, using a dragline excavator with $1\frac{1}{2}$-cu.yd. bucket, on Indian Creek throughout 1941.

Patchen Property. Charles Patchen, 80 Broadway, Placerville, holds 80 acres in sec. 18, T. 10 N., R. 9 E., M. D., containing $\frac{1}{4}$-mile of stream gravel 100 to 300 feet wide and 8 feet deep in Martell Ravine. He states that this pays by hand-work.

River Pine Mining Company, 2432 19th Avenue, San Francisco, operated its dragline dredge on North Fork of Cosumnes River in sec. 26, T. 9 N., R. 10 E., M. D., near Nashville, during the last half of 1941. The dredge was operated in Amador County during the first half of the year.

Setter property yielded 548 ounces of gold and 70 ounces of silver from 77,500 cubic yards of gravel washed in 1942.

Starbuck Property. Frank M. Starbuck of Rescue holds 120 acres in sec. 16, T. 10 N., R. 9 E., M. D., which was worked by placer miners in the early days. Two to 6 feet of gravel are covered by 6 feet of overburden. It is reworked from time to time in the rainy season and produces a little gold. Part of the tract near the mouth of Sweetwater Creek was worked recently by a dragline dredge. Marcus Starbuck holds 40 acres in the adjoining section 17 that may pay by working with a dragline dredge. Gravel is 250 feet wide and 10 to 15 feet deep. It was worked by hand in the early days. Bedrock is decomposed granite.

Two Channel Mine. This is a consolidation of several old mines held by W. S. Eaton, for whom Ed Green, Placerville, is agent. More than 2000 acres in secs. 9, 10, 15, 22, 27, 34, T. 13 N., R. 11 E., M. D., 8 miles northeast of Georgetown are included. A channel containing white quartz boulders, and another containing cemented gravel and andesitic boulders are found here. Both have been worked by hydraulic and drift mining. The cemented gravel was crushed in stamp mills. Drifting has been done for a distance of 2500 feet south from Otter Creek, which cuts the channels. According to Bert Bryan of Smiths Flat, 2 miles of channel remain unworked in the vicinity of Kentucky Flat. He was planning to prospect the ground in the summer of 1944. Descriptions of old workings are contained in the references given below under the names of Kates, Kenna, Kentucky Flat, Mississippi, Tiedemann and Two Channel.

Bibl.: State Mineralogist's Reports XII, pp. 114, 115, 117; XIII, p. 160; XV, p. 302; XXXIV, p. 251.

Van Dyke, Modrell, and Warner, Box 822, Ione, operated a dragline dredge with a $\frac{3}{4}$-cu.yd. bucket on the Emma Gordon property during part of 1941.

Ventura mine of 90 acres in sec. 20, T. 10 N., R. 12 E., M. D., is owned by Ed Christian, Placerville. A tunnel has been driven through a ridge containing a lava-capped channel; 1300 feet of this on the Christian property connected with workings in adjoining property. Christian states that at least 80 acres of the property contains a gravel deposit but that little is known of the gold-content. The same channel may exist in 100 acres held by F. H. Richardson of Placerville and in 160 acres in sec. 24, T. 10 N., R. 11 E., held by Mrs. Hedwig Bitzer and Alma L. Howard, Smiths Flat.

Wulff Property. W. C. Wulff of Rescue holds 35 acres of placer ground, which produces gold by small-scale methods, in sec. 8, T. 10 N., R. 9 E., M. D. The productive part is 30 to 36 inches deep. The gold has evidently come from eroded pockets in the immediate vicinity.

FRESNO COUNTY

Hopkins and Becker, 3231 Fernside boulevard, Alameda, operated a suction dredge invented by Becker on the San Joaquin River near Friant in 1941 and recovered 298 ounces of gold and 61 ounces of silver from 121,000 cubic yards of gravel. The dredge had a 6-inch centrifugal gravel pump, gasoline engine, and riffle tables mounted on a steel pontoon hull. The suction point could be lowered vertically 28 feet below water level.

Grant-Service Rock Company recovered 114 ounces of gold and 16 ounces of silver in preparing 339,955 tons of sand and gravel for concrete aggregate in 1942. In 1943, recovery was 36 ounces of gold and 4 ounces of silver in preparing 210,000 tons of sand gravel.

Griffith Company and Bent Company, 418 South Pecan Street, Los Angeles, which supplied gravel for building the Friant dam in 1940, recovered 443 ounces of gold and 73 ounces of silver from 400,000 tons of gravel. In 1941, the recovery was 4,990 ounces of gold and 747 ounces of silver from 3,935,620 tons of sand and gravel. In 1942, recovery was 205 ounces of gold and 29 ounces of silver from 121,700 cubic yards of sand and gravel.

Twin Bar Mining Corporation, 1537 College Avenue, Berkeley, operated a dredge on properties of Grant Pacific Rock Company and V. Roulard in 1942.

HUMBOLDT COUNTY

The eastern part of Humboldt County contains placer gravels on the Trinity and Klamath Rivers. A little gold is produced by small-scale methods from beach sand. A number of these operations and properties have been described in a recent report of the State Mineralogist.[1] The principal producer in recent years has been the Pearch mine described below.

Orick Placers, Inc., operated a washing plant in the Gold Bluff district in 1943 to recover gold from beach sand.

Pearch mine of 180 acres in sec. 32, T. 11 N., R. 6 E., H., formerly owned by P. L. Young, Orleans, has been sold to Roy McGain, Orleans, who is operating it (1940). The property is 1 mile northeast of Orleans on the dirt highway to Happy Camp. An old terrace of Klamath River with bedrock about 50 feet higher than the present river contains a placer deposit 80 feet thick at the face of the present hydraulic pit. Only the lower part of the deposit is gravel with boulders of a maximum diameter of 18 inches. Overlying this is a great thickness of fine overburden and soil. The slate and chloritic schist bedrock dips steeply in several directions.

During the water-season the mine is kept in operation for 24 hours per day by a crew of 10 men. A mile and one-half of flume from Pearch and Cheney Creeks has a capacity of 1,300 miner's inches with a head of 225 feet at the mine. Three giants are set up, one with 6-inch nozzle-opening, the other two with 5-inch openings. They are supplied with water by means of 840 feet of 36-inch and 30-inch pipe, 900 feet of 20-inch pipe, and 2,800 feet of 15-inch pipe. Gold is recovered by means of wood-block riffles in a 'Y' sluice reaching both ends of the pit and discharging to Klamath River. To reach the river from one end of the pit, 192 boxes, each 12 feet long, are required. The 'Y' branch to reach the other end of the pit contains 94 boxes. Grade is half an inch per foot. McGain estimates that he mines 1,000 cubic yards per day during a working-season of 5 to 6 months. Water wheels furnish power for compressed air and electric lights.

Production for 1939 is given by the U. S. Bureau of Mines[2] as follows: "Cleaning bedrock at the Pearch hydraulic mine on the Klamath River northeast of Orleans yielded 154 ounces of gold and 23 ounces of silver." Figures on production for 1940 are not available, but in 1941 the mine produced 266 ounces of gold and 38 ounces of silver from 128,500 cubic yards of gravel.

Bibl.: State Mineralogist's Reports XIV, p. 404; XXI, p. 315.

[1] Averill, C. V., Mineral resources of Humboldt County: California Jour. Mines and Geology, vol. 37, pp. 499-528, 1941.
[2] Merrill, C. W., and Gaylord, H. M., Gold, silver, copper, lead, and zinc in California: U. S. Bur. Mines, Minerals Yearbook 1939, p. 233, 1940.

Fig. 82. Pearch mine, Humboldt County. *Photo by courtesy of Roy McGain; reprinted from California Journal of Mines and Geology, October 1941, p. 513.*

IMPERIAL COUNTY

The placer gold production of Imperial County has been small in recent years. The following notes are extracts from a recent report by Sampson and Tucker,[1] contained in the California Journal of Mines and Geology for April 1942. The same issue contains a few notes on gold placers by Henshaw.[2]

Gold Delta placer mines are in sec. 20, T. 13 S., R. 19 E., S.B., 9 miles east of Glamis, and are owned by Gold Deltas Corporation, C. R. Zappone, president, 1001 Subway Terminal Building, Los Angeles.

Placer gravels in washes of present water courses from the Chocolate Mountains have been developed by a shaft sunk 300 feet through unconsolidated gravel to bedrock. Cuts and shafts indicate a body of gravel 2 miles long and half a mile wide. A 6-foot thickness of gravel at the bottom of the 300-foot shaft is said to contain $0.50 to $2.00 per cubic yard in gold.

Picacho Basin mine (Placer). The basin consists of low, rolling hills covered in part by washed gravel and detritus from the neighboring hills. These superficial deposits are gold-bearing and have been washed by Mexicans for years and recently, by dry washing machines. The gravels have been worked along the arroyo between the Picacho lode mine and the old village at Picacho. It is reported that the gold-bearing gravels have an average value of 35 cents per cubic yard. These gravel deposits are 25 miles north of Yuma. Idle.

Bibl.: State Mineralogist's Report XXII, p. 260; Bull. 92, p. 156.

[1] Sampson, R. J., and Tucker, W. B., Mineral resources of Imperial County: California Jour. Mines and Geology, vol. 38, pp. 105-145, 1942.
[2] Henshaw, Paul C., Geology and mineral deposits of the Cargo Muchacho Mountains, Imperial County, California: California Jour. Mines and Geology, vol. 38, p. 190, 1942.

Potholes (Placer). These dry placers are located 10 miles northeast of Yuma, Arizona, and several miles west of Laguna Dam; elevation 150 feet. The gold-bearing gravels of this region were extensively worked in the early days by Indians and Mexicans and are now worked out. Value of the gold produced from the Potholes placers is reported to have been $2,000,000.

Bibl.: State Mineralogist's Reports XII, p. 242; XIII, p. 344; XXII, p. 261; Bull. 92, p. 155.

KERN COUNTY

Holcomb Valley Placer Company, lessee of the Monarch Rand mine in the Randsburg district, produced gold in 1943 as a byproduct of a placer operation carried on primarily for scheelite. Gravel was delivered by a carry-all to a stationary washing plant.

Monarch Rand Mining Company operated a dry-land dredge in the Randsburg district intermittently during 6 months of 1942 for the recovery of scheelite and gold.

Placer Concentrators produced gold in 1943 as a byproduct from a placer scheelite operation on the Patsy and Victory No. 2 mines in the Randsburg district.

Rand Gold Dredging Associates, Russ Building, San Francisco, installed a connected-bucket-type dredge with eighty-two 3-cu.ft. buckets to work the Tungold mine $1\frac{1}{4}$ miles northwest of Randsburg in 1942. The dredge was moved from Shasta County, where it had been operated by Roaring River Dredging Company. It was reconstructed, and jigs were added in order to recover scheelite as well as gold. Operations started on November 1, 1942, and continued until June 10, 1943, when the dredge capsized. It was righted later in the year as described by Macaulay.[1]

LOS ANGELES COUNTY

Gold was discovered in Los Angeles County in 1834, and the placers of San Francisquito, Placerita, Casteca, and Santa Felicia canyons were worked between the years 1834 and 1838 by the priests of San Fernando and San Buenaventura Missions. The placers of San Gabriel Canyon were worked by the priests of San Gabriel Mission and also by the native Californians until the discovery of gold in northern California by Marshall in 1848 at Sutter's mill.[2] In recent years production of placer gold in this county has been small. Two producers of sand and gravel for the construction industry reported the recovery of small amounts of placer gold from the San Gabriel district in 1943.

[1] Macaulay, W. B., Righting capsized dredge takes 50 minutes: Eng. and Min. Jour., April 1944.
[2] Tucker, W. B., and Sampson, R. J., Mineral resources of Los Angeles County: California Jour. Mines and Geology, vol. 33, pp. 173-270, 1937.
Bradley, W. W., California mineral production. 1924: California Min. Bur., Bull. 96, p. 47, 1925.

MADERA COUNTY

H. A. Berg, Box 581, Madera, operated a suction dredge on the Fresno River in the Dennis district in 1941, and recovered 257 ounces of gold and 74 ounces of silver from 22,000 cubic yards of gravel.

Cassaurang Ranch. Two suction dredges were operated on this ranch in the Dennis district where the Fresno River passes through it, in 1941. Operators were E. J. Gibbons and Richard A. Cassaurang, Madera.

G. E. Noble & Sons operated a suction dredge on the J. T. Pierce ranch in the Potter Ridge district during nearly all of 1942. Recovery from 7790 cubic yards of gravel was 118 ounces of gold and 35 ounces of silver.

Polk Ranch. Suction-dredge operations on this ranch in the Potter Ridge district produced 148 ounces of gold and 32 ounces of silver from 9000 cubic yards of gravel in 1942.

MARIPOSA COUNTY

Although small placers yielded many millions of dollars to early miners in Mariposa County, the rich Tertiary channels found in counties farther north are lacking because the protective blanket of lavas and other volcanic ejecta did not extend this far south.[1] Hence the placers of Mariposa are now less important than those of other counties along the Mother Lode. Short descriptions of the many lode-gold mines of this county are contained in State Mineralogist's Reports XXIV and XXXI.[2]

Barker Corporation, Hornitos, operated a dragline dredge on Eldorado Creek 1 mile from Hornitos in 1940. In 1941, operations were continued here and on the following other properties: Givens, Trabucco, Turner, and Waltz in the Hunter Valley district, and the Adams, Explorers, Inc., Munn, Penrose, R. Williams and Stratton in the Mother Lode district. In 1942 there were 22,000 cubic yards of gravel washed at the Kehoe property, which yielded 112 ounces of gold and 32 ounces of silver; 109,000 cubic yards washed at the J. Lord property yielded 833 ounces of gold and 229 ounces of silver; and 163,500 cubic yards washed at the Trabucco property yielded 1016 ounces of gold and 280 ounces of silver.

Thurman and Wright, 960 Russ Building, San Francisco, operated a dragline dredge with 4½- and 6-cu.yd. buckets on property of the Crocker-Huffman Land and Water Co. during the last half of 1941. Operations were continued on Burns Creek during part of 1942.

[1] Julihn, C. E., and Horton, F. M., Mineral industries of the United States; California, Mines of the southern Mother Lode region Part II, Tuolumne and Mariposa Counties: U.S. Bur. Mines Bull. 424, pp. 1-179, 1940.
[2] Laizure, C. McK., current mining activities in the San Francisco district with special reference to gold: California Jour. Mines and Geology, vol. 31, pp. 27-46, 1935.
Laizure, C. McK., Mariposa County: California Div. Mines, Mining in California, State Mineralogist's Rept. 24, pp. 79-122, 1928.

Trebor Corporation, Box 51, Mariposa, with Robert D. Mueller in charge, operated a dragline dredge with 2-cu.ft. bucket on the Chase Ranch in the Hunter Valley district in 1940. Earlier operations were conducted at Mormon Bar and Agua Fria Creek. In 1941, operations were conducted on the following properties: Fretz, Gaskill, Machado, Trabucco, Turner, and Waltz. In 1942, operations for 3½ months on the C. C. Pierce property on Corbett Creek yielded 289 ounces of gold and 57 ounces of silver from 100,000 cubic yards of gravel.

MERCED COUNTY

P. H. Bottoms, Box 121, Merced Falls, operated a dragline dredge with 2-cu. yd. bucket in the Snelling district during part of 1940.

Merced Dredging Company, 1805 Mills Tower, San Francisco, operated an electric connected-bucket dredge with sixty 10-cu.ft. buckets half a mile south of Snelling in 1940, 1941, and part of 1942.

San Joaquin Mining Company, 1805 Mills Tower, San Francisco, operated a connected-bucket dredge with sixty-four 10-cu.ft. buckets 3 miles west of Snelling in 1940, 1941, and part of 1942.

Snelling Gold Dredging Company of Snelling operated two electric connected-bucket dredges on the Merced River between Merced Falls and Snelling throughout 1940 and 1941, and during part of 1942. The property is in secs. 10, 11, 12, T. 5 S., R. 14 E., M. D., and covers a deposit of uncemental gravel and sand 10 to 20 feet deep with bedrock of volcanic ash. One of the dredges has seventy-two and the other sixty-six 7-cu.ft. buckets. Otherwise the dredges are much alike with steel hulls 42 by 96 feet and the usual riffle system.

Thurman and Wright, 960 Russ Building, San Francisco, operated a dragline dredge with a 6-cu.yd bucket on land of Crocker-Huffman Land and Water Company, in 1941. This operation was partly in Mariposa County and partly in Merced County.

Yuba Consolidated Gold Fields, 351 California Street, San Francisco, operated a connected-bucket dredge with seventy-two 9-cu.ft. buckets 4 miles east of Snelling in 1940 and part of 1941, in T. 5 S., R. 15 E., M. D. The dredge and part of the property were taken over from La Grange Gold Dredging Company in 1930. Gravel ranged in depth from 18 to 36 feet, and bedrock is slate near Merced Falls but is volcanic ash on the lower part of the tract. A little platinum was recovered with the gold.

NEVADA COUNTY

Some of the largest known reserves of placer gravels in California are in Nevada County. Long stretches of Tertiary river channels that were formerly the sites of great hydraulic mines remain unworked as a result of the injunction-decision of Judge L. B. Sawyer in 1884 against North Bloomfield Mining Company restraining that company from discharging debris into the streams. Federal legislation passed in 1893, called the Caminetti Act, created the California Debris Commission. This act is printed in full in the appendix of this report. For large hydraulic mines on the Sacramento-San Joaquin drainage, tailing must be restrained by dams. The Upper Narrows dam on the Yuba River was built for this purpose and was completed in 1939, but only a little hydraulic mining has been done above it to date (1945). One reason for this was Limitation Order L-208 of the War Production Board, which shut down nearly all the gold mines, remaining in effect from October 8, 1942 to July 1, 1945.

Estimates of yardages available for hydraulic mining above this and other dams have been made by Jarman [1] and Bradley.[2] A few recent operations have been described in a report by Logan [3] and older ones in a report by MacBoyle,[4] but MacBoyle's report is out of print.

Calaveras Gold Dredging Company operated a dragline dredge at Steep Hollow in the You Bet district from April 8 to August 20, 1940.

Clerkin property consists of 156 acres of patented land in sec. 24, T. 17 N., R. 7 E., M.D., near French Corral, owned by W. P. Clerkin of French Corral. Clerkin states the ground contains 30 to 40 acres of gold-bearing gravel, of which a 60- to 70-foot depth contains practically no gold. From 8 feet to 10 feet at the bottom is blue gravel, and the last foot contains most of the gold. Clerkin thinks that this deposit represents a high bench of the old Eocene channel about 100 feet wide and adjoins the bedrock tunnel claim of the French Corral placer. Clerkin works the property for about 2 months in the spring of each year with 750 miner's inches of water that flows from Bloody Run Creek from melting snow.

Dakin Company, 917 Sacramento Street, San Francisco, operated a dragline dredge at Champion Flat along Deer Creek from January 1 to March 1, 1940. The excavator was equipped with a 1½-cu.yd. bucket.

Esperance. See French Corral.

French Corral placer is a property of 1,700 acres in secs. 23, 24, 25, 35, T. 17 N., R. 7 E., M. D., including the Kate Hayes, Esperance, Fraser and Alexander, Bedrock Tunnel, and Milton placer mines, owned by G. M. Standifer, Balfour Building, San Francisco. The property contains approximately 1 mile of unworked length of the Eocene channel worked by hydraulic methods at North Bloomfield, North Columbia, Badger Hill, American Hill, North San Juan, Sweetland, Birchville, and French Corral.

---

[1] Jarman, Arthur, An investigation of "the feasibility of any plan or plans whereby hydraulic mining operations can be resumed in this state": California Min. Bur., Mining in California, State Mineralogist's Rept. 23, pp. 54-116, 1927.

[2] Bradley, W. W., Dams for hydraulic mining: California Jour. Mines and Geology, vol. 31, pp. 345-367, 1935.

[3] Logan, C. A., Mineral resources of Nevada County, gold placer mining: California Jour. Mines and Geology, vol. 37, pp. 431-436, 1941.

[4] MacBoyle, Errol, Mines and mineral resources of Nevada County: California Min. Bur., State Mineralogist's Rept. 16, pp. 91-108, 1919.

In November, 1940 a report was made on this property by L. A. Smith, who sank some shafts to sample the gravel and who had the results of some drilling that had been done about 1938. The following is abstracted from Smith's report: About one-third of a mile of channel on this property has been partly worked by the hydraulic method, and the resulting excavation is known as the Esperance pit. An average of about 65 feet of the top gravel has been piped off of an area roughly 600 by 1,500 feet. The bottom gravel remaining on bedrock measures roughly 200 yards wide by 500 yards long by 12 yards deep. Gold content of this 1,200,000 cubic yards should be higher than the channel average, and the deposit can be worked mechanically. Smith's sampling results in this bedrock gravel area turned out as follows:

```
Shaft No. 1   21.4 ft.   60.7¢ per cu.yd.   Bedrock
Shaft No. 2   31.5 ft.   41.6¢ per cu.yd.   Bedrock
Shaft No. 4   24.5 ft.   36.0¢ per cu.yd.   Bedrock
        Average of the above bedrock shafts—45.1¢
Shaft No. 3    8.0 ft.   23.6¢ per cu.yd.   No bedrock
        Average of the above four shafts—43.1¢
Cut No. 1   30.0 ft.   36.0¢ per cu.yd.   No bedrock
        (Extends above collar of shaft No. 1)
Cut No. 2    5.0 ft.   45.0¢ per cu.yd.   No bedrock
        (Extends above collar of shaft No. 3)
    Average value of all cuts and shafts—41.4¢ per cu.yd.
```

The gravel that was sampled was tough and uncemented in general, but it contained lenses of cemented gravel. It was washed through a trommel 4 feet long and 14 feet of sluice box. Smith estimates that 70 to 90 percent of the gold was recovered. He thought that a moderate amount of scrubbing in a large trommel would insure a good recovery of the gold.

The unstripped ground north of the Esperance pit was drilled about 1938 with the results given below. This work was done as exploratory work and all holes are therefore not in the channel. The lines were roughly 1,200 feet apart, and the number "2" was not used.

Line No. 1 has three holes in the channel—holes No. 7, No. 9, and No. 10. The average depth is 89 feet and average value 21.1 cents per cubic yard. The width of channel is about 600 feet.

Line No. 3 has four holes in the channel—holes No. 6, No. 7, No. 8 and No. 9. Using holes No. 7, No. 8 and No. 9, Smith figured a 500-foot width with average depth of 56 feet and average value of 14 cents per cubic yard.

Line No. 4 is made up of three very poor holes, the deepest of which (128 feet) did not reach bedrock. The channel at Line 4 is evidently comparatively narrow for an unknown distance or until it widens again in the Birchville pit.

The Line 1 drill holes appear to confirm the shaft samples in the shallow ground. Smith assumed 30 feet of 40-cent bottom gravel at Line 1, and 59 feet of 12-cent top gravel overlying it and thus arrived at the 89 feet of 21.1-cent gravel shown by the drill holes.

Line 3 drill holes do not stand up to the average indicated by Smith's shaft results. Thirty feet of 40-cent bottom gravel at Line 3 would show a 21-cent average for the 56-foot average depth without assuming any values in the top 26 feet.

Before these drill hole results are accepted, Smith believed they should be checked by sinking two or more shafts over selected drill holes. He did not sample the upper or overburden gravels, but pan tests that he made would indicate a value in the neighborhood of the recoveries reported by Schroder and others who hydraulicked top gravel from the Esperance pit. These reported recoveries run from 10 to 17 cents, old price, or an average of probably 20 cents per cubic yard at present price of gold. This figure is confirmed to some extent by one sample taken near his shaft No. 1. This sample began at approximately 30 feet above bedrock and extended to approximately 60 feet above bedrock, with an average gold content of 36 cents per cubic yard for the 30-foot sample.

Smith proposed to work by mechanical methods the gravel already stripped in the Esperance pit, about 1,200,000 cubic yards, then to prospect the unstripped ground and strip by the most feasible means, either mechanical methods or hydraulicking. He expected this work to expose an additional 2,400,000 cubic yards of gravel of a grade that would show a profit if worked by mechanical methods. In the fall of 1945, such a project had not yet been started.

Water for this property is obtained from Shady Creek, and is brought into the Pine Grove reservoir by means of a 4-mile ditch. The original capacity of the reservoir was 300 acre-feet. Applications have been made for 53 second-feet or 2,120 miner's inches of water.

Greenhorn Dredging Company operated a dragline dredge at Quaker Hill in the You Bet district from January 1 to December 31, 1940, using an excavator equipped with a 2-cu.yd. bucket.

Hall and French. See French Corral.

A. B. Innis of Nevada City operated a dragline dredge equipped with a 1½-cu.yd. bucket at the Malakoff mine from October 16 to December 31, 1941. The yield from 72,000 cubic yards of gravel was 333 ounces of gold and 33 ounces of silver. The operation was continued in 1942 and the yield from 146,000 cubic yards of gravel was 694 ounces of gold and 64 ounces of silver.

Kate Hayes. See French Corral.

Kaufield and McKinley of Lincoln, operated a non-floating washing plant on the Parker Ranch from June 27 until September 13, 1940. Kaufield and Danison of Nevada City operated a dragline dredge on Columbia Hill near North Bloomfield from March 1 to April 20, 1941.

M. K. Gibson Mining Company, Grass Valley, operated a dragline dredge on the Elder, Martel, Neirzert, and Thomas properties in 1941.

Milton. See French Corral.

Omega mine is in Sec. 16, 17, T. 17 N., R. 11 E., M. D., 3 miles southeast of the town of Washington. It is owned by South Yuba Mining and Development Company, Charles A. Kaas, secretary-manager, 420 Market Street, San Francisco. The last mining was done in the season 1941-42, ending in the summer of 1942, by Omega Company, of which Jack Little was manager.

A new 30-inch pipe line has been installed to take water directly from the ditch, giving a pressure of 175 pounds per square inch. The old pipe line from reservoirs is also a 30-inch line, but it gives only 90 pounds per square inch pressure. Three reservoirs hold about 35 acre-feet of water. The company takes 5000 miner's inches of water from the South Fork of the Yuba River under old water-rights. This amount is usually available to about the middle of July, then the amount gradually declines. The main ditch system is 12 miles long and starts just below Lake Spaulding. One-third of this 12 miles is flume 4 feet by 6 feet, including 4 miles at the start. The flume was repaired with 1200 M board feet of lumber in 1941. A lower ditch system, part of which is 3- by 4-foot flume, picks up water from various canyons to feed the reservoirs with 1200 miner's inches of water. It is 7 miles long. Other equipment includes a camp to house 75 men, and a sawmill with a capacity of 16 M to 20 M board feet of lumber per day operated by steam power. It includes two 60-inch saws, a four-saw edger, and a cutoff saw. Lumber is produced only for the use of the mining company.

Gold is recovered in a sluice 5 feet wide, $3\frac{1}{2}$ feet deep, and $\frac{1}{2}$ mile long, equipped with cross riffles of 30-pound and 45-pound steel rails. The grade is 6 inches in 16 feet. Duty of the water is about 3 cubic yards per miner's inch per day. Bedrock is a slaty schist.

According to Theodore Larsen of Nevada City, who was superintendent of the last operations, the company holds about 3000 acres of land containing 25,000,000 to 30,000,000 cubic yards of gravel. The face of the pit is carried 600 feet wide and 150 to 185 feet high in an Eocene channel 1500 feet wide.

The Omega Company began hydraulicking March 9, 1941, utilizing storage space for tailing in the recently completed Upper Narrows Debris Dam at Smartsville. During the season the company washed 429,637 cubic yards of gravel, which yielded 1302 ounces of gold and 49 ounces of silver. In 1942 the mine was operated by Omega Company, and lessees from January 17 to July 7. The yield from 818,175 cubic yards of gravel was 1749 ounces of gold and 54 ounces of silver. In 1943 bedrock was cleaned, yielding 64 ounces of gold and 2 ounces of silver.

Pilot Dredging Company, Cottonwood, operated intermittently in 1940 on the Coleburn property with a dragline dredge.

Relief Hill Mine. See Western Gold, Inc.

William Richter & Sons operated a dragline dredge at Scotts Flat from January 1 to October 13, 1940. In 1941 this firm operated a dragline dredge on the Donnelly and Johnson property.

San Juan Gold Company, F. L. Morris, president, Monadnock Building, San Francisco, holds 5484 acres of patented mining claims on San Juan Ridge principally in the following sections: Sec. 36, T. 18 N., R. 8 E.; Sec. 1, T. 17 N., R. 8 E.; Secs. 1, 2, 4, 5, 6, 7, 8, 9, 11, 12, T. 17 N., R. 9 E.; Secs. 35, 36, T. 18 N., R. 9 E.; Secs. 16, 21, 22, 31, T. 18 N., R. 10 E., M. D.

These holdings include the famous Malakoff hydraulic mine formerly operated by North Bloomfield Mining Company. According to officials of the present company, this old company recovered $2,830,000 during the period from 1870 to February 1884 from gravel that yielded 12 and a fraction cents per cubic yard with gold at $20.67 per ounce.

In the vicinity of North Columbia is a part of the same channel that has been stripped by old hydraulic mining to a depth of about 150 feet. Gravel still remains in this part of the channel to a depth of about 300 feet. Maps in the possession of F. L. Morris show that this has been drilled to determine the gold content for a distance of $2\frac{1}{2}$ miles,

FIG. 83. Badger Hill property of San Juan Gold Company.

on the Central, Consolidated, and Western placer mines. Rows of holes about 500 feet apart were drilled across the channel, and the distance between the holes is about 200 feet. Morris states that this sampling has indicated a gold content considerably higher than the recovery mentioned above for 100,0000,000 cubic yards in a strip down the center of the channel 600 feet in width. As bedrock was reached in only small areas of the old workings, gravel remaining in this stripped part of the channel should be better than the average gravel already mined because of the concentration of gold on bedrock.

Application for new water-rights has been made as follows: Humbug Creek, 20 cubic feet per second; Spring Creek, 15 cubic feet per second; Grizzly Canyon, 25 cubic feet per second; Bloody Run, 50 cubic feet per second; Middle Fork of Yuba, 300 cubic feet per second. Rights are held on Wolf Creek for 25 cubic feet per second and on Malakoff pit for 5 cubic feet per second. Old ditches are being retained and may be reconditioned to handle all water except that of the Middle Fork of Yuba River. Estimates have been made on the cost of a new system for this water of $800,000 for flume and dam only or $1,540,000 total for the entire system complete to the mine. This includes 22 miles of flume and 2 miles of ditch.

As the cost of installing this water-system is high, owners of the property are considering the installation of mechanical methods of mining instead of hydraulicking. If mechanical mining is used, storage of tailing on the property may prove feasible.

A. B. Innis operated a dragline dredge with a 1½-cu.yd. bucket at the Malakoff mine from October 16 to December 31, 1941. The yield from 72,000 cubic yards of gravel was 333 ounces of gold and 33 ounces of silver. In 1942, the yield from 146,000 cubic yards was 694 ounces of gold and 64 ounces of silver.

Western Gold Incorporated operated the *Relief Hill mine* by the hydraulic method in 1941 and was one of the first companies to utilize debris storage behind the Upper Narrows Dam on the Yuba River. W. H. Taylor, 942 Russ Building, San Francisco, is president; and Claude E. Clark, Graniteville Star Route, Nevada City, is manager. The mine is 3 miles east of North Bloomfield in secs. 4, 9, T. 17 N., R. 10 E., M. D.

FIG. 84. Western Gold, Inc., Relief Hill mine. *Photo by Thomas F. Ryder.*

Jarman[5] estimated that 6,000,000 cubic yards of gravel have already been washed at this mine and that 5,000,000 to 15,000,000 cubic yards were still available. He stated that owners estimated that 30,000,000 cubic yards were available at that time. Present owners estimate a still higher yardage available. Drift mining has been done at this property as well as hydraulic mining. During the war the mine has not been in operation because of shortage of labor and Limitation Order L-208 of the War Production Board. The operators are planning to open up the mine for hydraulicking on a large scale when labor and materials are available.

The following description of an earlier period of operation at this mine is from Gardner and Johnson:[6] The Relief Hill Mining Company began operations near Camptonville in the autumn of 1931 and worked 4 months in 1932 before the water supply failed. An old mine was being rejuvenated; the gravel was 200 feet thick. About 500 miner's inches of water was used during the season. A total of 1000 inches will be used when the mine is fully reopened. The old pit was cleaned and virgin gravel reached in 1932 just as the water played out. Tailings were impounded behind dams in a dry canyon. The ditch line is 7 miles long. The pipe line is 14 to 22 inches in diameter, and the effective head is 210 feet. The sluice boxes are 48 inches wide; riffles are wooden blocks. A duty of 3 cubic yards per 24 hours per miner's inch is expected. A crew of 15 men worked 120 days in 1932.

Wyandotte Dredging Company, Box 228, Nevada City, operated a dragline dredge, using a 2½-cu.yd. bucket on the Perrin and Pingree Ranches from October 18 until December 31, 1940. The yield from

---

[5] Jarman, Arthur, op. cit., p. 111.
[6] Gardner, E. D., and Johnson, C. H., Placer mining in the western United States, Part II, Hydraulicking, etc.: U.S. Bur. Mines Inf. Circ. 6787, p. 52, 1934.

87,000 cubic yards of gravel was 771 ounces of gold and 109 ounces of silver. The company also operated a dragline dredge on property of the Alpha Stores in the You Bet district during 1940. In 1941 this company did further work on the Perrin and Pingree properties. The yield from 130,000 cubic yards of gravel washed at the Perrin property was 1186 ounces of gold and 155 ounces of silver; the yield from 70,000 cubic yards washed at the Pingree property was 339 ounces of gold and 58 ounces of silver.

You Bet mines include many of the old hydraulic mines famous in the early days of that method of mining, such as the following: Palmyra, Newark, Arkansas and Greenhorn, Starr, Red Dog, Missouri Canyon, Gail Placer, Rose and Duryea, Emigrant, Smith and Powell, Chicken Point, Atkins and Taylor, You Bet, Brown Bros., Washington, Browns Hill, Niece and West, Birds Eye Canyon, Poverty, and Walloupa. The group includes 1150 acres of patented land and 120 acres of locations in secs. 25, 35, 36, T. 16 N., R. 9 E.; secs. 29, 30, 31, 32, T. 16 N., R. 10 E.; secs. 2, 11, 14, 15, T. 15 N., R. 9 E.; sec. 6, T. 15 N., R. 10 E., M. D.; also 151 acres of timberland in sec. 28, T. 16 N., R. 10 E.; also the following ditches: a 12-mile English ditch with the first right on 1500 miner's inches from Steep Hollow Creek; a 5-mile Star ditch on the south fork of Greenhorn Creek with the second right to 500 miner's inches of water, and 13/21 interest in the Irish ditch with the second water right on Steep Hollow Creek. This property is now owned by Alpha Stores, Ltd., Fred F. Cassidy, president, Nevada City, California.

Immense yardages of gravel were washed at these properties by the hydraulic method in the early days from one of the great Tertiary river channels, but large yardages still remain unworked. The Jarman[7] report of 1927 quotes F. A. Goodale as estimating that the property still contains 12,000,000 to 24,000,000 cubic yards of gravel containing enough gold to pay for hydraulic mining. Tom Brady, who lives at You Bet and who holds the adjoining Jupiter group of 100 acres in sec. 31, T. 16 N., R. 10 E., states that a part of the channel starting at a point east of You Bet and running northward has never been worked, except a little on the surface. Brady says that the best content of gold is found where the bedrock starts to rise on the west side of the channel and that this has not yet been mined in this particular section of the channel.

The following quotation is from the Colfax folio by Waldemar Lindgren:[8]

"At Red Dog and Hawkins Canyon, near You Bet, the deep gravel has again been exposed and is beyond doubt continuous between the two points. The gravel here is similar to that at Quaker Hill. The deepest gravel has been hydraulicked only in the places mentioned but considerable drifting by means of tunnels and inclines has been done from Niece and West's claims for 1½ miles northeast on the Steep Hollow side. The channel has very little fall, the average elevation being 2620 feet. It is estimated 47,000,000 yards have been removed, leaving over 100,000,000 yards available. Much of this would be difficult to wash on account of lack of grade. Reports of yield and grade of gravel are not available but the You Bet diggings have probably produced $3,000,000."

---

[7] Jarman, Arthur, op. cit., pp. 44-116.
[8] Lindgren, Waldemar, U.S..Geol. Survey Geol. Atlas, Colfax folio (no. 66), p. 9, 1900.

You Bet Mining Company did some hydraulic mining on this property in the winter of 1913-14 after building a debris dam and getting a permit from the California Debris Commission. Work was stopped in 1914 by a court injunction on the ground that the water below the dam was rendered unfit for domestic purposes on account of its turbidity. In 1915 many of the claims were leased to Chinese miners, who did some drift mining that is reported to have been very productive. Apparently the last hydraulic mining was done under the supervision of J. W. Scott about 1935. He used 3,000 miner's inches of water under a 280-foot head and recovered gold in a sluice that had a slope of 7 inches to 12 feet. Hydraulic mining was done on the Red Dog property with a bank 200 feet high, and at the Brown's Hill property near the town of You Bet with a bank 252 feet high. He was able to remove rim-gravel without blasting at the rate of 6,000 cubic yards per 24 hours or 2 cubic yards per miner's inch per day. Some of the bottom ground required blasting and was moved at the rate of 10,000 cubic yards per day or 3½ cubic yards per miner's inch per day, yielding 51 cents per cubic yard, according to his report.

Mining cost was 7¼ cents per cubic yard, and storage in the Combie Reservoir of the Nevada Irrigation District cost 3 cents per cubic yard. Tailing from the Red Dog property was discharged to Greenhorn Creek through a tunnel that starts 70 feet above the creek and is 22 feet down in the bedrock at the mine. Pipe lines in use during this period of operation included one line 705 feet long of pipe 26, 20, 18, and 15 inches in diameter. A second pipe line contained 1,350 feet of the same sizes of pipe. Nozzles used on the giants were 7 inches and 6 inches in diameter at the outlet. Because of difficulties with muddied water and consequent law-suits brought by Pacific Gas & Electric Company, coupled with the fact that the Nevada Irrigation District needs the remaining space in the Combie Reservoir for the storage of water, it is doubtful that any further storage of hydraulic tailing will be sold in this reservoir. However, some consideration has been given to raising this dam to provide additional space for the storage of hydraulic tailing.

In 1940 the Wyandotte Dredging Company, the San Carlos Gold Dredging Company, and the Pilot Dredging Company operated dragline dredges to handle tailing from former hydraulic operations on Greenhorn Creek. In 1945 the property was being operated under a lease and option held by Phil P. Fredericks, Manx Hotel, San Francisco.

PLACER COUNTY

Many important old placer mines are located at Dutch Flat, Gold Run, Michigan Bluff, Forest Hill, and other places in Placer County. Descriptions of many of them were published by Logan [1] in the California Journal of Mines and Geology for January, 1936. This publication contains a long table of placer mines with references to older reports on them, and the reader is referred to that table for references to literature on the older operations. The famous drift-mines of the Forest Hill Divide were described by Ross E. Browne [2] in State Mineralogist's Report X, for 1890, and a good map and vertical sections accompanied the report. It is out of print but may be consulted in many libraries.

Below are a few notes on recent operations and on a few properties that are believed to contain reserves of possible commercial grade.

H. J. Aalders and W. W. Prather of Lincoln operated a dragline dredge using a 1½-cu.yd. bucket on the Gladding Ranch 4½ miles north of Lincoln from January 1 to July 15, 1940.

C. N. Chittenden of Lincoln operated a non-floating washing plant on the Rizzi Ranch from January 1 to July 20, 1940, and moved it to the Mulligan Ranch on August 1 where operations were continued until the end of 1940. The yield from 75,000 cubic yards of gravel was 222 ounces of gold and 45 ounces of silver, and from 26,000 cubic yards of gravel was 132 ounces of gold and 31 ounces of silver from the respective properties. In 1941 Chittenden operated a non-floating washing plant on the Johnson Ranch in the Lincoln district. Gravel was delivered by dragline excavator with a ¾-cu.yd. bucket. The yield from 43,500 cubic yards of gravel was 282 ounces of gold and 51 ounces of silver.

Duffy property, in sec. 3, T. 13 N., R. 11 E., and sec. 24, T. 13 N., R. 9 E., M.D., and intermediate sections, is held by George L. Duffy of Forest Hill. Duffy states that he controls about 18 miles of mineral rights on the Middle Fork of American River either by options on patented ground or by special-use permits from U. S. Forest Service and the Federal Power Commission. The holdings are for a proposed dredging project and extend from the original Ruck-a-Chucky dam site to the mouth of the Rubicon River. The gravel has been sampled for a distance of 1½ miles from a point near the mouth of Volcano Canyon near the line between sec. 5 and sec. 6, T. 13 N., R. 11 E. and going down stream. Casing 11 inches in diameter was sunk at intervals of 100 to 250 feet. The holes were staggered, alternate ones being on opposite sides of the river. The depth ranged from 20 to 25 feet, and Duffy states that the average value in gold was 60 cents per cubic yard with gold figured at $35.00 per ounce. Width of the gravel is 350 to 400 feet. Eight miles at the lower end of the holding were sampled by means of sinking 70 caissons, 5 feet in diameter, by hand. Many of these did not reach bedrock, as it was possible to sink them to a depth of only 15 to 16 feet. Gravel was washed in a sluice and long tom, and according to Duffy gave a return of 32 cents per cubic yard. Gravel is 500 to 700 feet wide on this lower end of the holding. Duffy states that early work

---

[1] Logan, C. A., Gold mines of Placer County: California Jour. Mines and Geology, vol. 32, pp. 49-96, 1936.

[2] Browne, Ross E., The ancient river beds of the Forest Hill Divide: California Min. Bur., State Mineralogist's Rept. 10, pp. 435-465, 1890.

on the bars and in the river was done in circular pits. The pits were kept unwatered, and the gravel was removed by means of hydraulic elevators. Much of the gravel between these circular pits was not disturbed by this early method of mining.

A part of Duffy's holdings known as Horseshoe Bar in secs. 4, 5, T. 13 N., R. 11 E., was worked by W. D. Ingram of Gridley with a dragline dredge in 1941 and 1942. The washing plant was designed to serve a 5-cu.yd. dragline excavator, but the one actually in use was a Northwest dragline with a 2-cu.yd. bucket. Results from this operation are said to have been good, but it was shut down by Limitation Order L-208 of the War Production Board in 1942.

El Oro Dredging Company of Colfax operated a dragline dredge on Indian Canyon in the Iowa Hill district from February 4 to September 19, 1940. A second dredge was operated in Shirttail Canyon from August 6 until October 31, 1940.

Gold Placers, Inc., 320 Capital National Bank Building, Sacramento, operated a dragline dredge on the Robinson Ranch in the Ophir district from April 30 to August 30, 1941, and on the Leak Ranch from September 7 to December 20, 1941.

Gold Recoveries Corporation, Box 58, Auburn, operated a dragline dredge on the William Ayers and Anderson property in the Ophir district during 1941.

Hallstrom and Lindblad, Route 7, Box 4343A, Sacramento, operated a non-floating washing plant, to which gravel was delivered by a dragline excavator with a 1¼-cu.yd. bucket, in the Ophir district during 1940. The yield at the Baker Ranch 6 miles east of Roseville from 124,000 cubic yards of gravel was 355 ounces of gold and 9 ounces of silver. The yield at property of the Placer Realty Corporation from 165,000 cubic yards of gravel was 471 ounces of gold and 30 ounces of silver. This firm continued operations in 1941 on the Joseph Mooney, Mathilda Bahr, and Rogers Ranches and in Miners Ravine, all in the Ophir district.

W. D. Ingram (see Duffy property also), Box 225, Foresthill, operated a dragline dredge on Horseshoe Bar on the county line between El Dorado County and Placer County during 1941.

Innis Dredging Company, Nevada City, operated a dragline dredge on Dry Creek from January 1 to June 1, 1940.

Jasper-Stacy Company (Recalp Company) of Lincoln worked out its ground in Auburn Ravine 2 miles east of Lincoln in May 1940. A dragline excavator with a 2-cu.yd. bucket was used.

Kaufield and McKinley, Box 274, Lincoln, operated a non-floating washing plant, using a mechanical excavator, on the Love Ranch in the Ophir district from February 20 to June 28, 1940.

La Kamp Bros. of Dutch Flat operated a non-floating washing plant at the Mutual mine in the Dutch Flat district during 1941. Gravel was delivered with a bulldozer.

Lebanon Consolidated Mines, 200 Bush Street, San Francisco, worked the Occidental drift mine in the Iowa Hill district from January 1 to December 31, 1941. The yield from 3766 cubic yards of gravel was 536 ounces of gold and 63 ounces of silver.

Lost Camp mine is a hydraulic mine of 440 acres 2 miles by road from Blue Canyon, at an elevation of 4300 feet, in secs. 22, 23, T. 16 N., R. 11 E., M. D. The channel contains interbedded layers of soft rhyolite tuff and free-washing white quartz gravel. Work on the property has proceeded at intervals over a long period by three different methods, ground sluicing, hydraulicking, and drifting. Several million cubic yards of gravel probably remain unworked, but little is known about the gold content.

The present owner is the Robie Estate, Wendell T. Robie, manager, Auburn, California, but the most recent work was done by a California corporation called Lost Camp Mining Company from 1934 to December 1941. The property includes water rights on Monumental and Fulda Creeks.

In 1944 a case was pending in the Sacramento Superior Court (No. 60474, Department 4), involving a complaint of Carmichael Irrigation District about muddy water discharged with the tailing from this mine. Stipulations limited the hydraulic season to the period from November 15 to April 30 of each year. This was a temporary arrangement which may be modified later. In this case McGeachin Placer Gold Mining Company and Mayflower Gravel Mining Company contend that the United States Debris Commission is a party in interest, because the tailing is impounded by a debris dam on the North Fork of the American River, and that the case should be tried in a Federal court.

In an entirely different suit in the Superior Court at Auburn, Lost Camp Mining Company was ousted from the property and the title confirmed to the Robie Estate. This suit involved a claim by the California Debris Commission of $4009.30 for tailing storage by the debris dam, but the Superior Court at Auburn excluded this point from the decision.

Mayflower Gravel Mining Company, care of Richard Detert, Mills Tower, San Francisco, holds a very large mining property at Foresthill now containing 5800 acres in secs. 15, 22, 23, 24, 25, 26, T. 14 N., R. 10 E., and including the following mining claims: Texas, Sacramento, Washington, Garland Mill Slope, Excelsior, Hope, Uncle Sam, Green Spring, Live Oak, Small Hope, New Jersey (mineral rights), Brushy Slide, Dardanelles, Oro, and Adams pit.

The mine was first worked by hydraulicking. Later it was an important drift mine, working the same channel as the bottom lead in the adjoining Paragon property. The principal production, about $1,600,000, was made between 1888 and 1899. The cemented gravel was crushed in a 20-stamp mill of 850-pound stamps, dropping 100 times per minute. The battery screen was an iron plate punched with holes 0.2-inch in diameter, and the capacity was 6½ tons per stamp-day. The gold was recovered by amalgamation. John Hays Hammond[3] quotes the following figures: From December 11, 1888, to September 24, 1889, a total of 33,787 tons of gravel was mined from a length of 1620 feet of channel and yielded on crushing $272,616.50 or $8.06 a ton.

The Mayflower operators worked downstream on the bottom or main lead to the boundary of the Garland Mill Slope claim (then under

---

[3] Hammond, John Hays, The auriferous gravels of California: California Min. Bur., State Mineralogist's Rept. 9, p. 120, 1890.

other ownership), as well as upstream to the Paragon line. The May-flower ground contains an unworked segment of Big Blue Lead 1½ miles long including those parts in the Excelsior and Garland Mill Slope claims. The bedrock tunnel of the Mayflower has been driven ahead but must be driven an additional 1500 feet to connect with the Excelsior-Baltimore bedrock tunnel. When this connection is made the 1½-mile segment of channel will be drained. Water has been too troublesome in the past.

Old reports describe an upper lead in this same channel which was worked in the Paragon mine. George Duffy of Foresthill did about a mile of drifting in several directions in the Mayflower shaft, 150 feet above bedrock, about 1938. His gravel averaged $3.50 per ton with gold at $35 per ounce. The south drift was the best and yielded $6 to $7 per ton from cemented gravel. Nuggets of $1 to $1.50 were found.

He also did some prospecting in the old hydraulic-bank near the collar of the Mayflower shaft by sinking five shafts 125 feet apart, each about 20 feet deep. These shafts represent the lower 20 feet of a 90-foot depth of gravel and gave returns of $4.50 per cubic yard, according to Duffy. He believes that this 90-foot depth could be dredged. Bedrock is rhyolitic tuff. A segment of the Orono inter-volcanic channel, 2 miles long, extending from Mayflower bedrock tunnel southward is supposed to exist in this property.

Recent work has been done in the hydraulic pit at Smith's Point and the Albright claim (sec. 27) on a bank 450 feet high. A 22-mile ditch from Shirttail Canyon will carry 5000 miner's inches of water giving a head of 525 feet. Only 325 feet of head has been used recently. A giant supplied by a 15-inch pipe uses 7- and 9-inch nozzles.

McGeachin Placer Gold Mining Company has extensive holdings of placer gravel in secs. 2, 3, 4, 10, T. 14 N., R. 10 E., M. D., near Iowa Hill. Several of these have been described by Logan [4] under the names McGeachin Placer Gold Mining Company, Morning Star drift mine, and Long Point Mining Company (Jupiter). C. H. Dunn, Sacramento, is president; I. E. Rose, Iowa Hill, is manager. The property includes 1700 acres. Ground suitable for hydraulic mining is known as the Big Dipper, which includes the old Irish and Byrne and Horman claims. Other claims include the Jupiter, Hazelroth, Schwab, Weber, and Winchester.

Water supply is obtained from North Fork of Shirttail Canyon and includes three ditches and the Morning Star reservoir 10 miles from the mine, which holds 1800 acre-feet. The main or McKee ditch carries 3500 miner's inches, and would furnish 400 feet of head at the mine if a 2465-foot penstock of 36- and 34-inch pipe were built. This ditch heads in sec. 25, T. 15 N., R. 10 E. The Morning Star ditch heads in sec. 18, T. 15 N., R. 11 E.

In connection with plans that were being made to resume hydraulic mining, a report was made in December 1938 by F. H. Reynolds and Company, Consulting Engineers, Sacramento, and samples were taken by I. E. Rose. Sampling was done in old hydraulic banks roughly 50 feet in height and consisted of cuts 2 feet wide and 1 foot deep. Four of the samples were used in making calculations of gold-content, and a fifth was discarded because it had been taken near bedrock, and

---

[4] Logan, C. A., op. cit., 1936.

FIG. 85.   Sampling hydraulic bank 50 feet high, McGeachin Placer Gold Mining Company.   *Photo by courtesy of I. E. Rose.*

the gold-content was high.   Rose thinks that the ground contains 25,000,-000 to 30,000,000 cubic yards of gravel without overburden that will yield 20 cents per cubic yard in gold, in a roughly circular area 3500 feet long and more than 2000 feet wide.   Additional yardage is available with overburden.   Rose states that 1,500,000 cubic yards have already been worked on the Irish and Byrne claim, and 5,000,000 to 6,000,000 cubic yards on the Horman.   Banks range from 40 to 285 feet in height.   Rose states that drifting operations carried on for 2 years recently yielded gravel running from $6 to $9 per cubic yard in gold.

H. W. McKinely of Newcastle operated a dragline dredge on the Fisher Ranch in the Ophir district from June 17 to July 31, 1941.

Midland Company, Inc., Box 8, Sawyers Bar, moved its dragline dredge from the Lincoln district to the North Fork of Salmon River in Siskiyou County during 1940.   The dragline excavator had a 1¼-cu.yd. bucket.

Panob Gold Dredging Company, Box 896, Lincoln, operated two dry-land outfits on the Forsyth and Lewis and G. E. Stoll properties during 1940. From March to October 1941 this company operated a non-floating washing plant on the Ferrari property in the Ophir district. Gravel was delivered by a dragline excavator with a 1½-cu.yd. bucket to a washing plant equipped with Ainlay bowls. The operation at the Forsyth and Lewis property was continued during 1941.

Pantle Bros. of Lincoln operated a dry-land placer machine equipped with four Ainlay bowls on the Ahart, Ferreva, and Kaneko Ranches in the Lincoln district during 1940. The yield from 179,800 cubic yards of gravel was 632 ounces of gold and 111 ounces of silver. The gravel was delivered with a 1-cu.yd. dragline excavator.

Paragon mine, in secs. 13, 18, 19, 30, 31, T. 14 N., R. 11 E., and sec. 24, T. 14 N., R. 10 E., M. D., is one of the large hydraulic mines of the Forest Hill district 2 miles from Foresthill at a place formerly called Bath. Logan [5] published two pages of description of this mine in State Mineralogist's Report XXXII, and that description is up to date with a few exceptions, as follows: Paragon Mines, Inc., W. E. Wilson, president, Foresthill, has acquired the lease and option formerly held by Alanta Mines, Inc., of which King G. Gillette was president. The property now contains 1760 acres owned by the J. F. Thompson Estate, of which Charles H. Segerstrom, Sonora, is administrator. Wilson states that an old tunnel that was driven through the hydraulic bank to carry the water-ditch contained leaks that probably helped to cause a serious cave at the property in 1935. He has abandoned this tunnel and moved the forebay and pipe-lines to the east side of pit.

Roseville Gold Dredging Company, 351 California Street, San Francisco, operated a dredge in Strap Ravine 6 miles east of Roseville from January 23 to the end of 1940. The dredge was driven by electric power and had a bucketline of seventy-two 3-cu.ft. buckets. This operation was continued during 1941.

Stewart Gravel Mines, c/o J. D. Stewart, 138 Commercial Street, Auburn, controls about 3000 acres on the great Eocene river channel that passes through Dutch Flat and Gold Run. The property is in secs. 2, 3, 4, 9, 10, T. 15 N., R. 10 E., M.D., and covers a length along the channel of 2 miles.

It is conspicuous because highway no. 40, the main route from California to Truckee and Reno, runs for a mile at the base of one of the old hydraulic banks nearly 200 feet high. The main line of the Southern Pacific Railroad is at the top of this old hydraulic face. The channel is 1½ miles wide at this point and the gravel was originally 500 feet deep in the middle of the channel, so a depth of several hundred feet remains unworked. According to Stewart, a length of 11,500 feet south of the highway has not been drifted on bedrock.

The mine is provided with a 4000-foot bedrock tunnel in greenstone, which requires no timbering. The portal of this is 75 feet higher in elevation than Canyon Creek, and this creek empties into North Fork of American River. The tunnel is driven on a grade of three-quarters of an inch per foot, and during hydraulicking operations it was provided with a sluice and steel rails for riffles to recover the gold. Tailing was discharged to Canyon Creek.

[5] Logan, C. A., op. cit.

The south end of the property for a length of about half a mile and width in the bottom of 400 to 1200 feet has been worked to bedrock by means of hydraulic mining. The lower 60 to 80 feet of the gravel is cemented and required blasting in advance of hydraulicking. After hydraulic mining ceased the property was worked by drifting and a tunnel was driven 1800 feet on the bedrock beyond the 4000-foot tunnel, which is at a lower elevation, to give access for drift-mining. Logan [6] has published details on production during short periods from 1872 to 1879 in State Mineralogist's Report XXXII.

To the north of the railroad, on the part of the channel that drains toward Bear River, are many millions of cubic yards of unworked gravel in the Dutch Flat district. Extensive deposits in this area are owned by James L. Gould, Soda Springs P. O., Placer County, and Nichols Estate Company, c/o Arthur Nichols, 846 Mendocino Avenue, Berkeley. Individual members of the Nichols family hold tracts in this vicinity also.

Volcano Mining Company, Ltd., 1018 Mills Building, San Francisco, worked the Volcano drift mine in secs. 18, 19, 20, T. 14 N., R. 11 E., M.D. between Foresthill and Michigan Bluff in 1940. The operation was continued in 1941, and 4000 tons of gravel yielded 206 ounces of gold and 27 ounces of silver.

PLUMAS COUNTY

The mineral resources of Plumas County were described in the California Journal of Mines and Geology [7] for April, 1937, which contains a geologic map of the county showing the locations of the principal mines and a long table of mines giving references to earlier reports. Placer mining was not active in the county in 1937, and only a few such mines are described. Descriptions of many of the old drift and hydraulic mines are contained in the chapter on Plumas County [8] of State Mineralogist's Report XVI, which is still available at this time (1945). A map showing drift and hydraulic mines of La Porte district is on file at the San Francisco office of this division.

Baker and McCowan, Box 305, Chico, moved a dragline dredge from Butte County to Meadow Valley in the Quincy district and operated from August to December, 1940 and during 1941. The dragline excavator had a 1½-cu.yd. bucket.

Innis Dredging Company of Nevada City moved its dragline dredge from Nevada County to Lights Creek in the Lights Canyon district and resumed operations August 4, 1940. The dragline excavator had a 2-cu.yd. bucket. In 1941 this operation was continued from January 1 to September 22, and the yield from 250,000 cubic yards of gravel was 1653 ounces of gold and 130 ounces of silver.

Lobicassa Company, Box 812, Sacramento, operated a dragline dredge on Jamison Creek in the Johnsville district from August 20 to December 24, 1941.

[6] Logan, C. A., Gold mines of Placer County, Gold Run district: California Jour. Mines and Geology, vol. 32, pp. 62, 63.
See also: Dutch Flat District, pp. 56-58.
[7] Averill, C. V., Mineral resources of Plumas County: California Jour. Mines and Geology, vol. 33, pp. 79-143, 1937.
[8] MacBoyle, Errol, Mines and mineral resources of Plumas County: California Min. Bur., State Mineralogist's Rept. 16, p. 188, 1920.

SACRAMENTO COUNTY

Natomas Company

Natomas Company, Forum Building, Sacramento, was the largest producer of placer gold in California in both 1940 and 1941. During those years the company operated seven dredges in the Folsom district. Operations were curtailed in 1942, 1943, and 1944, but were being increased again in the summer of 1945. One of the dredges is equipped with buckets holding 17 cubic feet each, two with 16-cu.ft. buckets, three with 12-cu.ft. buckets, and one with 9-cu.ft. buckets. The depth which they can dig below the surface of the water ranges from 30 feet to 70 feet. Bedrock is volcanic ash. More gravel exists below the volcanic ash in parts of the dredging field, but it is not gold-bearing. Thomas McCormack of Rio Vista is president, and R. G. Smith of Natomas is general manager of gold dredging. This company designs and builds its own dredges, and the following notes on recent practices have been supplied by R. G. Smith:

Main Drive. The main drive of the bucket line has been changed to direct current (d-c) motors with Ward-Leonard controls. The drive is through reduction gears which are entirely enclosed in cases. The reason for this is the more efficient variable speed and the greater ease of control. Because of this, the company has been able to increase bucket-line speeds as much as 50 percent where the gravel is not too hard. To supply the direct current a motor generator set is installed in the bottom of the hull. Many of the controls for other machinery are installed in the same room. This helps to maintain the center of gravity of the dredge at a low point. Other new drives for main bucket line used by different companies consist of two-motor alternating current drives with either constant or variable speeds through V-belt drives to the main drive shaft; also a single two-speed motor drive through V-belts to the single main drive shaft of the old-type conventional drive.

Ladder Hoist. Where motors have been installed for the main drive, it has become necessary to have a separate ladder hoist. These are driven through enclosed gears and are generally operated with regenerative braking. The hoist is equipped with magnetic brakes and post or gravity brakes operated pneumatically; also with Lilly over-speed and over-haul control. Greatly increased hoisting speed has been adopted.

Swing Winch. Where motor generator sets have been installed for d-c main drives, individual bow-line winches driven by d-c motors have been installed. They are driven through enclosed reduction gears and are equipped with Ward-Leonard controls. This gives efficient variable speed and maximum ease of control for the side-swing. Line travel on both sides of the dredge is made the same. Variations of this practice have consisted of using a-c drive as heretofore, isolating the bow-line drums to operate without going through a chain of gears on the side-line winch. The side-line winch is used as drive for one bow-line and a separate winch and motor are used for the other bow-line. Most of these installations are equipped with pneumatic control for both shifting of frictions and for brakes. Brakes are usually of the gravity type, which are released by the pressure of compressed air.

Side-Line Winch. Pneumatic control of both brakes and frictions has become general practice, and the winches are driven by enclosed

FIG 86.   Natomas Company dredge.

reduction gears.   The use of pneumatic control for the winches has done away with the great number of operating levers which formerly had to be placed to one side of the bucket-line.   Now the winch room can be placed on the longitudinal center-line of the dredge, and this gives the operator a much better chance to see what he is doing.   Pneumatic control reduces the manual effort required of the operator to a minimum.

Screen Drive.   Screens are now driven either directly through enclosed reduction gearing set in the same line with screen or with V-belt drive with motor set on a horizontal plane.

Gold Saving.   The tendency has been to increase the total width of gold-saving devices per cubic yard handled rather than to increase the total area.   The total width of riffle-tables has been divided into narrower sluices, and this tends to increase the effective riffle area.   The reason is that a tilting of the dredge in the longitudinal direction giving a fore and aft pitch, tends to throw all the fluid in the sluices to one side of the sluices.   If the sluice is kept narrow, this tilting is not likely to expose the riffles on the high side of the sluice.   Natomas Company has used jigs on some of its dredges since 1914.   Where a jig 42 inches wide is used, the riffle sluices are made 21 inches wide.   This is in contrast to a 30-inch width formerly used.   On the Natomas dredges on which jigs are used only a launder is placed between the distributor box beneath the screen and the jig, that is, no riffles are used ahead of the jig.   The main consideration for the use of jigs is that the gold is difficult to amalgamate.   Two of the dredges of Natomas Company are not equipped with jigs because the gold which they recover is not tarnished.   Jigs may be able to save fine gold a little better than riffle sluices as ordinarily handled.   Natomas uses both Bendelari and Pan-American[1] jigs.   On the Bendelari jig the agitation is effected by an eccentric which actuates a circular rubber diaphram in the hutch.   On the Pan-American jig the whole bottom of the jig is moved up and down by an eccentric, and the joint between the stationary and the moving part of the hutch is a circular rubber part shaped like a tire-casing.   Jigs need a thicker pulp than riffles.   Too much water will carry the gold past the jigs.   Hence only the high-pressure water goes over the jigs, and an equal amount

[1] Made by Pan-American Engineering Company, 820 Parker Street, Berkeley, California.

of low-pressure water is added below the jigs. On a dredge handling 500 to 600 cubic yards per hour, of which 40 percent goes to the gold-saving devices, 6000 gallons per minute of high-pressure water is used and 6000 gallons per minute of low-pressure water. An arrangement that is quite common is to use a single-cell jig as a rougher, with riffle sluices below the jig to take the overflow and to serve as emergency gold-saving equipment. The concentrates are amalgamated on riffles and then put over a cleaner jig, from which final concentrates flow to a ball-mill for scouring the rusty gold. A variation consists of some riffle tables ahead of the jigs and also of amalgamation of the jig concentrate in a continuous barrel amalgamator. Another arrangement consists of two-cell jigs with no riffles on the jig overflow. Concentrates are either subjected to riffle amalgamation or fed directly to cleaner jigs. The final product is then amalgamated.

Hull Construction. Although some of the recently-built dredges have been equipped with the conventional riveted steel hull, the tendency toward welded construction has been evident. Pontoon-type hulls have met with considerable favor even for dredges of 9 cubic feet and under, and for digging to depths up to 70 feet. The pontoons are of welded construction and are bolted together during assembly in the field. Hulls with welded construction have been in service for several years without any evidence of failure.

Buckets. Rivetless bucket lips have been coming into greater favor rapidly. Perfection of small construction details and the operating advantages gained have been contributing factors to popularity. Low lips that are welded to the bucket are coming into use in some places, but not yet in California.

R. G. Smith points out that a dredge should be tailor-made to fit the particular tract of ground that is to be worked. The design of the dredge must take into consideration each of the following:

1. Depth of the gravel.
2. Whether the gravel is tight or loose.
3. Whether large boulders are present.
4. Whether there is a high percentage of sand and silt.
5. Length of time that the dredge is expected to last.

On one Natomas dredge the fines from the riffle sluices go to a sump containing a revolving sand-wheel. Buckets around the circumference of this wheel pick up the sand from the sump and discharge it to the same stacker that carries the oversize from the trommel. Silt overflows from the first sump to a second sump, where it is picked up by a pump and discharged through a pipeline that runs along the side of the stacker.

Other Operators

Capital Dredging Company, 351 California Street, San Francisco, operated two connected-bucket dredges on its property 5 miles south of Folsom throughout 1940, 1941, and until October 15, 1942. One dredge had 88 and the other 100 buckets of 18-cu.ft. capacity. They were both electrically driven.

Carson Creek Dredging Company, 216 Pine Street, San Francisco, washed gravel on the Martin Quinn Estate from September 11 until the end of 1940, using a dragline excavator with a 1¾-cu.yd. bucket. The operation was continued from January 1 until February 5, 1941, and was then taken over by Northwest Development Company.

Climax Dredging Company of Folsom operated a dragline dredge on the J. Vincent property in the Folsom district from January 1 to April 8, 1941.

Cosumnes Gold Dredging Company, 351 California Street, San Francisco, operated a connected-bucket dredge in the Cosumnes River district, 7 miles southwest of Sloughhouse during 1940, 1941, and until October 20, 1942. The dredge is equipped with 63 buckets of 12-cu.ft. capacity, and is electrically driven.

Cutter and Mueller recovered gold as a by-product in the operation of a commercial sand and gravel plant at Fair Oaks in 1942. The recovery from 40,320 cubic yards of gravel was 183 ounces of gold and 15 ounces of silver.

General Dredging Corporation of Natoma or 811 W. 7th Street, Los Angeles, operated a dragline dredge on the American River in the Folsom district during nearly all of 1940. This corporation was dissolved September 30, 1941, but continued to operate as General Dredging Company, a partnership. Dredge no. 1, equipped with a dragline excavator having a 5-cu.yd. bucket, was operated on American River. Dredge no. 2, operating on the ancient river channel in the same district, used a dragline excavator with a 2-cu.yd. bucket. Dredge no. 4, working gravel along the American River near Fair Oaks, also used a dragline excavator with a 2-cu.yd. bucket. Operations were continued in 1942.

Hoosier Gulch Placers, 1015 25th Street, Sacramento, operated a dragline dredge on Katesville Gulch and on the Logtown property in the Cosumnes River district during 1941. The dragline excavator had a 2-cu.yd. bucket. A second dragline dredge served by a 2½-cu.yd. bucket was operated on the Hutchison property from January 5 to October 31, 1940. This company operated boat no. 1 on the Biggs Ranch and boat no. 2 on the Rossi property throughout 1941 and during part of 1942.

Humphreys Gold Corporation, 910 First National Bank Building, Denver, Colorado, operated a dragline dredge in the Cosumnes River district on the Fassett-Parker-Hanlon property from January 1 to December 5, 1940. The equipment, which included a dragline excavator with a 2½-cu.yd. bucket, was then moved to the Hutchison property where operations were continued during all of 1941 and from January 1 to April 6, 1942. In 1942 the floating washing plant was served by four dragline excavators, each equipped with a 2½-cu.yd. bucket. These were used to strip overburden as well as to deliver auriferous gravel to the plant.

Lancha Plana Gold Dredging Company, La Lomita Rancho, Lockeford, operated a connected-bucket dredge at Sailors Bar, American River, from April 20, 1940 until November 11, 1942. The dredge is equipped with 84 buckets of 6-cu.ft. capacity and is electrically driven.

Lobicassa Company, Box 812, Sacramento, operated a dragline dredge, with a 1½-cu.yd. bucket on the Mahon property from June 5 to October 17, 1940, when the property was worked out.

McQueen and Downing, 1040 38th Street, Sacramento, operated a dragline dredge on Deer Creek in the Folsom district from September 10 until December 17, 1940 and from January 1 to February 14, 1941.

Natomas Company. See page 278.

Pacific Coast Aggregates, Inc., 1401 42nd Street, Sacramento, recovered 111 ounces of gold and 11 ounces of silver as a by-product in the operation of a commercial sand and gravel plant at Fair Oaks in 1942. In 1943 this company, together with Fair Oaks Gravel Company, recovered 249 ounces of gold and 21 ounces of silver.

SAN BERNARDINO COUNTY

The mineral resources of San Bernardino County have been described by Tucker and Sampson [1] in the California Journal of Mines and Geology for October 1943. Although this county contains a wide variety of mineral deposits of commercial importance, and although the report cited contains descriptions of many lode-gold mines, little placer mining for gold has been done in the county recently.

Hoefling Bros., Box 768, Sacramento, recovered placer gold as a by-product of an operation carried on throughout 1943 principally for scheelite at the Spud Patch mine in the Randsburg district. Gravel amounting to 597,139 cubic yards was mined with a dragline excavator, of which 134,934 cubic yards were trucked to a stationary washing plant. The by-product production of precious metals was 210 ounces of gold and 45 ounces of silver.

Holcomb Valley Placer Company, 973 North Main Street, Los Angeles, operated a non-floating washing plant in the Holcomb Valley district, to which gravel was delivered by tractor and scraper from July 7 to November 16, 1941. The yield from 16,265 cubic yards of gravel was 204 ounces of gold and 10 ounces of silver.

SAN JOAQUIN COUNTY

California Gold Dredging Company, 351 California Street, San Francisco, operated an electric connected-bucket dredge with 81 buckets of 6-cu.ft. capacity during 1940 and 1941, in the Jenny Lind district.

Gold Hill Dredging Company, 311 California Street, San Francisco, operated two electric connected-bucket dredges on the Jenny Lucas, Alex Perie, Putnam, Thorne, and Osterman properties in the Camanche district during 1941. Some work was done in San Joaquin County during 1940 also. One dredge had 66 buckets of $7\frac{3}{4}$ cu.ft. capacity and the other had 87 buckets of $8\frac{1}{2}$-cu.ft. capacity. Operations were continued in 1942 on some of the properties mentioned above and also on property of California Lands, Inc. and Central Bank of Calaveras during most of 1942. The company also operated one dredge at a time during parts of 1943.

Gold Valley Dredging Company operated a dry-land washing plant to which gravel was delivered by gasoline dragline excavator with a $\frac{1}{2}$-cu.yd. bucket at the Murdock Ranch in the Camanche district from February 25 to May 24, 1942. The yield from 9,316 cubic yards of gravel was 83 ounces of gold and 6 ounces of silver.

Lobicassa Company, Box 812, Sacramento, operated a dragline dredge using a $1\frac{1}{2}$-cu.yd. bucket on the Foster Ranch in the Camanche district from January 1 to May 14, 1942. The property was then abandoned as worked out.

Mokelumne Sand and Gravel Company, 527 East Lodi Avenue, Lodi, produced a small quantity of gold as a by-product in preparing sand and gravel for concrete aggregate during 1943.

---

[1] Tucker, W. B., and Sampson, R. J., Mineral resources of San Bernardino County: California Jour. Mines and Geology, vol. 39, pp. 427-549, 1943.

San Gruco Company and C. E. Gruwell of Angels Camp operated dragline dredges on the McGurk property in the Bellota (Linden) district in 1940. Each operator used a dragline excavator with a 1½-cu.yd. bucket.

Smith-Notterman Company, 245 West Rose Street, San Francisco, operated a dragline dredge with a 1¾-cu.yd. bucket on the Elmer Cady and Lewallen Ranches in the Jenny Lind (Bellota, Linden) district during 1941. The operation was continued in 1942 from January to March 26, and the yield from 93,105 cubic yards of gravel was 344 ounces of gold and 14 ounces of silver.

A. G. Watkins & Sons of Linden operated a dragline dredge equipped with a 2-cu.yd. bucket during 1940 and parts of 1941 on the Calaveras River.

SHASTA COUNTY

A report on the mineral resources of Shasta County was published in the California Journal of Mines and Geology for April 1939,[1] and a map of the county showing locations of mines was included; also a table of mines giving references to older reports. All of the placer operations mentioned below were started after that report was written. A few earlier operations of similar nature are described in the April 1939 Journal.

B. H. K. Mines, Box 325, Orland, operated a dragline dredge equipped with a 1¼-cu.yd. bucket on the R. C. Connelly and Robert Litsch properties on Clear Creek from November 15 to December 31, 1941. The yield from 54,400 cubic yards of gravel was 339 ounces of gold and 48 ounces of silver. The operation was continued from January 1 to May 15, 1942, and the yield from 120,000 cubic yards of gravel was 760 ounces of gold and 104 ounces of silver.

J. P. Brennan, 1343 Butte Street, Redding, operated a dragline dredge on Tadpole Creek from January 1 until October 7, 1940, then moved the equipment to Champion Gulch and continued the operation until June 1941. Both of these operations were in the Igo district. The dragline excavator was equipped with a 1¼-cu.yd. bucket.

Clear Creek Dredging Company, Box 598, Redding, operated a dragline dredge using a 1½-cu.yd. bucket on Clear Creek in 1940. In 1941 a second dredge with a 2½-cu.yd. bucket was added. This company operated in 1942 also.

Columbia Construction Company, 1522 Latham Square Building, Oakland, prepared 1,781,466 tons of gravel to be used in the construction of Shasta Dam from a point upon the Sacramento River near Redding in 1940 and recovered a substantial quantity of gold. In 1941 this company prepared 4,038,167 tons of gravel and recovered 2810 ounces of gold and 301 ounces of silver as a by-product. Two dragline excavators were used, one with a 5-cu.yd. bucket and the other with an 8-cu.yd. bucket. In 1943 the company produced 1,500,000 cubic yards of sand and gravel and recovered 1555 ounces of gold and 166 ounces of silver as a by-product.

Crow Creek Dredging Company, Box 558, Redding, operated a dragline dredge on Crow Creek in the Igo district intermittently during

[1] Averill, C. V., Mineral resources of Shasta County: California Jour. Mines and Geology, vol. 35, pp. 108-191, 1939.
See also Averill, C. V., Gold dredging in Shasta, Siskiyou, and Trinity Counties: California Jour. Mines and Geology, vol. 34, pp. 96-125, 1938.

1940.   During 1941, 220,000 cubic yards of gravel were delivered by a dragline excavator with a 1½-cu.yd. bucket and were washed.   In 1942 operations continued on Cottonwood Creek from January 1 to April 13, and 100,000 cubic yards of gravel yielded 580 ounces of gold and 20 ounces of silver.

DeKarr and Herbert of Redding operated a dragline dredge, using a ¾-cu.yd. bucket, on the Fred Kohle property on North Cow Creek from January 16 to March 17, 1941.   The yield from 23,800 cubic yards of gravel was 297 ounces of gold and 46 ounces of silver.

Dobbin Gulch Dredging Company, Box 539, Redding, operated a dragline dredge with a 1¼-cu.yd. bucket, on the Montgomery property on Flat Creek from January 1 to May 30, 1941.   The yield from 142,160 cubic yards of gravel was 853 ounces of gold and 62 ounces of silver. In 1942 operations were conducted on the Roaring River from March 3 to June 2.   The yield from 70,500 cubic yards of gravel was 169 ounces of gold and 11 ounces of silver.

French Gulch Dredging Company, 2404 Russ Building, San Francisco, installed a connected-bucket dredge equipped with 76 buckets of 4½-cu.ft. capacity on Clear Creek near French Gulch and began operations on September 2, 1940.   The operation was continued during 1941.

C. E. Gruwell, Hotel Redding, Redding, operated a dragline dredge on the Fish, Forschler, Rais, and Russell Ranches in the Igo district during 1941.

Lincoln Gold Dredging Company of Lincoln operated a dragline dredge having a 2½-cu.yd. bucket on the Brady property in the Igo district in 1942.   The yield from 37,234 cubic yards of gravel was 162 ounces of gold and 26 ounces of silver.

R. S. Olson, 1178 Walnut Avenue, Redding, operated a dragline dredge from January 1 to March 10, 1940, on China Gulch and from June 18 to December 5 on Daly Gulch.   Both are in the Igo district. Operations on Daly Gulch were continued in 1941.

Pioneer Dredging Company, Box 305, Redding, operated a dragline dredge equipped with a 3-cu.yd. bucket, in Happy Valley from January 1 to August 21, 1940.

FIG. 87.   Thurman Gold Dredging Company. *Photo by courtesy of Yuba Manufacturing Company.*

San Gruco Company, Redding, moved its dragline dredge equipped with a 1½-cu.yd. bucket to property of the Happy Valley Land and Water Company and operated from November 1 until the end of 1940; also during 1941.

Tehama Dredging Company, Box 727, Anderson, operated a dragline dredge, which had a ¾-cu.yd. bucket, at the Gold Acres mine near Gas Point from March 20 to June 30, 1941. The yield from 48,860 cubic yards of gravel was 242 ounces of gold and 17 ounces of silver.

Thurman Gold Dredging Company, 235 Montgomery Street, San Francisco, installed a connected-bucket dredge equipped with 72 buckets of 9-cu.ft. capacity on Clear Creek and began operations on December 1, 1940. Operations continued during 1941 and until October 14, 1942.

SIERRA COUNTY

Depot Hill hydraulic mine owned by F. J. Joubert of Camptonville has been operated practically every season for many years. Storage for the tailing is available behind the Bullards Bar dam of the Pacific Gas & Electric Company. The mine is 5 miles north of Camptonville on the state highway running to Downieville in sec. 19, T. 19 N., R. 9 E., M. D. More than a page of additional details about this mine is contained in the California Journal of Mines and Geology for January 1942.[1]

Indian Hill mine was described by Gardner and Johnson[2] as follows:

B. F. Dyer operated the old Indian Hill mine near Camptonville during 1931 and 1932. The material washed up to the end of 1932 consisted mainly of slides from the faces of the old workings. The gravel deposit was 35 feet thick; the grade of bedrock was 1 inch to the foot.

---

[1] Averill, C. V., Mines and mineral resources of Sierra County: California Jour. Mines and Geology, vol. 38, p. 29, 1942.
[2] Gardner, E. D., and Johnson, C. H., Placer mining in the western United States, Part II, Hydraulicking, etc.: U.S. Bur. Mines Inf. Circ. 6787, pp. 50-51, 1934.

FIG. 88.   Depot Hill hydraulic mine. *Reprinted from California Journal of Mines and Geology, January 1942, p. 25.*

FIG. 89.   Poverty Hill Properties dredge under construction.   *Reprinted from California Journal of Mines and Geology, January 1942, p. 36.*

FIG. 90.   Removing overburden ahead of dredging with carryall at Lancha Plana, Amador County.   *Photo by courtesy of Yuba Manufacturing Company.*

FIG. 91.   William Richter and Sons dragline dredge.   *Reprinted from California Journal of Mines and Geology, January 1942, p. 38.*

Water was brought to the mine through a 9-mile ditch and 3000 feet of 22-inch and 1500 feet of 15-inch pipe. The head was 130 feet. One No. 6 giant with a 4-, 4½-, or 6-inch nozzle was used. Boulders up to 14 inches in diameter were run through the sluice boxes. The sluiceway was down a narrow gulch and consisted of six sections of boxes (2 to 6 boxes to the section) and the rock bottom of the gulch between sections. There was a drop of 10 or 15 feet at the end of each section of boxes. The fall and cascading down the rocky gulch between each section broke up all cemented material and washed the gravel free of clay.

The boxes were 40 inches wide and 40 inches high; the grade was ½-inch to the foot. The upper five sections of boxes were paved with wooden blocks; the riffles in the lower section were of rock paving. Seventy percent of the gold was caught in the upper two boxes. Three undercurrents were used near the lower end of the line. The discharge of one undercurrent went into the main sluice before the next was taken out. The grizzly opening for an undercurrent in the bottom of the main sluice was 18 by 40 inches. The grizzlies were of 1¼- by 3-inch iron bars set on edge 2½ inches center to center. Additional top water was run over the undercurrents from an opening in the side of the sluice. The first undercurrent was 8 feet wide by 24 feet long. The riffles consisted of rows of pine blocks 6 inches thick by 6½ inches deep separated by 1½-inch crossboards. The second undercurrent was 8 feet wide and 20 feet long. The riffles consisted of 3½- by 3½-inch angle iron ⅜ inches thick and set crosswise on 5-inch centers. The third undercurrent at the end of the lowest box was 10 by 12 feet. The riffles were four angle irons 1 inch apart at the head of the undercurrent and rock paving from there down. A crew consisted of seven or eight men. About 100,000 cubic yards was washed during the 1932 season at a cost of 8 cents per cubic yard exclusive of construction work.

Loftus Blue Lead Mining Company, 801 Columbia Street, South Pasadena, operated a hydraulic mine in 1940 and 1941 on a large group of claims running from St. Louis to Howland Flat, a distance of 4 miles by road, in secs. 31, 32, T. 22 N., R. 10 E., M. D.; secs. 5, 6, 7, T. 21 N., R. 10 E.; and sec. 12, T. 21 N., R. 9 E. In 1940 the yield from 60,000 cubic yards of gravel was 352 ounces of gold and 25 ounces of silver. A few additional details about this mine are contained in the California Journal of Mines and Geology for January 1942.[3]

Pioneer Project mine in secs. 13, 14, 23, 24, T. 21 N., R. 9 E., M. D., is a consolidation of the old Pioneer with the adjoining Comet, Challenge, and Riffle claims. It was worked in 1942 and 1943 by A. J. Just, W. H. Pike, and A. J. Modglin of LaPorte. Hydraulic operations for a period of 10 days in 1942 yielded 71 ounces of gold and 4 ounces of silver from 17,000 cubic yards of gravel. A substantial quantity of gold was recovered during a 3-month period of operation in 1943.

Poverty Hill Properties, 974 Mills Building, San Francisco, is a partnership of which the general partners are A. J. Oyster, W. C. Van Fleet, Walter W. Johnson. Operations were conducted in 1940-41 on a part of the main La Porte channel, one of the old Eocene auriferous channels. The property consisted of 1100 acres in sec. 32, T. 21 N., R. 9 E., M. D., and sec. 5, T. 20 N., R. 9 E. The property was first worked by hydraulicking, but later a connected-bucket dredge with

---

[3] Averill, C. V., op. cit., p. 29.

FIG. 92. Ruby mine, surface plant. *Reprinted from California Journal of Mines and Geology, January 1942, p. 39.*

FIG. 93. Ruby mine, underground slusher hoist. *Photo by courtesy of L. L. Huelsdonk; reprinted from California Journal of Mines and Geology, January 1942, p. 40.*

82 buckets of 6-cu.ft. capacity was installed in a pond in one of the hydraulic pits. Overburden was stripped to a depth of 40 feet with caterpillar tractors and carryalls. A few additional details about this operation are contained in the California Journal of Mines and Geology for January 1942.[4]

William Richter & Sons, Route 2, Box 400, Oroville, operated a dragline dredge equipped with a $1\frac{1}{2}$-cu.yd. bucket on property owned by the Pacific Gas & Electric Company on the Yuba River in secs. 16, 17, T. 19 N., R. 9 E., M. D., in 1941 and 1942. In 1941 operations from June 1 to December 31 yielded 1403 ounces of gold and 179 ounces of silver from 280,000 cubic yards of gravel. In 1942 operations from January 1 to April 30 yielded 243 ounces of gold and 29 ounces of silver from 58,000 cubic yards of gravel.

Ruby mine has been operated during recent years by C. L. Best, 800 Davis Street, San Leandro. In 1943 operations were on a maintenance basis only because of Limitation Order L-208 of the War Production Board. The yield from 500 cubic yards of gravel was 323 ounces of gold and 12 ounces of silver. Because of the newly developed methods of drift mining used at this mine, a description of the operations in 1941 is reprinted below from the California Journal of Mines and Geology for January 1942:[5]

---

[4] Averill, C. V., op. cit., pp. 35-37.
[5] Averill, C. V., op. cit., pp. 38-42.

FIG. 94.   Ruby mine, timbering.   Slusher scraper at left.   *Photo by courtesy of L. L. Huelsdonk; reprinted from California Journal of Mines and Geology, January 1942, p. 41.*

19

FIG. 95. Gold nuggets from the collection of C. L. Best, mined at the Ruby drift gravel mine, Sierra County, California. Total weight, 899.15 troy ounces; total value $30,037.65. *Photo by courtesy of L. L. Huelsdonk, superintendent, Ruby mine.*

The Ruby mine was being operated in 1941 by C. L. Best, Caterpillar Tractor Company, San Leandro, with L. L. Huelsdonk, Goodyears Bar, in charge. Including a lease on the Mott property, there are 1,150 acres, of which 300 are patented, in secs. 10, 11, 14, 15, T. 19 N., R. 10 E., M. D. The mine is reached by 15 miles of road from Downieville, mostly steep mountain road, dirt surface.

The Bald Mountain Extension channel, one of the oldest Tertiary channels, branches from the Bald Mountain channel at a point north of Forest. Bald Mountain channel is the same as the Ruby and City-of-Six channels. The last two named are simply continuations of the Bald Mountain channel to the north. The Bald Mountain Extension channel was worked in the Ruby mine in the nineties from an adit level driven from the side of the mountain on which the town of Forest is located. Present work is on the opposite side of the mountain. An old adit level (portal elevation 4,707 feet) was utilized for a distance of 1,800 feet. Work beyond that point is new. The adit is in the Tightner formation for 3,320 feet, then in serpentine for 520 feet, then in gabbro and schist for 610 feet, then passes into a second belt of serpentine. A point in the adit is 1,850 feet south of the common corner of secs. 10, 11, 14, 15, T. 19 N., R. 10 E. The contact of the second belt of serpentine and the gabbro-schist is 200 feet east of the point in the adit just described. The adit then continues in a general southeasterly direction to a point where a raise was put up to the intervolcanic channel. Distance from the portal to this raise is 5,850 feet and the raise is 109 feet high. From the top of the raise 400 feet of drifting was done in a southerly direction on the channel and 4,000 feet in north and northeasterly directions on the channel. From this point the channel winds considerably, and 700 feet more of driving will be needed to connect with the Larry shaft, of which the collar elevation is 5,163 feet and the bottom elevation is 4,954 feet. Several thousand feet of additional exploratory work have been driven on the channel, and a connection for air, involving 3,000 feet of work, has been made to the Golden Bear shaft.

The intervolcanic channel that is being worked is 200 feet lower than the Bald Mountain Extension channel and cut off the Bald Mountain Extension channel. Apparently much of the gold in the intervolcanic channel was derived from the older channel. The intervolcanic channel varies from 60 feet to 160 feet in width and is breasted to a height of 6 to 8 feet. Channels are capped by as much as 900 feet of lava, which is mostly andesite, but basalt is found on top of the andesite in places. Large boulders are stored underground. The finer gravel is moved by slusher scrapers to raise-chutes and hauled in trains by storage battery locomotives to the washing plant at the portal of the main adit level. Timbering comprises stulls and caps specially designed with a mortise and tenon and handled by one man.

In the summer of 1941, the crew comprised 18 men, and 80 to 100 tons of gravel were treated per day, but when the crew was 43 men, 200 tons were treated per day and a maximum of 250 tons was reached. The reason for the small crew in 1941 was that many men had left to engage in defense activities. Gravel passes from storage bin over Hungarian riffles of alloy steel 50 inches wide by 1¼ inches deep; then to a vibrating screen, which is a double screen. The upper screen is of 2-inch square openings, and rods are half an inch in diameter. The screen which is below is four-mesh of No. 12 wire. Undersize goes to a six-unit Huelsdonk table 20 feet long by 7 feet wide. The washing plant will treat 500 tons of gravel per 24 hours. The screen mentioned above is vibrated by an eccentric and 20-hp. motor with a 3¾-inch stroke at the rate of 200 vibrations per minute. Undersize goes to the Huelsdonk table mentioned above, which is vibrated with a ¾-inch to 1-inch stroke at a rate of 200 vibrations per minute. The end of the table farthest from the vibrating screen is set three-quarters of an inch lower than the end near the screen. Recovery amounting to 10 to 20 percent of the total is made on this table as fine gold. The remainder is made on the first riffle and the screen about equally divided. Steel bars are placed across the screen to hold it down and these have a tendency to act as riffles. Below the screen additional riffles are provided in the sluice that carries away the oversize, but little gold is recovered from these. Nuggets as big as 52.33 ounces valued at $1,758 have been recovered. C. L. Best is saving all nuggets above $100 in value for exhibit purposes and in 1941 had a collection of 123 that had been recovered since 1937. Gold is 940 to 950 in fineness. Most of the tailing is stacked on the property by means of a belt conveyor.

The second serpentine belt mentioned above is 490 feet wide, then the workings pass into the Blue Canyon slate, which is the bedrock of the channel being worked. On the contact of the second serpentine belt and the Blue Canyon formation is a fault called the Independence, on which is a 6-foot quartz vein. Another quartz vein 4 feet in width was cut 110 feet farther ahead in the adit in the Blue Canyon formation. A third vein known as the Wolf vein strikes north and dips 65° W. It is 6 inches to 12 feet in width. This vein was worked in the years 1935, 1936, and 1937 to a depth of 200 feet below the main adit level. Drifts were run north on the 200-foot level for 600 feet and the ore was stoped through to the main adit level. This work on the quartz vein had been discontinued and the workings are now full of water. Ore was treated in a stamp mill of 30 tons daily capacity, and treatment was amalgamation on plates followed by tables and flotation. The quartz averaged $5.80 per ton in the mill but additional gold was recovered as high-grade. This vein is in the Tightner formation and was found at a distance of 2,120 feet from the portal of the main adit.

Camp facilities are provided for a crew of 40 men, and the property is well equipped with repair shops, drill sharpeners, air compressors, and other modern machinery. Electric power is supplied by Pacific Gas and Electric Company.

Tennessee Manxman drift mine was worked throughout 1942 by C. S. Poor. The yield from 1,200 cubic yards of gravel was 142 ounces of gold and 16 ounces of silver. In 1943 operations from January 1 to November 1 yielded 74 ounces of gold and 8 ounces of silver from 680 cubic yards of gravel.

SISKIYOU COUNTY

Beaver Dredging Company, 615 F Street, Marysville, worked a dragline dredge with a 5-cu.yd. bucket on Indian Creek 6 miles west of Fort Jones from April 16 to December 31, 1941.

C. & E. Dredging Company, 1002 Pacific Building, Portland, Oregon, operated a dragline dredge using a 2-cu.yd. bucket on McAdams and Cherry Creeks in the Deadwood district from May 9 to December 31, 1941.

Cal Oro Dredging Company, 681 Market Street, San Francisco, operated a connected-bucket dredge on the Lange property in the Greenhorn district from January 28 to September 22, 1940.

Etna Gold Dredging Company, 1730 Franklin Street, Oakland, operated a connected-bucket dredge with 80 buckets of 3-cu.ft. capacity on Wildcat Creek 2 miles north of Callahan in 1940. The yield from 800,000 cubic yards of gravel was 4299 ounces of gold and 642 ounces of silver. The operation was continued until October 1941.

Farnsworth mine in the Liberty district was operated by the hydraulic method by E. A. McBroom from March 1 to May 10, 1943. The yield from 500 cubic yards of gravel was 22 ounces of gold and 2 ounces of silver. See Salmon River Mining Company also.

The Gallia Placer Mining Company[1] operated the Gallia mine on the North Fork of the Salmon River near Sawyers Bar during the 1932 season. The gravel was 30 to 35 feet thick and contained some large boulders. Water was brought to the mine under a 265-foot head through a 2000-foot line of 36- to 15-inch pipe. Enough grade was not available for the disposal of tailings, and the gravel as mined contained too many boulders for the successful operation of a hydraulic elevator. The gravel was cut and swept to a Ruble elevator by a giant with a 3½- or 4-inch nozzle. It was then put through the elevator by another giant with a 4½-inch nozzle. The Ruble was 4 feet wide and elevated the oversize 25 feet. The Grizzly consisted of 90-pound rails 2½ inches apart set lengthwise on 10- by 10-inch stringers. The undersize from the grizzly went through a 24-inch sluice with riffles consisting of angle iron and rails placed crosswise in the boxes.

The sluice discharged into a hydraulic elevator with a 20-inch intake. A 4-inch nozzle was used in the high-pressure jet; the material was elevated 30 feet. The elevator discharged into a second sluice.

The capacity of the plant was limited by the quantity of material that could be run over the Ruble elevator. The cutting and driving giant was used for a few hours and then shut off until the accumulated material could be handled in the Ruble. The giant at the Ruble operated continuously. About a week with the full crew was required to move the Ruble; it had to be moved every 3 weeks, as the dump room behind it was exhausted.

Boulders too large to go up the Ruble were moved back by a derrick. Large boulders uncovered in cutting were dragged from the pit by means of a donkey engine. Water was used not only for the giants in the pit and the jet of the hydraulic elevator but also for operating the derrick and running a dynamo for operating a sawmill and an air compressor.

---

[1] Gardner, E. D., and Johnson, C. H., Placer mining in the western United States, Part II, Hydraulicking, etc.: U.S. Bur. Mines Inf. Circ. 6787, 1934.

The working crew consisted of two men in the pit and one man on the ditch line on each of two 12-hour shifts. Although 50 to 60 cubic yards per hour could be cut and swept to the Ruble by the cutting-giant, the average capacity, including the time for moving the Ruble, was 200 cubic yards per day. The labor cost, assuming $4 per man-shift, would be 11 cents per cubic yard. The cost of supplies would be about 2 cents and supervision 4 cents, making the operating cost 17 cents.

Happy Camp Dredging Company, Happy Camp, operated a dragline dredge on the Allen property from May 1 to May 31, 1940. The dredge was then moved to property owned by Grant Smith and was operated there from September 1 to November 30.

Horton Gulch Mine was described by Gardner and Johnson[2] as follows:

J. O. McBroom operated the Horton Gulch placers on the South Fork of the Salmon River near Cecilville. The 1932 season extended from January 1 to April 11. The gravel was fairly tight. The grade of bedrock was 1 inch to the foot. One giant with a 5-inch nozzle, working under a 65-foot head, was used for both cutting and sweeping the gravel into the sluice boxes. About 30 inches additional by-wash water was used for moving the gravel through the sluice which consisted of three 12-foot boxes 24 inches wide. The riffles were hard boulders hand-shaped to make a pavement 7 to 10 inches thick. An undercurrent was used for 1 month and then discarded; about 1 ounce of gold was cleaned up from the undercurrent during the month's run. All large boulders were blasted. An average of 80 cubic yards per day was washed during the 1932 season. Two men were employed. At $3.50 per shift the labor cost would be 9 cents; supplies would amount to about 2 cents per cubic yard, making a total of 11 cents.

Joubert mine in the Liberty district was worked by lessees, H. J. Dickinson, Stanley Czerwinski, and others of Sawyers Bar in 1940 and 1941. Hydraulic operations in 1940 yielded 354 ounces of gold and 55 ounces of silver. In 1941 the yield from 27,900 cubic yards of gravel was 385 ounces of gold and 58 ounces of silver.

Larsen Bros. and Harms Bros., Route 4, Box 2220, Sacramento, operated dragline dredges on the Klamath River and on Horse Creek during 1940 and 1941. In 1943, 114 ounces of gold and 18 ounces of silver were recovered from concentrates accumulated at the Moccasin dredge before it was closed because of War Production Board Limitation Order L-208. The following description of operations in 1940 are reprinted from the California Journal of Mines and Geology for April 1941:[3]

Scandia mine, in secs. 7, 8, 9, 15, T. 46 N., R. 10 W., M. D., on Horse Creek near the Klamath River, was being operated in 1940 by Larsen Bros. and Harms Bros., Route 4, Box 2220, Sacramento. Emmet Miles, Horse Creek, Siskiyou County, was in charge at the mine. The property is reached by means of 2 miles of dirt road turning from the graveled state highway along the Klamath River.

2 Op. cit.
3 Averill, C. V., Dragline dredging in Siskiyou County: California Jour. Mines and Geology, vol. 37, pp. 328-331, 1941.

FIG. 96.   Dragline dredge at Scandia mine.

Low bars of Horse Creek for a total length along the creek of 6 miles were being dredged.   The width was about 1,000 feet in the lower part of the tract but less above.   Depth of gravel was 12 feet to 18 feet and practically all the gold was in the lowest 2 feet.   From a tract of 100 acres about 2,000,000 cubic yards had already been dredged in the fall of 1940.

The washing plant is of Bodinson make and is similar in all respects to the ones described in the chapter on dragline dredging (*ante*).

It is of 3,600 bank-yards capacity per 24 hours and is built to serve a 3-cu.yd. dragline excavator.   The bucket in use is a $2\frac{1}{4}$-yard Esco.   The trommel is punched with $\frac{3}{8}$-inch to $\frac{3}{4}$-inch holes, and undersize goes to standard-dredge-type Hungarian riffles.   Murphy 6-cylinder diesel engines rated at 160 hp. each supply the power on both the excavator and the washing plant.

The following figures on cost of equipment and cost of operation were furnished by Emmet Miles: 95 Northwest dragline with 60-foot boom, $58,000; washing plant, $40,000; D7 tractor with bulldozer, $7,000; D8 tractor with carryall, $10,000; truck, welder, pickup, $1,900; miscellaneous lighting plants and pumps, $1,000; air compressor, $200; spare generator for washing plant, $900.   The last item mentioned furnishes power to an electric motor on the upper end of the belt stacker.   The total crew comprises 14 men.

Gravel actually washed on boat cost 5 to 6 cents per cubic yard to handle.   Cost of removing and leveling overburden was 7 cents per cubic yard.   No depreciation was included in these costs.   To cover the cost of mining, stripping overburden and leveling, including depreciation but excluding profit, it was necessary for the gravel to run 10 cents per cubic yard.

Restoration of the land in order to make it available for farming appears to be an accomplished fact at this mine.   A tract of 100 acres is being so restored, and 60 acres were to be planted to new crops, alfalfa, rye, and sweet clover in the fall of 1940.   Some of the fields were already green with the new crops in October, 1940.

The land along the creek averages 1,000 feet in width. It carries an overburden of soil 5 to 6 feet in depth. For dredging, it is divided into strips roughly 300 feet in width. From the first strip the soil over- burden is removed to wasteland along the outside of the tract. This is done with a D8 Caterpillar tractor and a carryall of 12 cubic yards capacity. At times a second outfit of 16 cubic yards capacity has been in use. After the soil has been removed from the 300-foot strip, the gravel beneath is dredged with the dragline to a depth of 12 to .18 feet. The belt-stacker on the dredge leaves the gravel behind in conical piles about 20 feet high. Next a Caterpillar tractor with bulldozer attachment levels these piles, and the running back and forth of the heavy machine packs the gravel considerably. Then the soil of the second 300-foot strip is removed by the carryalls and placed on the leveled gravel of the first strip, and is spread out and graded so that the land is ready for farming. The process is continued to the last strip, where the gravel is spread in the form of a dike to hold the stream in a prepared channel. Previously the stream ran near the center of the tract and gave trouble from flooding and washing the land. Hence, the operators believe that the land is left in better condition for farming than it was originally.

Efforts have been made in the past to level the gravel from a dredge and to throw out the fines directly from the dredge on top of the gravel. Disturbing of the gravel by the dredge results in an increase of voids between the various boulders and cobbles, and the fines go into these voids to a great extent. Even if soil is left on top, it may gradually sink to fill the voids, leaving gravel and boulders exposed on the surface. The present method overcomes the difficulty in two ways: the heavy machine's pack the gravel, and the five to six feet of soil put on top gives a wide margin of safety.

Unfortunately this method can not be applied to all dredging-land. First, the average tract does not contain such a thick layer of good soil. Second, the method is expensive, and the average tract does not contain enough gold to pay for it. The operators figure that the extra cost of restoring the land was $20,000 for a tract that is worth less than $10,000 as a farm. As an example of the expense involved, consider the fact that one tractor worth about $8,000 had already been worn out in roughly

FIG. 97.   Dragline dredge at Moccasin mine.

a year and a half, and in the fall of 1940 a second was well on the way to being worn out.

However, the land along Horse Creek contained enough gold to pay for the restoration and a profit besides. The original restored tract was used to induce the owner of an adjoining tract to allow his land to be worked by the same method, so the whole operation is profitable.

Moccasin mine, in sec. 14, T. 46 N., R. 10 W., M.D., was also operated by Larsen Bros. and Harms Bros., Route 4, Box 2220, Sacramento, in 1940. The property is on the Klamath River about 1½ miles up the river from Horse Creek. It is reached from the bridge at Horse Creek by means of a road up the north bank of the river. A river bar about 2,000 feet wide carrying gravel to a depth of 18 feet to 35 feet was being dredged. Total depth in some places including 10 feet of fine overburden was 45 feet.

The Moccasin outfit includes the largest dragline excavator used for dredging in northern California. The only larger dragline used for any purpose is the one that excavates gravel from the Sacramento River at Redding for aggregate to build Shasta Dam. The 5-yard Monighan at Moccasin mine walks around much as a man walks and is equipped with a large foot on each side for that purpose instead of the Caterpillar treads commonly used on smaller draglines. It has a 100-foot boom and is capable of digging to a depth of 45 feet. The bucket in use is a 4½-cu.yd. heavy-duty Esco.

The washing plant is of Bodinson make and has a capacity of 6,000 cubic yards per 24 hours. The trommel is 47 feet by 72 inches and the largest holes are three-quarters of an inch. AR (abrasion resistant) steel used on the first section of the trommel had given 7½ months of service in the fall of 1940. The pump is a 14-inch United Iron Works pump driven by 100-hp. General Electric motor. The 85-foot stacker carries a 42-inch belt. The barge is 40 feet wide by 64 feet long by 54 inches deep. Power is furnished by a 300-hp. Fairbanks Morse diesel engine on each machine, the dragline and the washing plant. The engine on the washing plant drives a 200-kva. Fairbanks Morse alternator, 480 volts. Boil-boxes similar to those used by Lincoln Gold Dredging Company in Trinity County are used under the trommel. A manifold supplies jets with water under pressure in the various compartments beneath the trommel and thus the sands are kept in agitation.

The overburden, 10 feet in depth, is removed and piled to one side by means of a D8 Caterpillar tractor and a 12-cu.yd. carryall. No attempt is made at resoiling as at the Moccasin mine. The river was to be turned into a channel prepared for it on the north side of the tract during the winter, 1940-41, so that the present channel of the Klamath River could be mined. Near the river the gravel is free of overburden and is 18 feet in depth. The width of the pond being carried in 1940 was 500 feet.

Costs of principal items of equipment set up on the job were furnished by R. H. Wallace, superintendent, as follows: Monighan, $75,000; washing plant, $70,000; D8 tractor and carryall, $16,000. The total cost of operating the equipment per cubic yard of gravel was stated to be 9 cents.

Lincoln Gold Dredging Company, Lincoln, operated a dragline dredge equipped with excavator having a 1¼-cu.yd. bucket on the Calkins property 1 mile east of Yreka from July 7 to December 2, 1941. The yield from 93,742 cubic yards of gravel was 556 ounces of gold and 78 ounces of silver. In addition, E. A. Kinkle recovered a small quantity of gold by using a dry-land plant. The same two operators also worked the Rose property. In 1942 the company operated a dragline dredge on several properties as follows: from property owned by the City of Yreka the yield was 304 ounces of gold and 44 ounces of silver from 43,029 cubic yards of gravel; from General Dredge property the yield was 1844 ounces of gold and 263 ounces of silver from 272,423 cubic yards of gravel; and from the Nunes property the yield was 259 ounces of gold and 37 ounces of silver from 31,128 cubic yards of gravel. All of these properties are in the Greenhorn district.

McQueen and Downing, 125 Dexter Street, Yreka, operated a dragline dredge on the Neville and Silva properties in the Klamath River district during 1941.

Midland Company, Inc., 1112 Pearl Street, Alameda, operated a dragline dredge which had a 1½-cu.yd. bucket on the North Fork of the Salmon River in Liberty district throughout 1941. The yield from 350,000 cubic yards of gravel was 1950 ounces of gold and 284 ounces of silver.

Northern Dredging Company, 310 Kearny Street, San Francisco, operated a dragline dredge on the Allen and the Collins properties in the Klamath River district from January to May 1941, when the company was dissolved. A dragline excavator with a 2-cu.yd. bucket was used.

Okoro Mines, Inc. of Callahan operated a dragline dredge intermittently from January to June 1940 and washed 45,000 cubic yards of gravel, from which 218 ounces of gold and 31 ounces of silver were recovered. In 1941 operations conducted from July 11 to December 31 with a dragline dredge equipped with a 2½-cu.yd. bucket yielded 771 ounces of gold and 101 ounces of silver from 245,000 cubic yards of gravel. In 1942 operations with the dragline dredge on the Hayden property during the month of January yielded 103 ounces of gold and 10 ounces of silver from 60,000 cubic yards of gravel.

Oro Trinity Dredging Company, Box 212, Oroville, operated a dragline dredge which had a 1½-cu.yd. bucket on Scott River during 1940 and until May 31, 1941.

P. D. Sacchi, E. L. Spellenberg, and F. Kubli, of Arcata, operated a dragline dredge using a 1½-cu.yd. bucket at forks of Salmon intermittently during 1940. They recovered 365 ounces of gold and 53 ounces of silver.

Salmon River Gold Dredging Company, 310 Kearny Street, San Francisco, operated a dragline dredge, using a 3-cu.yd. bucket on several properties in the Salmon River district during 1941.

Salmon River Mining Company property was described by Gardner and Johnson[4] as follows: The Farnsworth brothers operated the mine of the Salmon River Mining Company on the South Fork of the Salmon River near Cecilville during the 1932 season. The gravel consisted of 6 feet of pay dirt overlain with 11 feet of overburden. The grade of the bedrock was three-fourths of an inch to the foot. Water under a 225-

---

[4] Op. cit.

foot head was brought to the mine through a 3-mile ditch and 1 mile of pipe line. The diameter of the first 3000 feet of the pipe line was 22 inches. This was reduced to 18 and then to 15 inches at the pit. The branch line on the floor of the pit to the different giants was of 11-inch pipe. Four giants with 6-inch nozzles were set up in the pit, but only two were used at a time. A fifth giant with a 5-inch nozzle was set up at the lower end of the sluice. Usually one giant cut the bank and one of the same size swept the gravel to the head of the sluice. The river ran alongside of the gravel being washed, and the sluice box emptied into it. The river water carried the sand and fine gravel downstream; coarse material, however, piled up in the stream. The dump giant was used 1½ to 2 hours during the working shift to stack the coarse material at the end of the box. At the end of the washing shift this giant was set with an automatic control so that water played on the boulders until the next morning. A windrow of boulders 30 feet high along the opposite bank of the river had been made by the giant. The largest boulders were washed to the top of the pile. The stream played in a vertical arc; it was depressed slowly and went up faster. About 1 minute was consumed in each cycle. The giant was overbalanced so that the stream was elevated when free. It was pulled downward by means of a 2-inch hydraulic cylinder fed through a hose from the pipe line. At the end of the stroke a trip turned a valve which shut off the water to the cylinder; at the top of the upward swing another trip opened the water valve. Each morning the river bed at the end of the sluice was free of boulders.

The sluice boxes were 36 inches wide by 30 inches high and were set on a grade of 7 inches to each 12-foot box. One setting of the sluice-way was sufficient for a season's work. The head boxes were protected by parallel rows of 6-inch poles placed horizontally on either side of the box. The poles were laid on an earth fill, the surface of which slanted upward at an angle of 25° from the edge of the boxes. At the end of the season the poles were removed and the underlying gravel was washed into the boxes. The riffles in the sluice consisted of rock paving. Diorite boulders with one flat side were selected from the washed gravel in the pit. These stones were dressed by hand to make a rigid paving with a fairly smooth upper surface. Formerly wooden blocks were used, but they had to be replaced every 60 to 70 days. The sluice was cleaned up at the end of the washing season. An undercurrent was used at the lower end of the sluiceway. The screen consisted of ¾-inch round steel rods 15 inches long, placed ⅛ inch apart lengthwise with the sluice. The undercurrent table was 5 feet wide and 11 feet long; wooden Hungarian riffles were used. Quicksilver was used in the sluice box; some reached the undercurrent where it was caught in the riffles.

Boulders up to 18 inches in diameter were put through the sluice. A hand derrick with a 25-foot mast and two 30-foot booms was set at the head of the sluice to remove any oversize boulders that were washed to this point. A derrick hoist was used for dragging stumps and large boulders from the main part of the pit. The hoist pulled over a 25-foot mast guyed with 4 lines; apparently, however, 5 lines should have been used. The cable (1½ inches in diameter) was pulled out by hand; the range was 400 feet from the hoist set-up. The hoist was double-geared and was run by an undershot water wheel driven by a 1½-inch nozzle. A stream from a 1-inch nozzle was used on top of the water wheel for braking. No explosives were used in the mine. The overburden was washed from the top of the gravel and run directly into the river. This work was done during low-water periods when enough water for only one giant was available.

Lumber cost $30 per M. An average of 223 cubic yards was worked per day during the 1932 season. The operating cost was 7 cents per cubic yard with labor at 6 cents.

Shasta Dredging Company (Thompson dredge), 737 North Central Avenue, Stockton, operated a dragline dredge on Brasswire Gulch 1 mile southwest of Hornbrook from May 12 to August 16, 1941, after moving the equipment from the Jenny Lind district in Calaveras County. The dragline excavator was equipped with a 2½-cu.yd. bucket.

Surveyor's Mistake mine on Vesa Creek in the Klamath River district was operated by Henry Beauman of Klamath River Post Office, during 1941. He used a non-floating washing plant to which gravel was delivered by mechanical means.

Von der Hellen and Webber, Box 217, Yreka, operated a dragline dredge using a 2-cu.yd. bucket on Humbug Creek throughout 1940. The operation was continued from January 1 to October 1, 1941.

William von der Hellen Mining Company, Box 1026, Medford, Oregon, operated a dragline dredge with excavator equipped with a 2½-cu.yd. bucket on the Klamath River throughout 1940. In 1941 the operation was continued, and the yield from 773,700 cubic yards of gravel was 6113 ounces of gold and 928 ounces of silver. The operation was continued at McConnell Bar from January 1 to September 27, 1942, with a 3-cu.yd. bucket and the yield from 596,800 cubic yards of gravel was 3594 ounces of gold and 554 ounces of silver.

Yreka Gold Dredging Company, 351 California Street, San Francisco, completed its operation 2 miles north of Yreka in 1940 and then moved its dredge to Seiad Valley, where operations were resumed on September 14, 1940. The dredge was equipped with 67 buckets of 6-cu.ft. capacity. The following description is reprinted from the California Journal of Mines and Geology [5] for April 1938:

Yreka Gold Dredging Company built a new dredge in 1937 to work in sec. 14, T. 45 N., R. 7 W., M. D., and adjoining sections along 2 miles of Yreka Creek just north of Yreka. Ethredge Walker is president and Albert Schubach is secretary, Balfour Building, San Francisco. Eric Peterson is dredge-master at Yreka. The dredge was built by Walter W. Johnson Company, Balfour Building, San Francisco, and the following details are furnished through the courtesy of that company.

The hull is approximately 82 by 42 by 7 feet, and is made of 19 pontoons about 20 by 10 by 7 feet, weighing 6 tons to 7 tons each. Exposed walls are made of $\frac{9}{16}$-inch steel and inside walls adjacent to other pontoons are of $\frac{3}{16}$-inch steel. The pontoons and all structural parts, the digging and stacking ladders, frame for revolving screen, distributors, and 10-ton spud are of electric-welded construction, which has proved very satisfactory.

The bucket-line carries buckets of 6-cu.ft. capacity each, to dig to a depth of 25 feet. Buckets are of the new rivetless-lip, bowl-shaped

<hr />

[5] Averill, C. V., Gold dredging in Shasta, Siskiyou, and Trinity Counties: California Jour. Mines and Geology, vol. 34, pp. 123-125, 1938.

design, and are made of manganese steel by American Manganese Steel Company, Oakland. Lower tumbler is made of manganese steel and is round; upper tumbler is of high-carbon steel, six-sided, and cast integral with shaft. The hopper-chute is lined with manganese steel bars. A special feature of this is a removable back plate for discharging boulders too large for the revolving screen. The boulders are dumped, without stopping the bucket-line, on a fork made of heavy bars. These are swung by a heavy shaft operated by a compressed-air cylinder to dump the boulder into a steel-lined chute which discharges into the pond. Dumping is regulated by a gate in the chute, so that the boulder can be placed in some part of the pond where it will be out of the way.

The revolving screen is 34 feet long by 6 feet in diameter, and is lined with manganese steel plates. Perforations are $\frac{3}{8}$-inch to $\frac{1}{2}$-inch and $\frac{5}{8}$-inch to $\frac{3}{4}$-inch in the sections of screen except the last, which has $\frac{3}{4}$- by $\frac{1}{2}$-inch slots for recovery of nuggets. Several feet at each end of the screen are not perforated. Undersize from the screen is treated on 1600 square feet of riffle-tables. Riffles are of angle-iron, $1\frac{3}{16}$ inches by $1\frac{3}{16}$ inches spaced at 1 inch; also of wood, some shod with steel, some with rubber. They are $1\frac{3}{16}$ inches deep spaced at 1 inch. Oversize from the screen is stacked by a stacker 90 feet long carrying a 36-inch American Rubber Company rib-stacker belt.

Water is pumped from the pond by Byron Jackson pumps of 82 percent efficiency. The 10-inch high-pressure pump furnishes 3200 gallons per minute at 65 feet head to the revolving screen. The 8-inch low-pressure pump furnishes 1800 gpm. to the riffle-tables. A 4-inch pump is provided for cleanups, washing decks, and fire-protection.

The winch is a combination ladder-hoist, swing-line and spud-line winch controlled entirely by compressed air. This method of control adds to the efficiency of the dredge. A two-speed, specially designed motor delivers 55 hp. at 1200 rpm. or 35 hp. at 600 rpm. At the higher speed, it provides ample power for raising the digging-ladder, raising the spud, and swinging the dredge when stepping ahead. The low speed is used for swinging during regular digging.

Other electric motors are as follows: 100-hp. variable-speed on the bucket-line, 60-hp. on the high-pressure pump, 15-hp. on the low-pressure pump, 40-hp. with reduction gearing on the revolving screen, 25-hp. with reduction gearing on the stacker, and 3-hp. on the fire-pump. Power is transmitted by the bucket-line and winch motors to the driven pulleys with multiple V-belt drives.

Power is taken on the dredge at 2400 volts and is stepped down by three 100-kva. transformers to 440 volts. A 5-kva. transformer is provided for lights.

The dredge is operated 24 hours per day by one dredge-master, three winchmen, three oilers, two shore-men, one tractor driver, and one cleanup man. The direct operating cost is 4.3 cents per cubic yard to which should be added $\frac{1}{2}$ cent per yard for management and shipment of bullion. No depreciation, no land-cost and no royalty are included. The capacity at Yreka is 140,000 cubic yards to 150,000 cubic yards per month. The same dredge would handle 210,000 cubic yards in easier ground. It cost approximately $160,000 including some miscellaneous pumping equipment for pumping muddy water out of the

Fig. 98. Steel hull of dredge of Yreka Gold Dredging Company. *Reprinted from California Journal of Mines and Geology, April 1938, p. 124.*

Fig. 99. Yreka Gold Dredging Company, dredge under construction. *Reprinted from California Journal of Mines and Geology, April 1938, p. 124.*

pond, but not including the 55-hp. Caterpillar tractor with diesel engine and bulldozer.

Yuba Consolidated Gold Fields (Siskiyou Unit), 351 California Street, San Francisco, was the leading gold producer in Siskiyou County in 1940. The company operated a Yuba type connected-bucket dredge with 72 buckets of 9-cu.ft. capacity near Callahan. The operation was continued throughout 1941.

The following description of the dredge is reprinted from the California Journal of Mines and Geology [6] for April, 1938:

Yuba Consolidated Gold Fields built a new dredge near Callahan, Siskiyou County, in 1936, in sec. 8, T. 40 N., R. 8 W., M.D. From a point near the confluence of Wildcat Creek and Scott River, it will work for several miles up the river. F. C. Van Deinse, 351 California Street, San Francisco, is vice-president and general manager. H. C. Perring is field-superintendent.

The dredge is No. 116 of Yuba Manufacturing Company, and is built on a steel hull not of the pontoon type, 122 feet 8 inches by 56 feet by 10 feet. It will now dig to a depth of 35 feet below water line, but is designed so that extensions can be put on both the hull and the digging-ladder; and it will then dig to a depth of 50 feet or 60 feet. To cope with very difficult digging, this dredge was equipped with machinery of sizes ordinarily used on dredges with 18-cu.ft. buckets, while its buckets are of 9-cu.ft. size. Concentric ladder suspension is used, that is the ladder and the bucket-chain turn on the same axis.

Gravel is screened in a trommel 8 feet in diameter by 48 feet long, of which 34 feet are perforated with $\frac{1}{2}$-inch to $\frac{5}{8}$-inch and $\frac{5}{8}$-inch to $\frac{1}{2}$-inch holes. It turns at 7 rpm. The trommel is lined with $\frac{1}{4}$-inch plates of "abrasion resisting steel," a high-carbon, high-manganese steel supplied by United States Steel Corporation. It costs more per pound than ordinary steels but less per cubic yard dredged. Undersize from the trommel is treated on 3500 square feet of riffle-tables in a double-deck arrangement. They are provided with wooden riffles shod with steel. For washing, 10,000 gallons per minute of water are pumped from the pond. The total connected load is 750 hp., which includes an extra-heavy digging motor about midway in size between those customarily used in 18-cu.ft. dredges and 9-cu.ft. dredges.

The dredge is operated for 24 hours per day by a total crew of 24 men including a man in the office. The actual capacity is 210,000 cubic yards per month in ground that is hard to dig.

---

[6] Averill, C. V., op. cit.

STANISLAUS COUNTY

C & E Dredging Company, 1002 Pacific Building, Portland, Oregon, operated a dragline dredge with excavator having a 1½-cu.yd. bucket on Littlejohn Creek two miles northwest of Knights Ferry intermittently between September 20 and December 17, 1940. The same equipment was used in dredging the adjoining Jack Welsh Ranch.

California Gold Dredging Company, 351 California Street, San Francisco, operated a connected-bucket dredge in the Jenny Lind district on the Stanislaus side of the county line during 1940.

Germain, A. G., operated a dragline dredge using a ½-cu.yd. bucket in the Knights Ferry district intermittently from July 15 to December 23, 1943. The washing of 655 cubic yards of gravel yielded 12 ounces of gold and 1 ounce of silver. The equipment was designed to be handled by one workman.

La Grange Gold Dredging Company, 1805 Mills Building, San Francisco, operated a connected-bucket dredge on the Tuolumne River in the La Grange district throughout 1940 and 1941. The dredge was equipped with 62 buckets of 10-cu.ft. capacity.

Placer Properties Company, Box 532, Oakdale, operated a dragline dredge on the Stanislaus River nine miles east of Oakdale throughout 1940. A 6½- and a 7½-cu.yd. bucket were tried on a 5-cu.yd. dragline excavator at various times during the year. The washing plant used a shaking screen in place of a trommel. In 1941 operations were continued with a 6-cu.yd. bucket at a point eight miles east of Oakdale. In 1942 operations were continued during most of the year with two dragline excavators, one with a 6-cu.yd. bucket and the other with a 2½-cu.yd. bucket. Operations were continued until December 12, 1943.

Tuolumne Gold Dredging Corporation, 1 Montgomery Street, San Francisco, operated a connected-bucket dredge from February 22, 1940 until April 13, 1941, when the dredge capsized. It was equipped with 100 buckets of 12-cu.ft. capacity. Operations were carried on throughout 1943 on a one-shift basis.

Vanciel, C. F., Route 2, Oakdale, operated a dragline dredge, employing an excavator with a 1½-cu.yd. bucket on the Anderson, Higginbotham, and Kaasa property in the Knights Ferry district from May 13 until December 13, 1941. The yield from 628,400 cubic yards of gravel was 2,198 ounces of gold and 179 ounces of silver.

Yuba Consolidated Gold Fields, 351 California Street, San Francisco, started operations with a connected-bucket dredge, electrically driven, in the La Grange district on December 15, 1941.

TRINITY COUNTY

The mineral resources of Trinity County have been described in the California Journal of Mines and Geology for January, 1941.[1] Further details about many of the placer mines mentioned below are contained in this report, as well as descriptions of lode mines and of mineral deposits other than gold. The report contains a long table of mines with references to earlier reports.

Arbuckle mine is a hydraulic mine near Weaverville that was operated by Arbuckle Bros. of Weaverville during 3 months of 1940.

B. H. K. Mines, Box 325, Orland, operated a dragline dredge equipped with a 1½-cu.yd. bucket on Littlejohn Creek in the Weaverville district from July 1 to the end of 1940. The yield from 184,000 cubic yards of gravel was 789 ounces of gold and 40 ounces of silver. In 1941 operations were continued on Little Browns Creek at several properties with the following results: at the Rehberger property operations from January 1 to May 2 yielded 751 ounces of gold and 41 ounces of silver from 176,000 cubic yards of gravel; at the M. K. Brown property operations from May 3 to July 1 yielded 405 ounces of gold and 24 ounces of silver from 95,000 cubic yards of gravel; at the Scharr property operations from July 20 to September 12 yielded 349 ounces of gold and 28 ounces of silver from 81,500 cubic yards of gravel; and at the Tye property operations from September 13 to October 22 yielded 150 ounces of gold and 10 ounces of silver from 55,000 cubic yards of gravel.

O. R. Batham, Box 325, Concord, operated a dragline dredge on the Bazet Estate property on the East Fork of Stuarts Fork from August 10, 1941 to the end of the year. The recovery from 205,550 cubic yards of gravel was 626 ounces of gold and 50 ounces of silver. Batham also carried on smaller operations at the Hook and Ladder and Nugget Bar properties.

J. P. Brennan, 1343 Butte Street, Redding, operated a dragline dredge using a ¾-cu.yd. bucket on Brown's Creek in the Weaverville district from July 17 to December 31, 1941.

Canyon Placers on Canyon Creek was worked by the hydraulic method by G. H. Bergin of Junction City in 1940 and 1941. More than a page of additional information about this property is contained in California Journal of Mines and Geology for January, 1941.[2]

Carrville Gold Company, 351 California Street, San Francisco, or 807 Lonsdale Building, Duluth, Minnesota, operated its dredge on the Trinity River about 3 miles north of Trinity Center throughout 1940 and 1941. Operations were conducted through the company's agent, Yuba Consolidated Gold Fields. The connected-bucket dredge has 75 buckets of 12-cu.ft. capacity.

Cinco Mineros Company, First National Bank Building, Oroville, operated a dragline dredge using a 1½-cu.yd. bucket near Hayfork throughout 1940. The operation was continued in 1941 on the Albiez, Crews, Parmenter, Ross, and Trimble properties.

[1] Averill, C. V., Mineral resources of Trinity County: California Jour. Mines and Geology, vol. 37, pp. 8-89, 1941.
[2] Averill, C. V., op. cit., pp. 29-30.

FIG. 100.   Goldfield Consolidated Mines Company, hydraulic mine.   *Reprinted from California Journal of Mines and Geology, January 1941, p. 37.*

FIG. 101.   Dredge of Junction City Mining Company.   *Reprinted from California Journal of Mines and Geology, January 1938, p. 117.*

Dobbin Gulch Dredging Company of Redding operated a dragline dredge equipped with a 1¼-cu.yd. bucket on the M. A. Brady property in the Weaverville district from June 13 to December 24, 1941. The yield from 213,800 cubic yards of gravel was 926 ounces of gold and 80 ounces of silver. In 1942 operations at the Brady property and the Sunshine mine were conducted from January 3 to October 19. At the Sunshine mine 107,300 cubic yards of gravel yielded 348 ounces of gold and 33 ounces of silver.

Golden Gravels Mining Company of Junction City operated at the Red Hill mine of Goldfield Consolidated Mines, near Junction City in 1941.

Goldfield Consolidated Mines, 1 Montgomery Street, San Francisco, operated its Red Hill hydraulic mine near Junction City in 1941 and produced a substantial quantity of gold. The mine was also operated during 1943.

Havilah Gravels, Inc. of Lewiston operated a dragline dredge which had an excavator with a 2-cu.yd. bucket on Eastman Gulch from November 23 to December 31, 1941. The yield from 7860 cubic yards of gravel was 338 ounces of gold and 48 ounces of silver. A nonfloating washing plant operated by J. W. Martin and R. W. Setzer on the same property from January 1 to August 1, 1941, recovered 163 ounces of gold and 19 ounces of silver from 20,000 cubic yards of gravel.

Interstate Mines, Inc., Box 14, Weaverville, operated a dragline dredge on the Lowden Ranch from January 4 to July 14, 1940.

Junction City Mining Company, 685 Sixth Street, San Francisco, operated a connected-bucket dredge near Junction City in 1940, 1941, and until October 28, 1942. In 1942 the yield from 2,077,000 cubic yards of gravel was 7878 ounces of gold and 735 ounces of silver. The following additional details about this dredge are reprinted from the California Journal of Mines and Geology for January 1941.[3]

Junction City Mining Company started a modern steel bucket-ladder dredge in sec. 18 and adjoining sections, T. 33 N., R. 10 W., M. D., near Junction City, on January 10, 1936, and has been operating continuously since that time. The company controls 8 miles of the river, the lower (northerly) end of the property being in sec. 1, T. 33 N., R. 11 W. Harvey Sorensen, 685 Sixth Street, San Francisco, is president; C. M. Derby, Mills Tower, San Francisco, is consulting engineer; and D. B. Wilson is superintendent at Junction City.

The hull of the dredge is new and of the late pontoon design, being no. 113 of Yuba Manufacturing Company. Transportation over mountain roads was one reason for adopting this design. The hull is 120 feet long by 52 feet wide by 8 feet 1 inch deep, and is made of 31 pontoons. These are designed and arranged so that the inside walls strengthen the hull at critical points. The largest pontoon weighs 24,000 pounds and the smallest 4800 pounds. Most of them weigh from 10,000 to 16,000 pounds. When assembled they form a rigid structure owing to the beam-effect of the side-walls. Some of the machinery from the old Madrona dredge was used.

---

[3] Averill, C. V., op. cit., pp. 40-42.

The bucket-chain contains 79 buckets of 9½-cu.ft. capacity each, and the dredge is capable of digging to a depth of 45 feet below waterline. A maximum depth of 58 feet has been reached by carrying part of the gravel as bank. Average depth of dredging is 28 feet. Bedrock varies from soft to hard but is decomposed enough so that a few inches of it can be taken up. The dredge is held in digging position by a single spud of 32 tons. The trommel is 7 feet in diameter and is perforated with ⅜-inch to ½-inch holes, but one section of 2-inch mesh is provided for recovery of nuggets. Riffles are of the Hungarian dredge type shod on top with ⅛-inch strap iron. The stacker for coarse tailing is 135 feet long and carries a 36-inch belt. The operating crew averages 24 men.

Electric motors are as follows: 50 hp. on a high-pressure 10-inch pump, 50 hp. on a low-pressure 10-inch pump, 50 hp. on an auxiliary 10-inch pump, 25 hp. on a 4-inch pump, 35 hp. on the winch, 35 hp. on the screen, 50 hp. on the stacker, and a 200-hp. digging motor.

The following figures on operation are furnished through the courtesy of C. M. Derby, consulting engineer. For the fiscal year ending June, 1937, the operating cost under rather severe conditions averaged 4.98 cents per cubic yard. This includes labor, material, power, ordinary taxes, and general expense. No land-cost, no royalty, and no depreciation are included. The average monthly yardage was 240,000 cubic yards. The approximate cost of the dredge was $250,000.

When the dredge was operated near the old Chapman mine, recovery of platinum group metals was as high as 2 ounces per week, and pieces weighing as much as half an ounce were recovered. In other locations the recovery was about one-half ounce per week. Analysis of a shipment follows: waste or sand, 38%; gold, 1.89%; platinum, 25.78%; iridium, 10.51%; osmium, 16.21%; ruthenium, 7.71%; palladium, 0.27%. Recovery of platinum from sands removed from the riffles at time of clean-up is made on a long tom on the dredge. The last cleaning is done by panning; and before the last panning, the concentrate is ground in a small ball mill and then is allowed to stand over night in nitric acid. Among the minerals contained in the concentrate is native cinnabar. The richest sand is found, and the poorest recovery is made in passing through ground that has already been mined. The channel of the river was mined shortly after 1850 in the days of the gold-rush. A dime that looked new, which carried the date 1838, was recently recovered.

C. L. Kalbaugh, operated a suction dredge and tractor on the Thursday No. 1 mine on Crow Creek from May 1 to September 15, 1942. The yield from 1000 cubic yards of gravel was 103 ounces of gold.

La Grange Placer Mines, Ltd., Box 141, Weaverville, operated its hydraulic mine between Junction City and Weaverville during parts of 1940, 1941, and 1942. In 1941 operations lasted from January 1 to July 1 and from December 16 to 31. The yield from 113,100 cubic yards of gravel was 757 ounces of gold and 84 ounces of silver. Operations conducted from January 1 to July 1, 1942, yielded 548 ounces of gold and 52 ounces of silver from 250,000 cubic yards of gravel.

Lewiston Placers of Lewiston operated its hydraulic mine near Lewiston from January 27 to July 1 and from December 6 to 31, 1941. In 1942 operations from January 1 to June 30 yielded 188 ounces of gold and 24 ounces of silver from 75,000 cubic yards of gravel.

Fig. 102.   Lincoln Gold Dredging Company, dragline dredge. *Reprinted from California Journal of Mines and Geology, January 1941, p. 46.*

Lincoln Gold Dredging Company of Lincoln operated dragline dredges in the Lewiston district in 1941 with the following results: from the Clark-Jansen property the yield from 109,139 cubic yards of gravel was 430 ounces of gold and 67 ounces of silver; from the Costa property the yield from 26,432 cubic yards of gravel was 134 ounces of gold and 9 ounces of silver; from the Dickerson property the yield from 65,856 cubic yards of gravel was 149 ounces of gold and 16 ounces of silver; from the Fancelli property the yield from 28,170 cubic yards of gravel was 141 ounces of gold and 19 ounces of silver; from the Froloff property the yield from 562,732 cubic yards of gravel was 2453 ounces of gold and 158 ounces of silver; and from the Phillips property the yield from 194,876 cubic yards of gravel was 1134 ounces of gold and 161 ounces of silver.   One of the dragline excavators was equipped with a 2½-cu.yd. bucket and the other with a 1½-cu.yd. bucket.   In addition to this production from dragline operations small amounts of gold and silver were recovered by hydraulic operations at the Costa and Phillips properties.

In 1942 operations were continued from January 1 to September 12 on the Costa property on Rush Creek with a dragline excavator having a 2½-cu.yd. bucket.   The yield from 271,744 cubic yards of gravel was 1406 ounces of gold and 95 ounces of silver.

As this dredge has several features that are different in design from those commonly used, the following description is reprinted from the California Journal of Mines and Geology for January 1941[4]:

Lincoln Gold Dredging Company is a partnership of E. M. Clark, French Gulch, and W. K. Jensen, Lincoln, California.   Late in 1939, a dragline dredge was installed on a tract of 60 acres held by leases, covering bars on Trinity River, at a point 4 miles west of Lewiston, in sec. 27(?), T. 33 N., R. 9 W., M. D.   It was planned to dredge this

[4] Averill, C. V., op. cit., pp. 45-47.

tract to a depth of 6 feet to 23 feet. The washing-plant is similar to those made by Bodinson, which have been described in some detail in a preceding chapter, but it is of Clark's own make. It has a capacity of 3500 cubic yards per day, and is driven by a D-13000 Caterpillar diesel engine. A 25-kw. electric generator is provided for lights and for one or two small motors as needed. Main drive is from engine to countershaft by multiple V-belt.

Several improvements on older designs have been incorporated. Disc-wheels, 36 inches in diameter, are attached to hand-winches instead of cranks. This is a safety measure to prevent the breaking of the operator's arm by a sudden strain which might reverse the direction of rotation of the crank. Beneath the trommel is the usual depressed trough containing baffles to regulate the flow of sand and water to each sluice. In each compartment of this, beneath the surface of the fluid, a jet of water supplied from a manifold impinges against a horizontal plate. Thus the contents of each compartment are kept in a state of agitation, giving the gold a chance to settle out. Clark says that most of his gold is recovered in these traps, and that it is not necessary to clean up the sluices so often. Sluices are equipped entirely with expanded metal lath over coconut matting, no riffles. Near the trommel a few strips of plate for amalgamation (silvered copper), 1½ inches wide and as long as the width of the sluice, are placed beneath the expanded metal lath. The pump-screen is in the form of a revolving drum to keep it free of floating trash. Several gates made of heavy steel bars, placed above the tailing stacker, are arranged to open upward only. Boulders traveling up the belt in the normal way pass through readily; but if a round boulder starts to roll back down the belt, it is stopped by one of the gates, and is given a new start in the proper direction. The dragline is a model 85 Northwest with a bucket of 2-cu.yd. capacity. The outfit includes also a Caterpillar tractor equipped with bulldozer.

North Fork Placer Mining Company of Helena operated the North Fork hydraulic mine 1 mile from Helena from January 1 to June 30, 1941. The recovery from 53,500 cubic yards of gravel was 277 ounces of gold and 30 ounces of silver. Operations at this mine in 1939 are described in the California Journal of Mines and Geology [5] for January 1941. The following description of an earlier period of operation is from Gardner and Johnson [6]: The North Fork placers on Trinity River at Helena were worked under a leasehold during the 1932 season by F. M. Reynolds, W. O. Kunman, and E. C. Mathews. A fourth man was employed. The mine was operated two 9-hour shifts with two men on a shift. The gravel deposit consisted of an old channel cutting through a ridge. The lower 15 feet of gravel was very tight and partly cemented. It was broken down by first cutting the bedrock from underneath it. After being broken down considerable piping was necessary to disintegrate the cemented fragments. The top gravel washed easily.

Water was brought to the mine from two sources in different flume lines. The lower flume emptied into a reservoir which supplied a giant

[5] Averill, C. V., op. cit., pp. 52-53.
[6] Gardner, E. D., and Johnson, C. H., Placer mining in the western United States, Part II, Hydraulicking, etc.: U.S. Bur. Mines Inf. Circ. 6787, p. 53, 1934.

with a 5-inch nozzle for about 5 hours' piping a day. A pipe line to the upper flume supplied one giant with a 7-inch nozzle steadily. Two bad breaks in the flumes during the season materially increased the cost per cubic yard washed. As the water supply decreased the diameters of the nozzles were reduced from 7 to 6 inches and finally to 5 inches.

Two sluices, consisting of seven 12-foot boxes 48 inches wide were used. One sluice emptied out of one end of the pit through a bedrock cut varying up to 30 feet in depth; the other box went out the opposite end. The riffles consisted of heavy rails placed crosswise in the boxes on top of 4- by 4-inch timber. An undercurrent was used at the end of the sluice that carried away most of the material. The undercurrent table was 12 by 20 feet and was decked with the type of Hungarian riffles used on dredges. Between 800 and 1,000 cubic yards was handled in 18 hours with a full head of water. The average daily yardage handled for the season was 770 cubic yards. One hundred and fifty thousand cubic yards was washed during the season (December 15 to June 30). The labor cost was 3 cents per cubic yard; supplies were estimated at $1\frac{1}{2}$ cents, making a total operating cost of $4\frac{1}{2}$ cents. The lessees had no supervision or general costs. The indicated costs do not include depreciation, interest on investment, or amortization.

Oro Trinity Dredging Company, Box 212, Oroville, operated a diesel-powered dragline dredge equipped with excavator using a $1\frac{1}{2}$-cu.yd. bucket near Weaverville from January 1 to June 18, 1940. The equipment then was moved to the Scott River district, Siskiyou County, where operations were resumed on August 10.

Placer Exploration Company. See Viking Dredging Company.

Red Hill mine is one of the mines operated by Goldfield Consolidated Mines Company and is mentioned above under that heading.

Reddings Creek Placer, Ltd., installed new equipment at the Wallace Bros. mine in sec. 33, T. 32 N., R. 9 W., M.D. The operation is of interest because a Ruble elevator was used. It was described in the California Journal of Mines and Geology [7] for January-April, 1933, but this is out of print. The following description is from Gardner and Johnson:[8] Placer operations on Redding Creek near Douglas City were begun in the spring of 1932; 56,600 cubic yards of gravel was washed by the time the water supply failed. The gravel bed, which was 9 feet deep and 120 feet wide, lay in a creek bottom. The fall of the creek was so slight (one-tenth inch to the foot) that enough grade could not be obtained for sluice boxes. A Ruble elevator was used for elevating the gravel and boulders and sorting out everything over 2 inches in diameter. Water under a 300-foot head was brought to the pit through a 24-inch pipe 3,000 feet long. The Y's in the pit were of 15-inch pipe. The gravel was cut and swept to near the entrance of the Ruble by a giant with a 5- or 6-inch nozzle, then the material was washed up the Ruble by means of a second giant with a 5-inch nozzle. A third giant with a 3-inch nozzle was used intermittently to level off the tailings piles.

The Ruble was 8 feet wide by 60 feet long and elevated the oversize 25 feet. It was lined with sheet steel. The grizzlies were of 3- by 6-inch timber set on edge; the top edge was steel-clad. They were placed crosswise on 3- by 6-inch sills laid lengthwise on the sheet-iron bottom of the

---

[7] Averill, C. V., Gold deposits of the Redding and Weaverville quadrangles: California Jour. Mines and Geology, vol. 29, pp. 68-69, 1933.
[8] Op. cit.

FIG. 103.   Weaver Dredging Company, dragline dredge.   *Reprinted from California Journal of Mines and Geology, January 1941, p. 63.*

chute. The plus 2-inch material was washed up through the elevator by the giant; the undersize dropped through the grizzly and ran down the bottom of the chute to four 12-foot boxes, 48 inches wide, set at right angles to the elevator. As the gravel was only 9 feet deep, the Ruble had to be moved three times during the season.   With 7 men and a Caterpillar tractor a week was required to move the elevator to a new location. A second elevator was planned next season to allow continuous production. Boulders were bulldozed; 2,000 pounds of 40-percent-strength gelatin dynamite was used for this purpose during the 1932 season. An average of 540 cubic yards per day was washed during the 1932 season. The operating cost of washing the gravel was 19 cents, of which three-fourths was for labor. The cost did not include ditch work (other than the ditch tender), construction costs, interest, depreciation, or amortization.

H. S. Smith, R. A. Smith, and R. I. Smith, operated a dragline dredge using a 3-cu.yd. bucket on the High Channel mine in the Hayfork district for 30 days in August and September, 1941. The yield from 100,000 cubic yards of gravel was 300 ounces of gold and 40 ounces of silver. In addition, a small quantity of gold was recovered at this property by hydraulic mining.

Swanson Mining Corporation of Salyer hydraulicked a small yardage of very high-grade gravel at the Salyer mine between February 9 and May 24, 1940. Operations were continued in 1941. Further details about this property are given under two headings, Salyer Consolidated Mines Company and Swanson Mining Corporation in the California Journal of Mines and Geology for January, 1941.[9] At this mine one ounce of platinum-group metals was recovered for each 20 ounces of gold and an analysis of this is given in the publication cited.

Trinity Dredge was operated on the Trinity River at a point about 4 miles north of Lewiston by C. R. and T. D. Harris of Lewiston from

---

[9] Averill, C. V., op. cit., pp. 59-62.

January 1 to November 15, 1940, when the deposit was exhausted. This dredge is described in the California Journal of Mines and Geology for January, 1941.[10]

Viking Dredging Company, Box 498, Chico, operated a dragline dredge equipped with a 2-cu.yd. bucket throughout 1940 on Filibuster Flat, Shanahan Bar, and Hidden Channel near the confluence of Redding Creek and Trinity River. The company operated the Hidden Channel, Tout, and Gasper properties in the Weaverville district from January 1 to February 28, 1941. Then the operation and equipment were taken over by Placer Exploration Company of Douglas City, which continued operations until December 2. The dragline dredge was equipped with a 2-cu.yd bucket.

Weaver Dredging Company, Box 216, Weaverville, operated a dragline dredge using a 1-cu.yd. bucket on East Weaver Creek from January 1 to June 12, 1940. Then the equipment was shipped to Montana. The company operated a second dragline dredge equipped with a 2½-cu.yd. bucket on La Grange property during parts of 1940. This operation was continued from January 1 to May 19, 1941, and the yield from 231,124 cubic yards of gravel was 976 ounces of gold and 89 ounces of silver.

W. E. Woodbury, hydraulicked 20,000 cubic yards of gravel at the Rex mine east of Weaver Creek, near Weaverville, during 1941.

TUOLUMNE COUNTY

Barker Corporation, Hornitos, operated a dragline dredge on Tuolumne River near Jacksonville from January 1 until November 18, 1940. Then the equipment was moved to Hornitos, Mariposa County.

Jackass property in the East Belt district was worked by L. R. Harris of Merced with a dragline dredge from April 20 to May 22, 1940.

E. A. Kent, 260 California Street, San Francisco, operated a dragline dredge with a 1¾-cu.yd. bucket on Six Bit Gulch near Chinese Camp in December 1940. A second dragline dredge equipped with a 2½-cu.yd. bucket was operated on Sanguinetti Ridge near Chinese Camp from June 29 until the end of 1940. Operations with these two dredges were continued in 1941 on the two properties mentioned above and on the Rosasco property. In 1942 operations with the two dredges were continued from January 2 to February 19 with the following results: on the Lyons Ranch 53,000 cubic yards of gravel yielded 224 ounces of gold and 15 ounces of silver; on the Rosasco Ranch 52,500 cubic yards of gravel yielded 308 ounces of gold and 21 ounces of silver.

La Bienvenita mine was operated during 1940 by E. Z. Bowman, Box 6, Chinese Camp, who used a nonfloating washing plant.

Menke-Hess property near Chinese Camp was operated in 1941 by Rio Development Company and McMillan & Company of Jamestown, who used a nonfloating washing plant.

Mullin & Company, Sonora, washed 55,700 cubic yards of gravel by dragline dredge on Sullivan Creek in 1940 and recovered 419 ounces of gold and 41 ounces of silver.

Mullin-Hampton Dredging Company of Sonora operated a dragline dredge which had an excavator with 1½-cu.yd. bucket on the Kaplan

---

[10] Averill, C. V., op. cit., pp. 62-63.

FIG. 104. Double stacker dredges. These two 18-cu. ft. Yuba dredges were especially equipped with double stackers and used in gold dredging operations by Yuba Consolidated Gold Fields near Hammonton, California. In cooperation with U. S. Army Engineers, these two dredges, about 1920, built two flood-control channels each 500 feet wide and about 5 miles long. *Cut by courtesy of Yuba Manufacturing Company.*

(Dondero) mine on Woods Creek 1 mile east of Columbia from January 29 to July 15, 1941. The yield from 85,000 cubic yards of gravel was 365 ounces of gold and 28 ounces of silver.

H. M. Richards did ground sluicing at Ohio Flat in the Challenge district and recovered 23 ounces of gold from 1000 cubic yards of gravel between January 1 and April 15, 1943.

YUBA COUNTY

Arundel Corporation, Box 951, Marysville, produced a substantial quantity of gold in the Smartsville district in 1940 in preparing gravel for concrete aggregate.

Dove Mining Company, Oregonhouse, operated a nonfloating washing plant on the Rose property in 1941.

Parks Bar Company, Box 932, Nevada City, operated a diesel-powered dragline dredge equipped with a 1½-cu.yd. bucket in Big Ravine in the Smartsville district from May 1 to October 31, 1940. The recovery from 95,000 cubic yards of gravel was 429 ounces of gold and 20 ounces of silver.

R. & M. Mining Company of La Porte operated a dragline dredge using an excavator with a 1¼-cu.yd. bucket at several properties on Slate Creek in the Strawberry Valley district in 1940 and 1941. In 1940 the Corley and Princess Pines properties were worked. In 1941 the Corley property yielded 423 ounces of gold and 36 ounces of silver from 134,000 cubic yards of gravel between April 15 and June 21; the Ophir property yielded 76 ounces of gold and 7 ounces of silver from 15,000 cubic yards of gravel between June 21 and July 8; and the First Chance property yielded 691 ounces of gold and 60 ounces of silver from 99,000 cubic yards of gravel between July 21 and November 27.

Sunmar Dredging Company, Box 228, Oroville, operated a dragline dredge on property of Mammoth Mining Company in the Smartsville district during 1940. The dragline excavator had a 2-cu.yd. bucket.

Williams Bar Dredging Company, 232 Montgomery Street, San Francisco, or Box 575, Marysville, operated a connected-bucket dredge on the Yuba River 4 miles northwest of Smartsville throughout 1940, 1941, and 1942. The operation was suspended January 13, 1943, under Limitation Order L-208 of the War Production Board. The dredge was equipped with 84 buckets of 6-cu.ft. capacity. In 1942 the yield from 2,872,327 cubic yards of gravel was 6,354 ounces of gold and 447 ounces of silver.

Yuba Consolidated Goldfields, 351 California Street, San Francisco, operated a fleet of six dredges at its property in the Yuba River basin near Hammonton in 1941. All the dredges were equipped with 18-cu.ft. buckets and electric power. Two had 87 buckets each; two had 100 buckets each; one had 126 buckets; one had 135 buckets. In 1943 the company was allowed to operate two of the dredges, the one with 126 buckets and the one with 135 buckets.

The following article, *Deep Gravels Dredged Successfully*, describes one of these dredges.

FIG. 105.   Yuba No. 20 dredge under construction.   *Photo by courtesy of Yuba Manufacturing Company.*

DEEP GRAVELS DREDGED SUCCESSFULLY

By Herbert Sawin [*]

Modern dredges, as operated by Yuba Consolidated Gold Fields, overcome obstacles today that seemed insurmountable only a few years ago. Early California dredge men considered a digging depth of 60 feet below water level the maximum range of economical bucket-line dredging. Later, owing to new designs and improved materials, 80 feet, then 112 feet, and now 124 feet, below water level are profitable operating ranges. Based on experience with Yuba 17, a 3,500-ton gold dredge built in 1934 and operated at Hammonton, California, which dredged abrasive and tightly packed gravel at depths to 112 feet below water level, Yuba 20 was designed and built for the same field, starting operations on May 1, 1939. This newest addition to a fleet of six 18-cu.ft. dredges digs to a depth of 124 feet below water level, at times against a bank of 50 feet. The contract for Yuba 20 was signed August 4, 1938. The first hull plates were laid in November, and the hull was launched 30 days later. The job was completed, and the dredge operated its first full day, on May 1, 1939. Considering the weight of 3,700 tons, the period of less than eleven months from contract date to starting of the dredge is noteworthy.

Electrically operated and displacing about 3,700 tons, the steel hull, superstructure, and gantries of the dredge alone weigh 1,500 tons. The digging units weigh 860 tons exclusive of gravel and suspersion parts. Its steadiness in the water while digging is noticeable at once to a visitor acquainted with placer mining dredges. The hull measures 250 feet, 8 inches by 80 feet by 11 feet. The digging ladder is 216 feet long and 13 feet ¾ inch deep; the main stacker measures 225 feet between pulley centers, and carries a 44-inch rubber conveyor belt. With the ladder raised to 30 feet above water, over-all length of the dredge is about 540 feet. These weights and dimensions clearly indicate the huge size of the dredge. Perhaps it is easier to visualize a dredge as long as an average city block and with its topmost point, the stern gantry, ten stories above the pond surface.

Describing the dredge briefly from "stem to stern," principal parts include the following:

1. Manganese-steel, two-piece lower tumbler with nickel-steel shaft.

2. 135 manganese-steel, rivetless-lip buckets.

3. Forged nickel-chromium steel bucket pins.

4. One-piece cast high-carbon-chromium steel upper tumbler having shaft cast integral.

5. Forged nickel-chromium steel tumbler wearing plates.

6. Perry bucket idler mounted on under side of digging ladder.

7. Bucket idler in well.

8. Packed lower ladder suspension blocks.

9. Cast-steel upper and lower block sheaves, 60-inch diameter.

10. Ladder hoist lines, 2-inch wire rope.

11. Monitor on bow to knock down high banks.

* Sales Engineer, Yuba Manufacturing Company, San Francisco, California. This article was published in Engineering and Mining Journal, vol. 144, no. 7, for July 1943, and is reprinted by permission of that journal.

12. Revolving screen 50 feet 6 inches long and 9 feet diameter, using ¾-inch Yuba ARS screen plates, and with friction drive at lower end.

13. Winch room on the center line of the dredge with flying bridge extending to both sides.

14. Two-drum ladder hoist winch on port side.

15. Eight-drum swing winch on starboard side, having separate drums for each of the port and starboard bow lines.

16. Auxiliary stacker 48 feet long with 44-inch belt to carry over-size material from the screen to main stacker or to rock chutes.

17. Main stacker about 18 degrees for normal work, but this can be changed as desired.

18. Double-drum stacker hoist winch driven by a single motor with worm drives.

19. Two spuds, box type, 37 inches by 60 inches by 70 feet.

20. Table area, 6,000 square feet, single-bank, double-decked with molded rubber Hungarian-type riffles.

21. Hinged top deck sluices can be raised to clean up lower sluices.

22. Tail sluices extending about 30 feet aft of stern.

23. Auxiliary spill chute aft of screen, permitting rock tailings to be discharged through a well in the stern.

24. Two sand wheels discharging to 30-inch belt conveyors which carry excess sand to the main stacker.

25. Conveyor idlers, troughing and return, of the anti-friction type.

Pumps include the following Yuba centrifugal units: one 14-inch, 70-foot head, 5,500 gpm.; one 14-inch, 52-foot head, 5,500 gpm.; one 6-inch dual, 116-foot head, 1,100 gpm.; one 4-inch auxiliary, 65-foot head, 450 gpm.; and a 2-inch service pump. A Yuba mud removal system using an 8-inch Byron-Jackson pump rated at 3,000 gpm., 245-foot head, also furnishes water to the monitor mentioned previously. Total installed load is 2,175 hp., all a.c. power.

Two variable-speed drive motors, each 300 hp., 600 rpm., are situated just aft of the upper tumbler, one on each side, with V-belts connecting them to the pulley shaft. This arrangement was first tried on Capital No. 4 dredge, built in 1937, where the multiple a.c. motor drive has been entirely satisfactory and resulting in a simple, trouble-free installation. The ladder winch is separate from the main drive, thus eliminating the large belt formerly needed. V-belt drives are also used on the swing winch, screen drive, and main and auxiliary stackers.

The main drive on Yuba 20 has proved to be highly successful. At Hammonton, about 90 feet below water level, there is a particularly hard (but not cemented) gravel stratum. The drive in question helps materi-ally in solving dredging problems associated with digging hard formations at great depths. The double V-belt main drive provides a flexible unit which acts as a safety link capable of absorbing severe shocks and pro-tecting the rest of the digging unit in the event of sudden or severe over-load. Bucket line speed is 21 per minute, based on the experience of several deep digging dredges in California.

In general, dredges digging 100 feet or more below water level have been found to excavate less material per day than units of the same bucket capacity digging at 80 feet or less. For example, Capital No. 4 dredge, digging to 82 feet with 18-cu.ft. buckets, was capable of turning out 127,-000 cubic yards of gravel per week, but its most economical rate was set at 100,000 cubic yards. For Yuba No. 20, on the other hand, the greatest weekly yardage has been about 90,000 cubic yards. Reasons for this are the longer time required on the larger dredge for unproductive operations such as raising the ladder, oiling, stepping, and moving, and the less sensitive control the winchman on the larger dredge has over the actual digging. Because of the greater weight of the dredge, he is not conscious of slight variations in digging depth as he would be on a smaller dredge, and if the buckets are not cutting deeply, several minutes will elapse before the half-empty buckets travel up into sight and the winchman drops the ladder to a correct digging position. With the larger dredge digging is necessarily slower in corners or at other points where caving might cause serious accidents. A further factor in the yardage dug by Yuba No. 20 is the tight formation mentioned in the foregoing.

Successful deep dredging owes much to the use of an idler to control the catenary of the bucket line in its return to the lower tumbler. A Perry patented idler, named for its inventor, O. B. Perry, was first used on Yuba 17 at Hammonton, California. Yuba 20 is also equipped with one. It is a cylindrical device, wheel-like in design, mounted on the underside of the digging ladder about midway between the tumblers. The buckets ride upside down over the idler, the contour of the face fitting the bucket lips. Renewable cast manganese-steel wearing plates are provided on the idler itself, and renewable forged nickel-chromium steel wearing bars protect the steel suspension unit. Both wearing plates and bars show little wear after many months of use, and because the long bucket line is better balanced, wear on bucket pins, bushings, and tumbler plates is reduced noticeably. The Perry idlers on Yuba's 17 and 20 make it possible to use buckets and bucket pins of the same design and metal sections as on shallower digging dredges in the same field, and buckets and pins are therefore interchangeable on all dredges in use in this field.

The Yuba mud removal system, also a help in deep dredging, provides suction behind the lower tumbler in the form of a pipe line inside the digging ladder with a flexible hose coupling to a pump on deck. On Yuba 20, a "Y" arrangement of gate valves permits discharge of mud pumped from the pond bottom either through a pipe line carried by the main stacker to a point far beyond the face of the rock tailings, or by a floating pipe line away from the dredge to a point several thousand feet distant, where the mud can be used for filling old holes or basins. In these basins muddy water can settle and be filtered through tailing piles to avoid stream discoloration. The mud sometimes reaches a depth of 30 feet on the bottom of the pond, crowding the lower tumbler and unless removed prevents free swinging of the dredge.

Auxiliary stacking and other stern-end equipment also attract favorable attention of experienced dredge operators. A well is provided on the center line of the hull at the stern, through which rock tailings can be discharged to make anchorage for the spuds, a departure made desirable by the length of the dredge. This is believed to be the first dredge so built. Part of the sand tailings are discharged to the main stacker belt to aid in binding rock tailings. The stacker was first used at an elevation of 23 degrees, the high elevation being necessary during

FIG. 106. Yuba No. 20 is an 18-cu. ft. dredge capable of digging 126 feet below water level against a bank 50 feet or more in height. It is owned by Yuba Consolidated Gold Fields and is operated near Hammonton, California. Total weight about 3,750 tons. Built by Yuba Manufacturing Company.   *Photo by courtesy of Yuba Manufacturing Company.*

early operations until the pond became deep enough to dispose of rock tailings without such high stacking.

Deep dredging methods must be developed to suit conditions that are decidedly different from those encountered in shallow dredging. An important change is in the design of the ladder hoist winch. On Yuba 20, this unit is separated from the main drive and occupies a deck space measuring 29 by 14 feet on the port side of the deck just inside the house. The total weight of the winch, including structural steel base, brake assembly, etc., is about 63 tons. The two ladder hoist lines are 2-inch wire rope on separate drums. Length of each is 2,300 feet. Each drum weighs 12 tons, and is driven through a pinion shaft which also carries the mechanical brake wheel. All shafts used on the winch assembly are nickel steel.

Power is supplied by a 500-hp., 1,160-rpm. type "CW" Westinghouse motor equipped with a Westinghouse Thrustor brake type HI-198, and operating through a Farrel speed reducer with a ratio of 7.181:1. All pinions and gears have herringbone cut teeth. The mechanical brake is of the post type, actuated by releasing pressure in a 10- by 10-inch Westinghouse air cylinder, counterweighted. Brakes are electrically operated and manually controlled from the pilot house. Manual control gives the winchman a finer "feel" and saves possible damage to equipment which might occur with full automatic brakes. The braking action, through the brake wheel to the drive pinions, slows down and stops both drums simultaneously. Immediately following the mechanical brake action, the Thrustor brake on the motor is applied automatically. A Lilly control is provided as a safety measure. Should the ladder be raised or lowered beyond safe limits or accidentally be dropped too fast, this unit would take control out of the winchman's hands and apply the brakes automatically. This type of control assures a longer life for winch parts. Internal expanding type clutches are used and are operated pneumatically from the pilot house. With drums revolving at a speed of 7.46 rpm., the raising and lowering speed for the digging ladder is about 12 vertical feet per minute at the lower tumbler.

The swing winch on Yuba 20 is on the starboard side just inside the front end of the house. The bow-line drums can be operated independently of the other drums on the winch, and a separate drum is used for each of the bow starboard swing lines. Other drums on the winch include two stern-line drums, two spud-line drums, and two spares. Clutches on all the drums are of internal expanding type controlled electrically through pneumatic cylinders mounted on the winch frame. The application of air cylinders to dredge equipment was originally developed and patented by Yuba about ten years ago for use on dredges to be operated in the tropics. Manually operated brakes are provided for the same reason as used on the ladder hoist winch. The winchman has better control in applying them, thus avoiding shocks to lines and thereby increasing the life of the wire rope.

The history of dredge mining proves that successful dredges are especially designed to suit conditions to be overcome in a particular field. There is practically no so-called "standard dredge." This applies in particular to deep-digging dredges, which present problems entirely different from those connected with shallow dredging. Smaller yardage with a given size of bucket naturally reduces the gross income.

21

The initial high cost of a dredge like Yuba 20 burdens the cost of operation, and properties sufficiently large to carry the load of such an investment are not found often.   Mechanically, the limit in size for mining dredges has not been reached.   Economic problems which affect the cost of dredging are the main factors limiting mechanical size at present.

An 18-cu.ft. dredge in California digging 80 feet below water level operates at a field cost of less than 3 cents per cubic yard.   Beyond 80-foot digging depth, the operating cost rises sharply, and for dredges digging 100 to 120 feet, the field cost is nearer to 5 cents per cubic yard. This rising cost must be given serious thought when considering a deep dredging venture, one reason being that areas large enough and deep enough to warrant the investment, probably would not have a high average value per yard.   For profitable deep dredging, the maximum dredging depth would be determined by an anticipated return commensurate with the extra operating costs.   It is probable that mechanical improvements and changes will be developed which may result in dredges of larger daily capacities with a given size of bucket, and do so economically, making otherwise worthless ground valuable mining property.

Operating data on Yuba 20 have been furnished by Yuba Consolidated Gold Fields, and with thanks to that company the following information is made available: Daily operating time (three shifts) has averaged 21 hours and 29 minutes.   This makes full allowance for shutdowns for all reasons, including moving, ladder inspections, clean-up, repairs, and greasing.   Gravel dug has averaged 12,260 cubic yards per day at a field cost of 4.32 cents per cubic yard.   The weekly power consumption has been about 145,000 kw.-hr.

On page 57 of this bulletin the gold mining tables as installed originally on Yuba No. 20 are described. Recent experience with Yuba No. 20 in digging through old rock piles and underlying sand beds from older operations pointed to a periodic excess of sand and greater quantities of fine gold. This condition was caused by concentrations from dredges operating in the same area at prior times which could not dig as deep as No. 20. To meet this condition jigs were proposed for use instead of riffles. Exhaustive studies were made to improve the recovery factor as a means of counteracting increased operating expense.

Before installing jigs, however, a thorough test was conducted. The extremely high volume of sand encountered at times produced difficult material handling conditions for jigs. Yuba jigs used experimentally proved to be capable of efficient operation. A full set of jigs designed and built by Yuba Manufacturing Company was installed on Yuba No. 20 in late 1946. The Yuba jig is of the horizontal thrust design with a small motor mounted between and operating two cells. The eccentric drives are completely enclosed and run in oil with heavy roller bearings used to reduce wear and to provide long operating life. Headroom requirements are kept to a minimum because of the horizontal thrust and this feature is of great advantage under dredge operating conditions.

There are twelve 4-cell rougher jigs and two 4-cell cleaner jigs on Yuba No. 20 and the complete circuit includes a ball mill and a small area of riffles over which jig concentrates are run. The gold recovery factor has been improved sufficiently to justify the change from riffles, even though an extra man per shift is needed to operate the jigs and auxiliary mechanical equipment installed on Yuba No. 20 to replace tables.

APPENDIX

LAWS AFFECTING PLACER MINING

A few laws that apply particularly to placer mining are brought together here for ready reference from a number of different publications. For detailed information on other phases of mining law, reference should be made to Bulletin 123, *American Mining Law*[1], and Bulletin 127, *Manner of Locating and Holding Mineral Claims in California*.[2]

---

[1] Ricketts, A. H., American mining law: California Div. Mines Bull. 123, pp. 1-1018, 1943.

[2] Ricketts, A. H., Manner of locating and holding mineral claims in California (with forms): California Div. Mines Bull. 127, pp. 1-35, with revisions by C. A. Logan, March 1944.

THE CAMINETTI LAW

An Act to create the California Debris Commission and regulate hydraulic mining in the State of California.

(Approved March 1, 1893.)

Be it enacted by the Senate and House of Representatives of the United States of America in Congress assembled, That a commission is hereby created, to be known as the California Debris Commission, consisting of three members. The President of the United States shall, by and with the advice and consent of the Senate, appoint the commission from officers of the Corps of Engineers, United States Army. Vacancies occurring therein shall be filled in like manner. It shall have the authority, and exercise the powers hereinafter set forth, under the supervision of the Chief of Engineers and direction of the Secretary of War.

SEC. 2. That said commission shall organize within thirty days after its appointment by the selection of such officers as may be required in the performance of its duties, the same to be selected from the members thereof. The members of said commission shall receive no greater compensation than is now allowed by law to each, respectively, as an officer of said Corps of Engineers. It shall also adopt rules and regulations not inconsistent with law, to govern its deliberations and prescribe the method of procedure under the provisions of this Act.

SEC. 3. That the jurisdiction of said commission, in so far as the same affects mining carried on by the hydraulic process, shall extend to all such mining in the territory drained by the Sacramento and San Joaquin river systems in the State of California. Hydraulic mining, as defined in section eight hereof, directly or indirectly injuring the navigability of said river systems, carried on in said territory other than as permitted under the provisions of this Act, is hereby prohibited and declared unlawful.

SEC. 4. That it shall be the duty of said commission to mature and adopt such plan or plans, from examinations and surveys already made and from such additional examinations and surveys as it may deem necessary, as will improve the navigability of all the rivers comprising said systems, deepen their channels and protect their banks. Such plan or plans shall be matured with a view of making the same effective as against the encroachment of and damage from debris resulting from mining operations, natural erosion, or other causes, with a view of restoring, as near as practicable and the necessities of commerce and navigation demand, the navigability of said rivers to the condition existing in eighteen hundred and sixty, and permitting mining by the hydraulic process, as the term is understood in said State, to be carried on, provided the same can be accomplished without injury to the navigability of said rivers or the lands adjacent thereto.

SEC. 5. That it shall further examine, survey, and determine the utility and practicability, for the purposes hereinafter indicated, of storage sites in the tributaries of said rivers and in the respective branches of said tributaries, or in the plains, basins, sloughs, and tule and swamp lands adjacent to or along the course of said rivers, for the storage of debris or water or as settling reservoirs, with the object of using the same by either or all of these methods to aid in the improvement and protection of said navigable rivers by preventing deposits therein of debris resulting from mining operations, natural erosion or other causes, or for affording relief thereto in flood-time and providing sufficient water to maintain scouring force therein in the summer season; and in connection therewith to investigate such hydraulic and other mines as are now or may have been worked by methods intended to restrain the debris and material moved in operating such mines by impounding dams, settling reservoirs, or otherwise, and in general to make such study of and researches in the hydraulic mining industry as science, experience, and engineering skill may suggest as practicable and useful in devising a method or methods whereby such mining may be carried on as aforesaid.

SEC. 6.   That the said commission shall from time to time note the conditions of the navigable channels of said river systems by cross-section surveys or otherwise, in order to ascertain the effect therein of such hydraulic mining operations as may be permitted by its orders and such as is caused by erosion, natural or otherwise.

SEC. 7.   That said commission shall submit to the Chief of Engineers, for the information of the Secretary of War, on or before the fifteenth day of November of each year, a report of its labors and transactions, with plans for the construction, completion, and preservation of the public works outlined in this Act, together with estimates of the cost thereof, stating what amounts can be profitably expended thereon each year.   The Secretary of War shall thereupon submit same to Congress on or before the meeting thereof.

SEC. 8.   That for the purposes of this Act "hydraulic mining" and "mining by the hydraulic process" are hereby declared to have the meaning and application given to said terms in said State.

SEC. 9.   That the individual proprietor or proprietors, or in case of a corporation its manager or agent appointed for that purpose, owning mining ground in the territory in the State of California mentioned in section three hereof, which it is desired to work by the hydraulic process, must file with said commission a verified petition, setting forth such facts as will comply with law and rules prescribed by said commission.

SEC. 10.   That said petition shall be accompanied by an instrument duly executed and acknowledged, as required by the law of the said State, whereby the owner or owners of such mine or mines surrender to the United States the right and privilege to regulate by law, as provided in this Act, or any law that may hereafter be enacted, or by such rules and regulations as may be prescribed by virtue thereof, the manner and method in which the debris resulting from the working of said mine or mines shall be restrained, and what amount shall be produced therefrom; it being understood that the surrender aforesaid shall not be construed as in any way affecting the right of such owner or owners to operate said mine or mines by any other process or method now in use in said State; *provided,* that they shall not interfere with the navigability of the aforesaid rivers.

SEC. 11.   That the owners of several mining claims situated so as to require a common dumping ground or dam or other restraining works for the debris issuing therefrom in one or more sites, may file a joint petition setting forth such facts, in addition to the requirements of section nine hereof; and where the owner of a hydraulic mine or owners of several such mines have and use common dumping sites for impounding debris or as settling reservoirs, which sites are located below the mine of an applicant not entitled to use same, such fact shall also be stated in said petition. Thereupon the same proceedings shall be had as provided for herein.

Sec. 12. A notice specifying briefly the contents of said petition, and fixing a time previous to which all proofs are to be submitted, shall be published by said commission in some newspaper or newpapers of general circulation in the communities interested in the matter set forth therein. If published in a daily paper, such publication shall continue for at least ten days; if in a weekly paper, in at least three issues of the same. Pending publication thereof said commission, or a committee thereof, shall examine the mine and premises described in such petition. On or before the time so fixed all parties interested, either as petitioners or contestants, whether miners or agriculturists, may file affidavits, plans, and maps in support of their respective claims. Further hearings, upon notice to all parties of record, may be granted by the commission when necessary.

Sec. 13. That in case a majority of the members of said commission within thirty days after the time so fixed, concur in a decision in favor of the petitioner or petitioners, the said commission shall thereupon make an order directing the methods and specifying in detail the manner in which operations shall proceed in such mine or mines; what restraining or impounding works, if facilities therefor can be found, shall be built, and maintained; how and of what material; where to be located; and in general set forth such further requirements and safeguards as will protect the public interests and prevent injury to the said navigable rivers, and the lands adjacent thereto; with such further conditions and limitations as will observe all the provisions of this Act in relation to the working thereof and the payment of taxes on the gross proceeds of the same; *provided*, that all expense incurred in complying with said order shall be borne by the owner or owners of such mine or mines.

Sec. 14. That such petitioner or petitioners must within a reasonable time present plans and specifications of all works required to be built in pursuance of said order, for examination, correction, and approval by said commission; and thereupon work may immediately commence thereon under the supervision of said commission or representative thereof attached thereto from said Corps of Engineers, who shall inspect same from time to time. Upon completion thereof, if found in every respect to meet the requirements of the said order and said approved plans and specifications, permission shall thereupon be granted to the owner or owners of such mine or mines to commence mining operations, subject to the conditions of said order and the provisions of this Act.

Sec. 15. That no permission granted to a mine owner or owners under this Act shall take effect, so far as regards the working of a mine, until all impounding dams or other restraining works, if any are prescribed by the order granting such permission, have been completed and until the impounding dams or other restraining works or settling reservoirs provided by said commission have reached such a stage as, in the opinion of said commission, it is safe to use the same; *provided, how-*

ever, that if said commission shall be of the opinion that the restraining and other works already constructed at the mine or mines shall be sufficient to protect navigable rivers of said systems and the work of said commission, then the owner or owners of such mine or mines may be permitted to commence operations.

SEC. 16.   That in case the joint petition referred to in section eleven hereof is granted, the commission shall fix the respective amounts to be paid by each owner of such mines toward providing and building necessary impounding dams or other restraining works.   In the event of a petition being filed after the entry of such order, or in case the impounding dam or dams or other restraining works have already been constructed and accepted by said commission, the commission shall fix such amount as may be reasonable for the privilege of dumping therein, which amount shall be divided between the original owners of such impounding dams or other restraining works in proportion to the amount respectively paid by each party owning the same. The expense of maintaining and protecting such joint dam or works shall be divided among mine owners using the same in such proportion as the commission shall determine.   In all cases where it is practicable, restraining and impounding works are· to be provided, constructed, and maintained by mine owners near or below the mine or mines before reaching the main tributaries of said navigable waters.

SEC. 17.   That at no time shall any more debris be permitted to be washed away from any hydraulic mine or mines situated on the tributaries of said rivers and the respective branches of each, worked under the provisions of this Act, than can be impounded within the restraining works erected.

SEC. 18.   That the said commission may at any time, when the condition of the navigable rivers, or when the capacities of all impounding and settling facilities erected by mine owners, or such as may be provided by Government authority, requires same, modify the order granting the privilege to mine by the hydraulic mining process so as to reduce amount thereof to meet the capacities of the facilities then in use, or if actually required in order to protect the navigable rivers from damage, may revoke same until the further notice of the commission.

SEC. 19.   That an intentional violation on the part of a mine owner or owners, company, or corporation, or the agents or employees of either, of the conditions of the order granted pursuant to section thirteen, or such modifications thereof as may have been made by said commission, shall work a forfeiture of the privileges thereby conferred, and upon notice being served by the order of said commission upon said owner or owners, company, or corporation, or agent in charge, work shall immediately cease.   Said commission shall take necessary steps to enforce its orders in case of the failure, neglect, or refusal of such owner or owners, company, or corporation, or agents thereof, to comply therewith, or in the event of any person or persons, company, or corporation working by said process in said territory contrary to law.

SEC. 20.   That said commission, or a committee therefrom, or officer of said corps assigned to duty under its orders, shall, whenever deemed necessary, visit said territory and all mines operating under the provisions of this Act.   A report of such examination shall be placed on file.

SEC. 21. That the said commission is hereby granted the right to use any of the public lands of the United States, or any rock, stone, timber, trees, brush, or material thereon or therein, for any of the purposes of this Act; that the Secretary of the Interior is hereby authorized and requested, after notice has been filed with the Commissioner of the General Land Office by said commission, setting forth what public lands are required by it under the authority of this section, that such land or lands shall be withdrawn from sale and entry under the laws of the United States.

SEC. 22. That any person or persons who willfully or maliciously injure, damage, or destroy, or attempt to injure, damage, or destroy, any dam or other work erected under the provisions of this Act for restraining, impounding, or settling purposes, or for use in connection therewith, shall be guilty of a misdemeanor, and upon conviction thereof shall be fined not to exceed the sum of five thousand dollars or be imprisoned not to exceed five years, or by both such fine and imprisonment, in the discretion of the court. And any person or persons, company, or corporation, their agents or employees, who shall mine by the hydraulic process, directly or indirectly injuring the navigable waters of the United States, in violation of the provisions of this Act, shall be guilty of a misdemeanor, and upon conviction thereof shall be punished by a fine not exceeding. five thousand dollars, or by imprisonment not exceeding one year, or by both such fine and imprisonment in the discretion of the court; *provided,* that this section shall take effect on the first day of May, eighteen hundred and ninety-three.

SEC. 23. That upon the construction by the said commission of dams or other works for the detention of debris from hydraulic mines and the issuing of the order provided for by this Act to any individual, company, or corporation to work any mine or mines by hydraulic process, the individual, company, or corporation operating thereunder working any mine or mines by hydraulic process, the debris from which flows into or is in whole or in part restrained by such dams or other works erected by said commission, shall pay a tax of three per centum on the gross proceeds of his, their, or its mine so worked, which tax of three per centum shall be ascertained and paid in accordance with regulations to be adopted by the Secretary of the Treasury, and the Treasurer of the United States is hereby authorized to receive the same. All sums of money paid into the Treasury under this section shall be set apart and credited to a fund to be known as the "Debris Fund," and shall be expended by said commission under the supervision of the Chief of Engineers and direction of the Secretary of War, in addition to the appropriations made by law, in the construction and maintenance of such restraining works and settling reservoirs as may be proper and necessary; *provided,* that said commission is hereby authorized to receive and pay into the Treasury from the owner or owners of mines worked by the hydraulic process, to whom permission may have been granted so to work under the provisions hereof, such money advances as may be offered to aid in the construction of such impounding dams or other restraining works, or settling reservoirs, or, sites therefor, as may be deemed necessary by said commission to protect the navigable channels of said river systems, on condition that all moneys so advanced shall be refunded as the said tax is paid into the said Debris Fund; *and provided further,* that in no event shall the Government of the United States be held liable to refund same except as directed by this section.

Sec. 24.   That for the purpose of securing harmony of action and economy in expenditures in the work to be done by the United States and the State of California, respectively, the former in its plans for the improvement and protection of the navigable streams, and to prevent the depositing of mining debris or other materials within the same, and the latter in its plans authorized by law for the reclamation, drainage, and protection of its lands, or relating to the working of hydraulic mines, the said commission is empowered to consult thereon with a commission of engineers of said State, if authorized by said State for said purpose, the result of such conference to be reported to the Chief of Engineers of the United States Army, and, if by him approved, shall be followed by said commission.

Sec. 25.   That said commission, in order that such material as is now or may hereafter be lodged in the tributaries of the Sacramento and San Joaquin river systems resulting from mining operations, natural erosion, or other causes, shall be prevented from injuring the said navigable rivers, or such of the tributaries of either as may be navigable, and the land adjacent thereto, is hereby directed and empowered, when appropriations are made therefor by law, or sufficient money is deposited for that purpose in said Debris Fund, to build at such points above the head of navigation in said rivers and on the main tributaries thereof, or branches of such tributaries, or at any place adjacent to the same, which, in the judgment of said commission, will effect said object (the same to be of such material as will insure safety and permanency), such restraining or impounding dams, and settling reservoirs, with such canals, locks, or other works adapted and required to complete the same.   The recommendations contained in Executive Document numbered two hundred and sixty-seven, Fifty-first Congress, second session, and Executive Document numbered ninety-eight, Forty-seventh Congress, first session, as far as they refer to impounding dams, or other restraining works, are hereby adopted, and the same are directed to be made the basis of operations.   The sum of fifteen thousand dollars is hereby appropriated from moneys in the Treasury not otherwise appropriated, to be immediately available to defray the expenses of said commission.

AMENDMENTS TO THE CAMINETTI ACT

Amendment to the 'Caminetti Act,' 1907

CHAP. 2077.   An Act To amend section thirteen of an Act of March first, eighteen hundred and ninety-three, entitled ''An Act to create the California Debris Commission and regulate hydraulic mining in the State of California.''

Be it enacted by the Senate and House of Representatives of the United States of America in Congress assembled, That section thirteen of an Act of March first, eighteen hundred and ninety-three, entitled ''An Act to create the California Debris Commission and to regulate mining in the State of California,'' is hereby amended so as to read as follows:

''SEC. 13.   That in case a majority of the members of said commission, within thirty days after the time so fixed, concur in the decision in favor of the petitioner or petitioners, the said Commission shall thereupon make an order directing the methods and specifying in detail the manner in which operations shall proceed in such mine or mines; what restraining or impounding works, if any, if facilities therefor can be found, shall be built and maintained; how and of what material; where to be located; and in general set forth such further requirements and safeguards as will protect the public interests and prevent injury to the said navigable rivers and the lands adjacent thereto, with such further conditions and limitations as will observe all the provisions of this Act in relation to the working thereof and the payment of taxes on the gross proceeds of the same: *Provided,* That all expenses incurred in complying with said order shall be borne by the owner or owners of such mine or mines: *And provided further,* That where it shall appear to said Commission that hydraulic mining may be carried on without injury to the navigation of said navigable rivers and the lands adjacent thereto, an order may be made authorizing such mining to be carried on without requiring the construction of any restraining or impounding works or any settling reservoirs: *And provided also,* That where such an order is made a license to mine, no taxes provided for herein on the gross proceeds of such mining operations shall be collected.''

Approved, February 27, 1907.

Amendment to the 'Caminetti Act,' 1934

An Act to amend the Act entitled ''An Act to create the California Debris Commission and regulate hydraulic mining in the State of California'', approved March 1, 1893, as amended.

Be it enacted by the Senate and House of Representatives of the United States of America in Congress assembled, That section 18 of the Act entitled ''An Act to create the California Debris Commission and regulate hydraulic mining in the State of California'' approved March 1, 1893, as amended (U. S. C., title 33, sec. 678), is amended to read as follows:

''SEC. 18.   The said commission may, at any time when the condition of the navigable rivers or when the capacities of all impounding and settling facilities erected by mine owners or such as may be provided by Government authority require same, modify the order granting the privilege to mine by the hydraulic mining process so as to reduce the amount thereof to meet the capacities of the facilities then in use; or, if actually required in order to protect the navigable rivers from damage or in case of failure to pay the tax prescribed by section 23 hereof within thirty days after same becomes due, may revoke same until the further notice of the commission.''

SEC. 2.   Section 23 of such Act as amended (U. S. C., title 33, sec. 683), is amended to read as follows:

''SEC. 23.   Upon the construction by the said commission of dams or other works for the detention of debris from hydraulic mines and the issuing of the order provided for by this Act to any individual, company, or corporation to work any mine or mines by hydraulic process, the individual company, or corporation operating thereunder working any mine or mines by hydraulic process, the debris from which flows into or is in whole or in part restrained by such dams or other works erected by said commission, shall pay for each cubic yard mined from the natural bank a tax equal to the total capital cost of the dam, reservoir, and rights of way divided by the total capacity of the reservoir for the restraint of debris, as determined in each case by the California Debris Commission, which tax shall be paid annually on a date fixed by said commission and in accordance with regulations to be adopted by the Secretary of the Treasury, and the Treasurer of the United States is hereby authorized to receive the same.   All sums of money paid into the Treasury under this section shall be set apart and credited to a fund to be known as the debris fund, and shall be expended by said commission under the supervision of the Chief of Engineers and direction of the Secretary of War, for repayment of any funds advanced by the Federal Government or other agency for the construction of restraining works and settling reservoirs, and for maintenance: *Provided*, That said commission is hereby authorized to receive and pay into the Treasury from the owner or owners of mines worked by the hydraulic process, to whom permission may have been granted so to work under the provisions thereof, such money advances as may be offered to aid in the construction of such impounding dams, or other restraining works, or settling reservoirs, or sites therefor, as may be deemed necessary by said commission to protect the navigable channels of said river systems, on condition that all moneys so advanced shall be refunded as the said tax is paid into the said debris fund: *And provided further*, That in no event shall the Government of the United States be held liable to refund same except as directed by this section.''

Approved, June 19, 1934.

Amendment to the 'Caminetti Act,' 1938

An act to amend section 23 of the Act to create the California Debris Commission, as amended.

Be it enacted by the Senate and House of Representatives of the United States of America in Congress assembled, That Section 23 of the Act approved March 1, 1893, entitled ''An Act to create the California Debris Commission and regulate hydraulic mining in the State of California'', as amended by the Act approved June 19, 1934, is hereby further amended by adding at the end thereof the following: ''The Secretary of War is authorized to enter into contracts to supply storage for water and use of outlet facilities from debris storage reservoirs, for domestic and irrigation purposes and power development upon such conditions of delivery, use, and payment as he may approve: *Provided*, That the moneys received from such contracts shall be deposited to the credit of the reservoir project from which the water is supplied, and the total capital cost of said reservoir, which is to be repaid by tax on mining operations as herein provided, shall be reduced in the amount so received.''

Approved, June 25, 1938.

DEFINITION OF HYDRAULIC MINING *

''Hydraulic mining'' is the process by which a bank of gold-bearing earth and rock is excavated by a jet of water, discharged through the converging nozzle of a pipe under a great pressure, the earth or debris being carried away by the same water, through sluices, and discharged on lower levels into the natural streams and water courses below; where the gravel or other material of the bank is cemented, or where the bank is composed of masses of pipe-clay, it is shattered by blasting with powder.

DEFINITION OF HYDRAULIC MINING FROM CALIFORNIA CIVIL CODE

1425. Meaning of hydraulic mining. Hydraulic mining within the meaning of this title, is mining by means of the application of water, under pressure, through a nozzle, against a natural bank.

---

* Ricketts, A. H., American Mining Law: California Div. Mines Bull. 123, p. 19, 1943.

TRINITY AND KLAMATH RIVER FISH AND GAME DISTRICT

The following are from the Fish and Game Code:

97.   Trinity and Klamath River district.   The following shall constitute the Trinity and Klamath River fish and game district: The Klamath River and the waters thereof, following its meanderings from the mouth of the Klamath River in Del Norte County to its confluence with the Salmon River, and also the Trinity River and the waters thereof, following its meanderings from its confluence with the Klamath River in the County of Humboldt to its confluence with the south fork of the said Trinity River.

482.   (a) It is unlawful to conduct any mining operations in the Trinity and Klamath River Fish and Game District between July 1st and November 30th, both dates inclusive, except when the debris, substances, tailings or other effluent from such operations do not and can not pass into the waters in said district.

(b) It is unlawful between July 1st and November 30th, both dates inclusive, to pollute, muddy, contaminate, or roil the waters of the Trinity and Klamath River Fish and Game District.   It is unlawful between said dates to deposit in or cause, suffer, or procure to be deposited in, permit to pass into or place where it can pass into said waters, any debris, substance or tailings from hydraulic, placer, milling or other mining operation affecting the clarity of said waters.   The clarity of said waters shall be deemed affected when said waters at a point a distance of one mile below the confluence of the Klamath River and the Salmon River or at a point a distance of one mile below the confluence of the South Fork of the Trinity River and the Trinity River contain fifty (50) parts per million, by weight, of suspended matter, not including vegetable matter in suspension and suspended matter occurring in said stream or streams due to an act of God.

(c) It is unlawful, between July 1st and November 30th, both dates inclusive, to carry on or operate any hydraulic mine of any kind on, along, or in any waters flowing into said Trinity and Klamath River District; provided, however, nothing herein contained shall prevent the operation of a hydraulic mine where the tailings, substance, or debris, or other effluent therefrom does not or will not pass into said waters of said Trinity and Klamath River Fish and Game District, between said dates, and provided further that any person, firm or corporation engaged in hydraulic mining shall have the right until the fifteenth day of July to use water for the purpose of cleaning up.

(d) Any structure or contrivance which causes or contributes, in whole or in part, to the condition, the causing of which is in this section prohibited, is a public nuisance, and any person, firm or corporation maintaining or permitting the same shall be guilty of maintaining a public nuisance, and it shall be the duty of the district attorney of the county where the condition occurs or the acts creating the public nuisance occur, to bring action to abate such public nuisance.

(e) Any person, firm, or corporation violating any of the provisions of this section is guilty of a misdemeanor.

(Amended by Ch. 760, Stats. 1939.)

PROTECTION OF DOMESTIC WATER SUPPLIES

The following is from the appendix of the Fish and Game Code, 33d edition, 1943-45, p. 239:

SECTION 1. Any person, firm or corporation, other than placer mine operators who hold permits from the California Debris Commission to operate, who has been engaged or who shall engage in the operation of a placer mine on any stream or on the watershed of any stream tributary directly or indirectly to the Sacramento or San Joaquin rivers shall record in the office of the county recorder of the county in which its mine is situate, within 60 days from and after the effective date.of this act, or, if operations are commenced after the effective date, then within 30 days after the commencement of such operations, a verified statement verified by the operator or by some one in behalf of the operator, showing:

(1) A description of the ground proposed to be mined by placer mining methods, described by United States Government subdivisions where possible.

(2) The names and addresses of the owners of the ground.

(3) The names and addresses of the operators of the mine.

(4) The proposed means or method of placer mining operation.

(5) The means which the operator proposes to use to prevent the pollution of any stream by the effluent from such operations.

In the event an owner or operator changes his address, or of a transfer of ownership or change of operator of any such mining property ,then within 10 days after any such transfer or change, a notice setting forth the names and addresses of the new owners or operators shall be filed in the office of the county recorder.

SEC. 2. No placer mining operator who does not hold a permit to operate from the California Debris Commission shall mine by placer process on any stream or on the watershed of any stream tributary directly or indirectly to the Sacramento or San Joaquin rivers without taking the following precautions to prevent pollution of the stream by the effluent from operations:

(1) Constructing a settling pond or ponds of sufficient size to permit the clarification of water used in the mining processes before the water is discharged into the stream.

(2) Mixing with the effluent from mining operations aluminum sulphate and lime, or an equivalent clarifying substance which will cause the solid material in the effluent to coagulate and thus avoid rendering the water in the stream unfit for domestic water supply purposes.

(3) Notwithstanding the provisions of Subdivision 2 of this section, any placer miner who is operating by dredging process and who desires to transport his dredger across a stream where the expense of constructing settling ponds in the stream itself would, in the opinion of the operator, be unduly heavy, shall have the right to conduct the dredger across the stream without constructing a settling pond under the following procedure: The operator shall give to the clerk or secretary, as the case may be, of each city or district owning or operating a domestic water supply the clarity of which is likely to be affected by the crossing operation, notice of the intent of the operator to cross the stream. Such notice shall be given at least seven days in advance of the date that the operator expects to cross the stream with his dredger. Upon the expiration of

the notice the operator may during the following 48 hours conduct his dredger across the stream even though some turbidity may be caused by the crossing operation. Having crossed the stream the operator shall thereupon and thereafter in its further operations observe the provisions of Subdivision 2 of this section.

Sec. 3. Any person, firm or corporation who fails to record the notice provided for in Section 1, or to install the protective devices or to give the notice provided for in Section 2, or both, shall be guilty of a misdemeanor. The operation of any placer mine on ground not covered by a permit to the operator from the California Debris Commission, without compliance with the provisions of both Sections 1 and 2 hereof, is hereby declared to be a public nuisance which may be enjoined upon suit brought by the district attorney of the county in which the operation has been conducted, or by any city or district whose domestic water supply is rendered unfit or dangerous for human consumption by the acts, or failure to act, of the operator. The superior court of the county in which the operation is conducted shall have jurisdiction to hear and determine the action and to award such relief as may be proper therein. Nothing in this act contained, however, shall be deemed or construed to deprive the State, any city, city and county, county, district, person, firm or corporation of any right to bring and maintain any action or proceeding, in any jurisdiction, which it was entitled to bring or maintain prior to the enactment of this act, or to receive or obtain in any such action any remedy accorded to it under existing law.

Sec. 4. If any section, subsection, sentence, clause or phrase of this act is for any reason held to be unconstitutional, such decision shall not affect the validity of the remaining portions of this act. The Legislature hereby declares that it would have passed this act and each section, subsection, sentence, clause and phrase thereof, irrespective of the fact that any one or more sections, subsections, sentences, clauses or phrases be declared unconstitutional.

(Added by Ch. 1215, Stats. 1941.)

PLACER MINING DISTRICTS

Sections 2401 to 2512 of the Public Resources Code, provide detailed procedure for the formation of placer mining districts. To give a general idea of the purpose, the first two sections are quoted below:

"2401. Districts may be formed in the manner provided by this chapter for the purpose of affording facilities for conducting placer mining without injury to property not owned by or included in the district.

"2402. Proceedings for the formation of a placer mining district shall be commenced by petition addressed to and filed with the board of supervisors of the county in which is located the largest proportion in value of the lands within the proposed district as shown by the last equalized county assessment roll. The petition shall be signed by twenty-five per cent of the owners of parcels of land subject to assessment for district purposes."

As other sections mentioned above are available in the Public Resources Code of California published both by Supervisor of Documents, 214 State Capitol, Sacramento 14, and by Bancroft-Whitney Company, 200 McAllister Street, San Francisco 2, they are not repeated here.

INDEX

A

B

E

F

N

O

P

Riffles—Continued
    steel, 118, 124, 125
    stone, 124
    stream-pebble, 30
    wooden-block, 30, 123-124 ; drawings of, 122, 123, 124
Rim Cam Gold Dredging Company, operations in Amador County, 232, 233
Rio Development Company, operations in Tuolumne County, 313
River Pine Mining Company, operations in Amador County, 232, 256 ; operations in
        El Dorado County, 256
Rivers, diagram showing down-faulting of bed, 205 ; diagrams showing action of erosion
        and suction eddies in, 177 ; magnetic methods of tracing channels, 227
Rizzi Ranch, Placer County, 271
Roaring River, Shasta County, 284 ; dredge, rubber substituted for iron in riffles, 42 ;
        Dredging Company, Shasta County, 260
Robie Estate, mining operations in Placer County, 273
    Ranch, Calaveras County, 253
Robinson Ranch, Placer County, 272
Rock Canyon Creek, El Dorado County, 255
Rocker, description of, 22-26, 29 ; drawing showing construction of knock-down, 24 ;
        drawing showing parts used in construction of, 25 ; photo showing use of, 14
Rogers Ranch, Placer County, 272
Romanowitz, Charles M., *Bucket-line dredging*, 51-60
Rosasco property, Tuolumne County, 313
Rose, I. E., photo by, 275
    mine, Nevada County, 269
    property, Siskiyou County, 298 ; Yuba County, 315
Roseville Gold Dredging Company, operations in Placer County, 276
Ross property, Trinity County, 305
Rossi property, Sacramento County, 281
Rottinger property, Butte County, 235
Rougher-jig, 68, 69, 73, 74 ; photo showing 4-cell block, 74 ; concentrates, necessity and
        methods for testing, 67 ; treatment of, 68, 69
Roughness coefficient *n*, table showing values of, 96
Roulard, V., property, Fresno County, 258
Ruble elevator, drawing showing, 108 ; photo showing, 110
Ruby channel, 291
    mine, Sierra County, 288, 289-292 ; cited, 81, 89 ; photo of, 288, 289 ; photo of gold
        nuggets from, 290
Rupley Ranch, Amador County, 232
Rush Creek, Trinity County, 309
Russell, Israel C., cited, 187
    Ranch, Shasta County, 284

S

Sacchi, Spellenberg, and Kubli, operations in Siskiyou County, 298
Sacramento County, placer mining in, 278-282
    mine, Placer County, 273
    River, recovery of gold from, 283
    -San Joaquin drainage, hydraulic mining on, 263
Sailors Bar, Sacramento County, 281
Salmon River, North Fork, placer mining on, 275, 293, 298 ; South Fork, placer mining
        on, 294, 298
    district, placer mining in, 298
    Gold Dredging Company, operations in Siskiyou County, 298
    Mining Company, operations in Siskiyou County, 293, 298-300
Salyer mine, Trinity County, 312 ; photo of hydraulic mining at, 166
Sampling, for hydraulic mining, 220 ; machine, photo of Bodinson, 30 ; placer deposits,
        31, 162, 219
Sampson, R. J., cited, 167, 259, 260, 282
San Andreas Gold Dredging Company, operations in Amador County, 232 ; operations
        in Calaveras County, 253
    Bernardino County, placer mining in, 282
    Buenaventura Mission, Los Angeles County, 260
    Carlos Gold Dredging Company, operations in Nevada County, 270
    Diego County, placer mining in, 157
    Domingo Creek, placer mining on, 254
    Fernando Mission, Los Angeles County, 260
    Francisquito Canyon, Los Angeles County, 260
    Gabriel Canyon, Los Angeles County, 260
        district, placer mining in, 260
        Mission, Los Angeles County, 260
    Gruco Company, operations in Shasta County, 285 ; and C. E. Gruwell, operations
        in San Joaquin County, 283
    Joaquin County, placer mining in, 282-283
        Mining Company, operations in Merced County, 262
        River, placer mining on, 257
        Valley, drawing showing delta formation in, 198
    Juan Gold Company, operations in Nevada County, 266-267 ; water-rights of, 267
        Ridge, Nevada County, 266
Sand-drag, photo showing, 76
Sanguinetti Ridge, Tuolumne County, 313

V

W

Y

Z

O

www.ingramcontent.com/pod-product-compliance
Lightning Source LLC
Chambersburg PA
CBHW031805190326
41518CB00006B/207